Aerospace and Aeronautical Engineering

Aerospace and Aeronautical Engineering

Edited by
Russell Mikel

WILLFORD PRESS

www.willfordpress.com

Published by Willford Press,
118-35 Queens Blvd., Suite 400,
Forest Hills, NY 11375, USA

ISBN: 978-1-68285-345-0

Cataloging-in-Publication Data

Aerospace and aeronautical engineering / edited by Russell Mikel.
p. cm.
Includes bibliographical references and index.
ISBN 978-1-68285-345-0
1. Aerospace engineering. 2. Aeronautics. I. Mikel, Russell.
TL545 .A37 2017
629.1--dc23

For information on all Willford Press publications
visit our website at www.willfordpress.com

WILLFORD PRESS

Printed in the United States of America.

Contents

Preface

This book on aerospace and aeronautical engineering discusses important innovations and theoretical breakthroughs that have occurred in the industry in this field. Aerospace engineering encompasses the entire process of aircraft and rocket design. Aeronautical engineering is a significant branch of aerospace engineering. This book includes some of the vital pieces of work being conducted across the world on various topics related to aerospace and aeronautical engineering. It seeks to provide comprehensive knowledge in this discipline. The various advancements in the field are glanced at and their applications as well as ramifications are looked at in detail. Most of the topics introduced in this book cover new techniques and the applications of aerospace and aeronautical engineering.

Various studies have approached the subject by analyzing it with a single perspective, but the present book provides diverse methodologies and techniques to address this field. This book contains theories and applications needed for understanding the subject from different perspectives. The aim is to keep the readers informed about the progress in the field; therefore, the contributions were carefully examined to compile novel researches by specialists from across the globe.

Indeed, the job of the editor is the most crucial and challenging in compiling all chapters into a single book. In the end, I would extend my sincere thanks to the chapter authors for their profound work. I am also thankful for the support provided by my family and colleagues during the compilation of this book.

Editor

Detection and Tele-replication of Human Hand Motions by a Robotic Hand

Lucian Milea[1], Monica Dascalu[2], Eduard Franti[2, 3], Suzana Cismas[4], Doina Moraru[1], Florin Lazo[2], Elteto Zoltan[2]

[1]Solaris Consult S.R.L., Bucharest, Romania
[2]Centre for New Electronic Architecture, Research Institute for Artificial Intelligence, Bucharest, Romania
[3]Micromachined Structures, Microwave Circuits and Devices Laboratory, National Institute for Research and Development in Microtechnologies (IMT), Bucharest, Romania
[4]Department of Biotechnology, University of Agricultural Sciences and Veterinary Medicine, Bucharest, Romania

Email addresses:

lucian@artsoc.ro (L. Milea), monicad@artsoc.ro (M. Dascalu), edif@artsoc.ro (E. Franti), suzanacismas@yahoo.com (S. Cismas) doinam@artsoc.ro (D. Moraru), florin_lazo@yahoo.com (F. Lazo), zelteto@gmail.com (E. Zoltan)

Abstract: This paper presents a tele-operated robotic hand controlled by replication of human hand motions and is focused on the description of technical solutions for detection and tele-replication of movements, in order to control a robotic hand. The purpose of such research is justified by the need of high precision human controlled operations in special environments. The system is based on a flex sensors set with processing units and has as effectors a robotic arm and an anthropomorphic hand. The current article displays the modality of achieving an anthropomorphic robotic arm capable of efficiently handling objects of different sizes. In order to implement and test the technical and computing solutions, the authors have used a commercial product as experimental platform and improved it both in its mechanical structure and in its command and control system. For implementing the motion algorithms of the robotic arm, a method was developed for decoding arm movements performed by a human operator. To this end, bending sensors placed at the human operator's joints (shoulder, elbow, wrist and fingers) were used. Signals collected from the sensors during the realization of these different movements by the human operator were decoded, processed and implemented in the drive system corresponding to the anthropomorphic robotic arm. In this regard, all sets of complex movements by the human arm operator were duplicated and implemented in the anthropomorphic robotic arm. The results obtained in handling various objects by means of using the anthropomorphic robotic arm have certified the effectiveness of this method.

Keywords: Robotic Hand, Robotic Arm, Flex Sensors, Motion Replication

1. Introduction

Tele-operation (distance command and control) of automatic and robotic systems is nowadays used in various domains, from automotive applications to medical high-precision systems, from aero-spatial vehicles to games. Many industries are now based on tele-operated factory automation and/or robotic systems. In particular, tele-operated robotic arms were developed and are now available as commercial products.

Mechanical arms, remotely operated by human workers, are nowadays used in more and more industries. They provide high precision in handling operations of different size objects. A special category is that of the robotic arms used for performing various operations in hazardous or toxic environments. Manipulating items under hazardous or toxic circumstances requires the human operator to wear special protective equipment so as to ensure personal safety; such equipment often makes movement and desired precision difficult or even drastically reduced.

Anthropomorphic mechanical arms have been made with a large number of degrees of freedom, similar to those of a healthy human arm. They allow the achievement of a high number of operations in manipulating objects. Teleoperation of a robotic arm can be done using a joystick or another type of teleoperation interface, or by replication of human motion (of the operator), or tele-replication [1]. Research and development reported in recent scientific literature show that currently, certain difficulties are faced in conceiving and developing algorithms that underlie the sets of complex movements for such anthropomorphic robotic arms.

In the case of tele-replication of human arm motion, the most common solution implies the use of an exoskeleton. The movements of the corresponding segments of the exoskeleton attached to the human arm are more easily put in correspondence with the artificial arm movement. For tele-replication of hand and finger motion, sensor-based systems like those implemented in virtual-reality gloves are preferred [2].

The system presented in this paper uses tele-replication of human motion as the method of remote control, using sensors for detecting arm and hand motions. The tele-operation algorithm is based on decoding the motion patterns for all the human operator's arm and hand joints, for complex sets of operations, and subsequently implementing them into an anthropomorphic robotic arm. Through this method, the anthropomorphic robotic arm is able to perform complex sets of movements which are useful in the manipulation of objects characterized by different sizes.

Although sophisticated devices like multisensory five-finger dexterous hand DLR/HIT hands are now available on the market, low-cost and high precision solutions are still needed (a performant robotic arm nowadays costs from $100,000 to $300,000).

2. The Sensor-Based Replication of Human Arm and Hand Motions in Robotic Systems

Replication of human movement by robotic arms and hands, so that the latter can be remotely controlled by human hand and arm movements, is an essential part of tele-operation. In order to achieve it effectively, the first necessary thing is to accurately detect (in point of quality and quantity) all the movements made by the human arm and hand (or only those joints that make up the segment of interest), followed by appropriate processing of the obtained information and the command of the robotic systems [3], [4].

Using of the most suitable sensors will be necessary in order to achieve accurate movements' detection, and the information obtained from them will be specifically processed to control the actuators of the robotic systems.

In doing so, we must consider that effector systems consisting of robotic arms and hands can have articulated structures and various drivers (not always similar to the human ones in point of cinematics and functionality), and they must be specifically ordered. Therefore, different systems often need necessary adjustment of the processing models used, or even completely new and specific approaches.

Detection of human hand movements can have several levels of complexity, depending on the monitored number of degrees of freedom and the required accuracy in various applications. In the general case, we can investigate the monitoring of all the degrees of freedom of the human hand and arm, from shoulder to fingers. Starting from the resulting information and in conjunction with the specific structure of the robotic arm to be controlled, suitable methods of processing the data and the control drives can be selected and / or designed.

In monitoring the position of each joint by its various degrees of freedom several sensors types may be used:

- strain bands (strain gauges) (of various types) for measuring the flexion of a joint

- flexion (or bending) sensors – they are the most commonly used in detecting human body movement. They are fixed on a jouint and fold with it. Typically they are resistive and the value of their electrical resistance is directly proportional to the bending angle and inversely proportional to the curvature radius [5]. They can be uni or bi-directional and some variants can even detect the bending direction. In terms of construction, flexion sensors may appear in many building variants:

- conductive ink on a flexible substrate - these are usually unipolar sensors, and, in order to perform detection in both directions, two such sensors should be used, placed in opposite directions.
- piezo resistive strips - they have a huge range of angles, they are bidirectional, but they have high costs.
- resistive polymer stacked between conductor layers - their resistance decreases when compressed, either directly or by bending; they have very low costs, but also a slow response, reduced accuracy and repeatability.

- optical sensors, in which the amount of light transmitted or reflected by a bent optical fiber is a measure of the bending degree thereof. They are made of plastic optical fiber, which at one end is fed from a light source, and, at the other end, has a photodetector; bending results in a loss in light transmission, which translates into a reduction in the detected voltage or an increase in resistance (for a photoresistor). Due to their operation, these sensors are unipolar.

- pairs of Hall sensors and magnets placed on the exoskeleton attached to the human hand.

- sensor networks for flexion and angle. Such an application enables the monitoring of all joints positions in the human body, so as to duplicate movements in a virtual space.

- displacement sensors placed in specific frames for measuring flexion / extension of the joints.

- 3D gyroscopes, accelerometers, magnetometers. Each can provide information about the absolute or relative position of the segments of the human body; their combined use can provide a complete and stable image on the current position of the human body, based on prior knowledge of the interconnection of the monitored segments. There are also inertial sensors for human motion tracking [6].

Important design decisions are: the type of sensor used, their number, their position on arm/joints. We choose the type of sensors to use – a bending sensor, and analyzing various properties of bending sensors [7], we decided to use

FS-L-0XXX-103-ST sensors (figure 1), considering the above mentioned data and the specifics of the application. Their main advantage is that they are easy to use and reasonably priced. They are uni-directional sensors and measure the bending degree with a reasonably linearity, and, thus, just as the sensor is bent, its electrical resistance increases in linear proportion.

Figure 1. The FS-L-00XX-XXX-ST bending sensor [8].

3. The Basic Structure of the Robotic System

The robotic system consists of a robotic arm with 5 degrees of freedom, at the end of which an anthropomorphic robot hand is fitted (five fingers, each with the same number of phalanges that mirror the human hand). This commercial product is controlled through a sophisticated software tele-operation interface that was further replaced with a customized control system.

The joints are powered by electric actuators which allow rotation between 30 and 180 degrees (depending on use and location). Of the 18 degrees of freedom of the system, only 16 are current. The actuators are pulse controlled by the PWM output of an Arduino Mega 2560 module, equipped with the ATMEL ATmega 2560 processor.

The metal fingers that originally equipped the arm had batracian fingers anatomical shape and were built from aluminum articulated modules that were operated by 5 micro actuators through a system of levers. The major drawback of the system was that it amplified the initial joints motion in proportion to the levers ratio (1: 5), making it imprecise to grip objects using the fingers. Meanwhile fingertips mobility was limited to the upper limit, making it impossible to squeeze a small object in the hand.

Because of these shortcomings in the initial system, the original fingers were replaced by modular translucent plexiglass that mimics the human fingers anatomical shape; the grabbing base ("the palm") was maintained together with the 5 micro actuators. The lever drive was replaced by a system of wires that connect to the last phalanx of each finger and are "pulled" by the 5 micro actuators, thereby arching the fingers up to the desired position.

Thus the accuracy of gripping various objects has improved. Regarding the drive system, the arm was equipped with multiple actuators to optimally drive each joint:
The shoulder - has two actuators: one for rotation around the 0Z axis (0-160 degrees) and another for lifting / lowering the arm vertically (0 - 90 degrees);
The elbow: 1 actuator 0-90 degrees vertical;
The wrist: 1 actuator, 0-160 degrees horizontally;

4. The Method of Processing, Command and Control

Based on the law of variation of the sensor output value depending on the input, and taking into account the type and range of controls specific to the actuators chosen, the solution of appropriate scaling the range of available values was chosen.

The sensors are connected into a resistive divider circuit, at the analog input of a a control module; thus, the bending variation will produce a variation in the internal resistance of the sensor, and implicitly the voltage on the analog input of the Arduino board; this voltage is converted into a numerical value in the range 0-1024.

The numerical value is scaled in accordance with the action range of the actuator on the arm corresponding to the respective joint (the map function is used = scales one range of linear values to another range or linear arithmetic values: map (value, fromLow, fromHigh, toLow, toHigh)).

5. Block Diagram of the Robotic System

The block diagram of the robotic system is presented in Figure 2. The robotic system used contains two main blocks: the block formed by the bending sensors, on the human operator's arm, together with the block formed by the mechanical arm, equipped with pressure sensors and actuators. Each of the two blocks has one control system Arduino 2560 type with AtMega 2560 microcontroller. The figure below illustrates a block diagram of the robotic system used and the flow of signals during its operation.

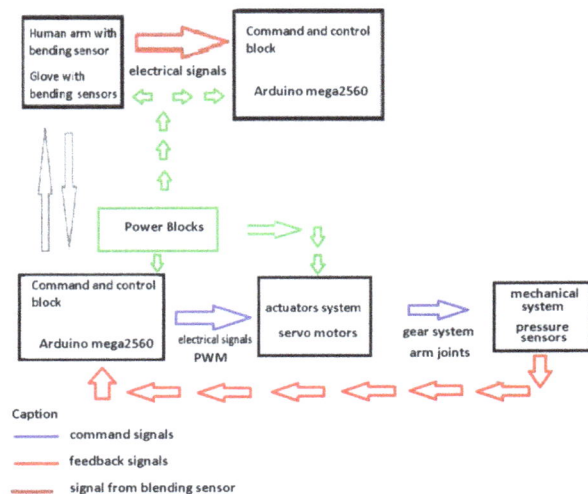

Figure 2. Block diagram of the robotic system.

6. Robotic System Implementation

In the implementation of the robotic system (figure 3), two boards Arduino mega 2560, equipped each with a microcontroller AtMega 2560, were used, as well as shoulder, elbow, wrist and fingers bending sensors placed on the human operator. For the shoulder, elbow and fingers, bending sensors 11cm long were used, and for the wrist a 5cm long sensor was used. The figure below schematically represents the robotic system implementation. In functional terms, signals from the human operator's bending sensors on the shoulder, elbow, wrist and fingers are taken on the analog inputs of microcontroller ATmega 2560 and converted into numerical values, which will be processed and then transmitted to the control unit the mechanical arm by bidirectional serial communication. Serial communication between the command blocks is bidirectional, with parametric 115200 baud rate, 8N1.

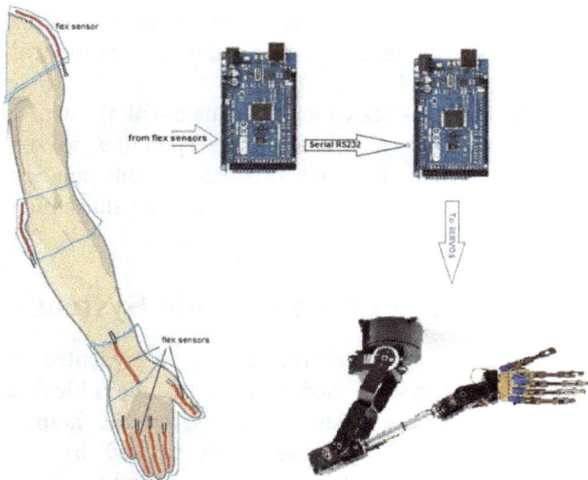

Figure 3. Implementation of the robotic system.

The data frame is 24 bytes (octets) in which an octet plays a role in error correction (CRC).

The data frame transmitted by the human arm and sensors glove control unit to the mechanical arm control block is the following:

Byte 0 = 0x20; - start frame identifier
Byte 1 = 0x0A; - start frame identifier
Byte 2 = 0x18; - dataframe length (0x18 = 24 bytes)
// fingers:
Byte 3 = finger 1 LSB byte
Byte 4 = finger 1 MSB byte
Byte 5 = finger 2 LSB byte
Byte 6 = finger 2 MSB byte
Byte 7 = finger 3 LSB byte
Byte 8 = finger 3 MSB byte
Byte 9 = finger 4 LSB byte
Byte 10 = finger 4 MSB byte
Byte 11 = finger 5 LSB byte
Byte 12 = finger 5 MSB byte // wrist:
Byte 13 = hand wrist flexion angle LSB byte
Byte 14 = hand wrist flexion angle MSB byte

Byte 15 = hand wrist rotation LSB byte
Byte 16 = hand wrist rotation MSB byte // elbow:
Byte 17 = wrist elbow angle LSB byte
Byte 18 = wrist elbow angle MSB byte // shoulder:
Byte 19 = shoulder lifting angle LSB byte
Byte 20 = shoulder lifting angle MSB byte
Byte 21 = shoulder rotation LSB byte
Byte 22 = shoulder rotation MSB byte
Byte 23 = CRC

7. Experiments and Results

By using this system, all necessary movement algorithms were detected and recorded using bend sensors included into support bandages and glove (see figure 4), so that the mechanical arm could perform the main sets of operations in remote-controlled interventions. These arrangements represents advances from past work, previously reported [9], [10]. Figure 5 is a collection of images from different experiments that involved grasping and manipulation of objects of different forms, materials and sizes.

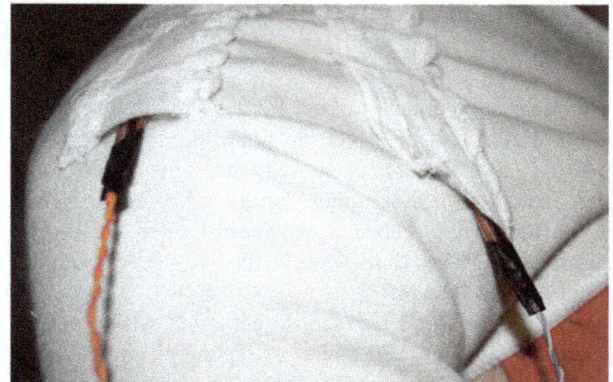

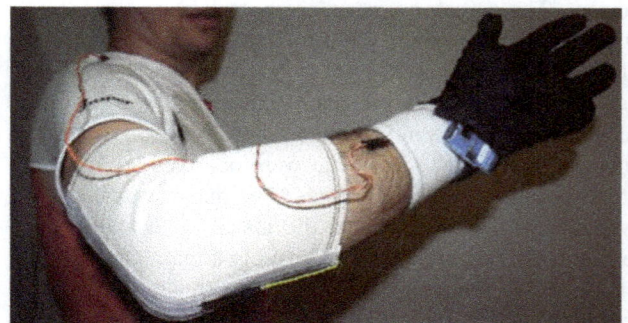

Figure 4. Detecting human arm and hand motions using bend sensors included into support bandages and glove.

A library of movement algorithms was thus created and organized under different categories of operations for remote-controlled interventions.

The figures below illustrate various types of moves performed using the mechanical arm and hand, which is remote-controlled by a human operator. Manipulation using a mechanical hand for different size objects is also presented in the figures below.

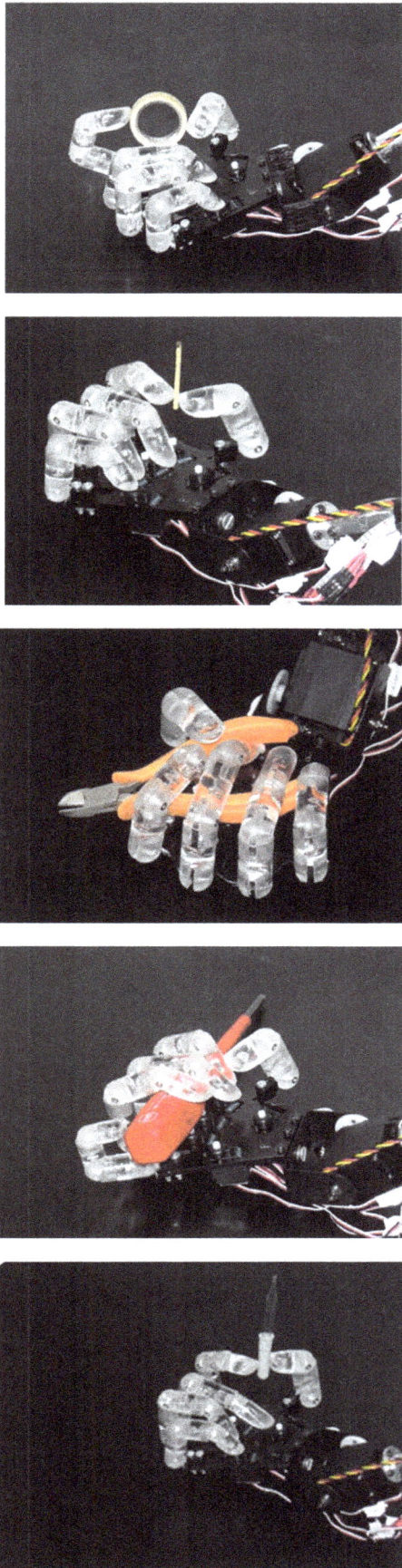

Figure 5. Handling different objects using the mechanical hand.

8. Conclusions

The robotic arm that was built enabled efficient handling of objects of different sizes. By the method presented in this paper we managed to decode the movement algorithms of the human operator's arm and to implement them into an anthropomorphic robotic arm.

Although a commercial product was used for offering an experimental basis, the system designed is original. The main novelty of the paper is the simple solution for replication of arm movement, in terms of sensors and processing algorithms. The obtained system is more versatile.

The results obtained in manipulating objects by means of the anthropomorphic robotic arm certify the effectiveness of the method used and clear the way for new experiments for achieving more complex operations with a higher degree of dynamism. In the next stages of the project, such operations of handling objects will be conducted, within certain sequences of predefined time.

Acknowledgement

This work was supported by a grant of the Romanian National Authority for Scientific Research, Programme for research - Space Technology and Advanced Research - STAR, project number 82/2013.

References

[1] A.D. Dragan, K.T. Lee, and S.S. Srinivasa. Teleoperation with intelligent and customizable interfaces, Journal of Human-Robot Interaction (JHRI), 2013.

[2] K.S. Hale, K.M. Stanney, Handbook of Virtual Environments: Design, Implementation, and Applications, 2nd Edition, CRC Press 2014, ISBN-10: 1466511842, ISBN-13: 978-1466511842.

[3] Leeper, A. E., Hsiao, K., Ciocarlie, M., Takayama, L., & Gossow, D. (2012), Strategies for human-in-the-loop robotic grasping, In Proceedings of the ieee/acm international conference on human-robot interaction, http://dx.doi.org/10.1145/2157689.2157691.

[4] Paravati, G., Sanna, A., Lamberti, F., & Celozzi, C. (2011, September), A reconfigurable multi-touch framework for teleoperation tasks, in Emerging technologies and factory automation, http://dx.doi.org/10.1109/ETFA.2011.6059219.

[5] Donghui Zhao, Xianfeng Chen, Kaiming Zhou, Lin Zhang, Ian Bennion, William N. MacPherson, James S. Barton, Julian D. C. Jones, Bend sensors with direction recognition based on long-period gratings written in D-shaped fiber, Applied Optics Vol. 43, Issue 29, pp. 5425-5428 (2004) doi: 10.1364/AO.43.005425.

[6] Daniel Roetenberg, Henk Luinge, and Per Slycke, Xsens MVN: Full 6DOF Human Motion Tracking Using Miniature Inertial Sensors, Xsens Motion Technologies BV (2009), https://www.xsens.com/wp-content/uploads/2013/12/MVN_white_paper1.pdf.

[7] L.E. Dunne, B. Smyth, B. Caulfield, A Comparative Evaluation of Bend Sensors for Wearable Applications, 2007 11th IEEE International Symposium on Wearable Computers, Boston MA USA, pp. 121-122, ISBN: 978-1-4244-1452-9, DOI: 10.1109/ISWC.2007.4373797.

[8] Digi-Key Electronics, Spectra Symbol Flex Sensors, media.digikey.com/pdf/Data%20Sheets/Spectra%20Symbol/F S%20Series%20Flex%20Sensor.pdf.

[9] E. Franti, G. Stefan, P. Schiopu, T. Boros, Anca Plavitu, Intelligent Control System for Artificial Arms Configuration, in Proceedings of The 5th EUROPEAN COMPUTING CONFERENCE (ECC '11), pag 312 - 316.

[10] E. Franti, G. Stefan, P. Schiopu, M. Teodorescu , Modular Software for for Artificial Arms Design, in Proceedingsof the International Conference on Automatic Control, Modelling & Simulation (ACMOS'11), Lanzarote, Spania, 27.05.2011 – 29.05.2011, pag. 387 – 391.

2

Design of mini wind tunnel based on coanda effect

Yassen El-Sayed Yassen, Ahmed Sharaf Abdelhamed

Mechanical Power Engineering, Faculty of Engineering, Port Said University, Port Said, Egypt

Email address:

y_yassen70 @yahoo.com (Y. El-S. Yassen), sh_ahmed99@yahoo.com (A. S. Abdelhamed)

Abstract: An experimental investigation and CFD treatment were employed to design mini-wind tunnel based on Coanda effect for model tests and basic research. The inlet source flow is efficiently creating smooth steady airflow with acceptable noise, achieving the possibility of placing the test target closer to the source of flow with reasonable estimates of turbulence intensity. The design aims at achieving flow uniformity in the working section midplane, preventing separation in the contraction and minimizing the boundary–layer thickness. Intensive measurements after construction demonstrate the significance of the design process and validate the CFD predictions. The results are represented in graphic form to indicate the aspects of the contraction ratio. The numerical and experimental results show the uniformity of velocity distribution inside the working section. Tracing of separation and backflow is crucial allowing a variety of realistic demonstrations to be performed. The numerical solution provides a powerful tool to demonstrate the rate of boundary–layer growth inside the working section and validate against the empirical correlations with insignificant wall–friction drag. Assessment study to address large–scale wind tunnel based on coanda effect would be considered.

Keywords: Separation, CFD, Coanda Effect, Mini–Wind–Tunnel, Boundary–Layer Growth

1. Introduction

A conventional wind-tunnel design is a complex field involving many fluid mechanics and engineering aspects. The first attempt in providing some guidelines for the complete design of low-speed wind tunnels was that due to [1]. However, recent experimental studies of flow through individual components of a wind tunnel [2–4] have led to increase understanding and design philosophy for most of the components of wind tunnel. More theoretical and experimental investigations have been written about this topic and e.g. [5–8] are useful references when designing and constructing conventional low-speed wind-tunnels. Typically, the air is moved through conventional tunnel using a fan or blower. The airflow created by the fan entering the tunnel is itself highly turbulent due to the fan blade's motion. The air moving through the tunnel needs to be relatively turbulence-free [9]. Therefore, the overall length of the tunnel increases to smooth out the turbulent airflow before reaching the subject of the testing. This design is less than ideal for a wind tunnel but it is still the prevalent design.

In the present study, both of computational treatment based on RNG turbulence model and experimental measurements are implemented to design an unconventional wind-tunnel (mini-wind tunnel) based on Coanda effect for model tests and laboratory teaching purpose. The inlet source flow is efficiently creating smooth steady airflow with acceptable noise, achieving the possibility of placing the test target closer to the source of flow with reasonable estimates of turbulence intensity. The design aims at achieving the flow uniformity in the working section midplane, without separation in the contraction and minimizing the boundary–layer thickness at entrance to the working section. Calibration of the proposed mini wind tunnel after construction is carried out. The boundary–layer growth inside the working section is determined using the empirical correlations and validated against the numerical results. Wall–friction drag is estimated. Both of CFD predictions and experimental results are validated against the uniformity of velocity distribution inside the working section. Also, tracing of separation and backflow through the tunnel is carried out for different values of contraction ratios. The overall length of mini wind tunnel is 160 cm with contraction ratio of 2.82 and the cross–sectional area of test section is 19×19 cm^2 with length 50 cm.

2. Experimental Design

2.1. Mechanism of Supply Air

The mechanism of air flow as shown in Fig.1 may be demonstrated in several steps which can all be explained by the Coanda effect, the Venturi effect, and Bernoulli's principle. First, the air is drawn in through the base of the machine, powered by what is called a mixed-flow impeller of 38 watt. It has nine fins with rows of tiny holes that reduce the friction caused by colliding high and low air pressure. The air is accelerated through an annular aperture and then passed over an airfoil-shaped ramp that channels (36 slits) its direction. By propelling air out of a 1.6 mm slit located on the inside of its ring, air flows across one side of the airfoil, the Coanda surface. This results in entrainment of surrounding air, just downstream of the airfoil. The Venturi effect is a factor in creating the optimal air velocity, resulting in the Coanda effect. As seen in the cross section of Fig.1, as the air exits the airfoil, it is funneled through a small slit. The fluid velocity greatly increases with this reduction in area, resulting in optimal entrainment of air. Also, the air is drawn in from behind, or induced, and this can be explained by Bernoulli's principle. As the air leaves the airfoil at a higher velocity than the surrounding air, it creates an area of low pressure. This pressure differential between the high velocity air and still air behind is what draws in the induced air. This combination of entrainment from the Coanda effect and inducement by Bernoulli's principle is what makes the air flow 15 times its air intake and an average turbulence level 0.024 %. According to the above explanation, the optimal Coanda surface profile for entrainment of air has found, one with an airfoil cross section with a 14° angle between the top and bottom surfaces and whose cord length is constant at approximately 10 cm. The construction of the airfoil was done by using fiberglass and a foam exoskeleton as can be seen in Fig.1.

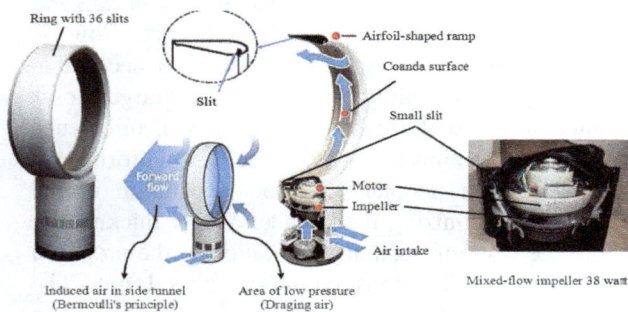

Fig 1. Air supply and construction of the airfoil.

2.2. Mini–Wind Tunnel Components

Manufacturing a mini-wind tunnel of card-board that was covered inside and outside with specific foil paper so as to obtain smooth surface of the inside tunnel. The general layout of the mini wind tunnel is shown in Fig. 2.

A mini-wind tunnel is manufactured as an open–circuit rig powered by a 38 watt induction motor with low average turbulence intensity 0.024 % and low average noise source

level (21.5 dB) compared with the allowable acoustics noise level in conventional small wind tunnels (70–90) dB [10,11]. The flow is accelerated by contraction (contraction ratio, CR =2.82) into a 19 cm square test section. The dimensions and specifications of the mini wind tunnel are shown in Fig. 3 and Table (1), respectively.

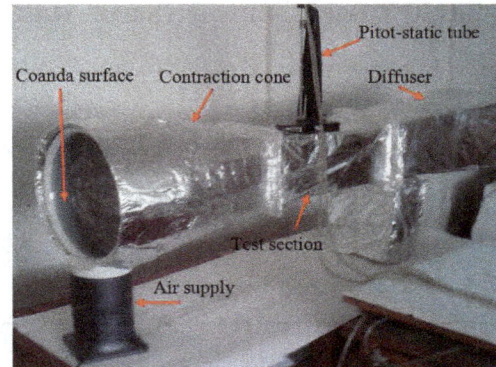

Fig 2. Test rig of mini-wind tunnel.

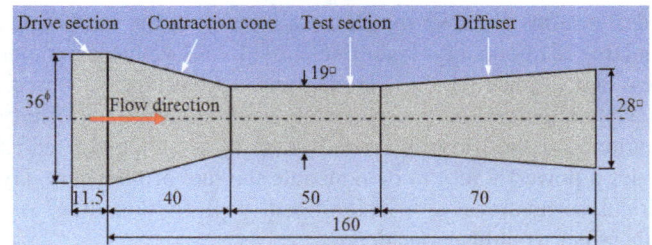

Fig 3. Mini-wind tunnel, dimensions in (cm).

Table 1. Specifications of Mini-wind tunnel.

Overall length (cm)	160
Entrance cone (cm)	36$^{\phi}$
Contraction ratio	2.82
Test section (cm)	19×19×50
Average turbulence leve at inlet tunnel	0.024 %
Air mass flow rate at inlet tunnel	0.008 to 0.0182 kg/s
Average noise source level at inlet tunnel	21.5 dB
Drive (motor)	38 watt

2.3. Experimental Measurements

The velocity profile through the mini-wind tunnel at eight sections (two in contraction section, four in test section, and two in diffuser section) is measured using Static-Pitot tube of 6 mm diameter. A simple slide probe traversing mechanism is used to locate the probe tip at every location (12 points, and 1.5 cm between each two successive points and five points, 5 mm between each near the wall). The collected measurements data at measuring station locations were averaged and used to calculate the mean velocity distribution of the air flow through the mini-wind tunnel. The uncertainty, using root-sum-square method in the measured mean velocity is ± 4.8 % within 95 % confidence. The measurement sections through the mini-wind tunnel (half of the tunnel) are shown in Figs. (4, 5).

3. Mesh-Independency

The quality of the grid plays a crucial role in the accuracy of CFD simulations and for achieving a fully converged solution. The body-fitted grid has the flexibility to model problems with irregular geometries. It offers modeling mini wind tunnel geometry accurately and produces fine grid resolution in regions of special interest. One of the important tasks in grid generation is to obtain grid-independent results and it is highly necessary to have a flow passage which is represented by as many grid points as necessary to give a realistic flow. The domain is meshed using an unstructured grid with different densities are tested. The refinement mesh with a factor of 2 is chosen around the airfoil for stability. However, the mesh of the final design shown in Fig. 6 has 61784 cells and average skew factor of 0.3481 produced best results.

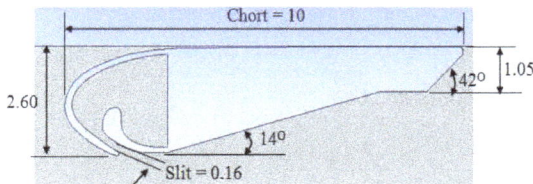

Fig. 4. Airfoil, dimensions in (cm)

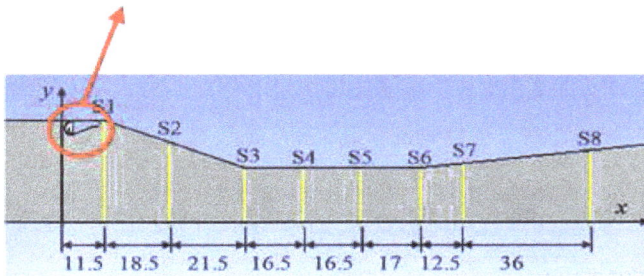

Fig. 5. Measurement Locations, dimensions in (cm).

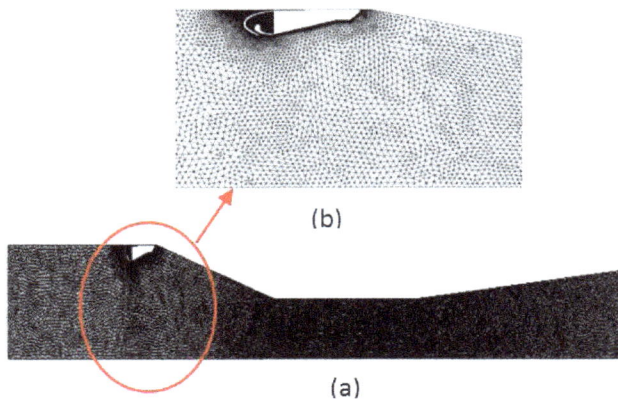

Fig. 6. (a) Grid construction (b) Grid details around the airfoil.

4. Results and Discussion

When investigating the results of the design and construction of an unconventional small–scale laboratory wind tunnel (mini–wind tunnel) based on Coanda. Effect, there are several key factors that have to be checked carefully. The most important factors concern mean–flow variations in time and space over the test–sectional cross section area, the turbulence fluctuation intensities, both in the streamwise and cross-stream directions. CFD simulation is performed using ANSYS FLUENT 14 in order to handle the axi-symmetric steady incompressible turbulent air flow using RNG turbulence model with enhanced wall treatment. In the present study, different contraction ratios (1.63, 2.54, 2.82, 3.14, and 5.19) are chosen for mini-wind tunnel based on Coanda effect with inlet mass flow rate 0.0182 kg/s, inlet diameter of the tunnel 0.36 m, and constant angle of the diffuser 3.5°. The inlet boundary conditions used in numerical solution are exactly the same as experiments.

Figure 7(a) shows the predictive results, for streamwise velocity profile at eight successive stations for contraction ratio CR=1.63. It can be seen that the velocity is maximum near the wall in sections 1 and 2 due to Venturi effect. The inlet velocity to the test section (section 3) is non-uniform, and the velocity profiles at sections 4, 5, and 6 are uniform. Also, the value of velocity in the test section is less than 2 m/s. However, it can be observed that no separation or back flow in the contraction cone, as shown in Figs. 7(b, c).

Figure 8(a) illustrates the numerical results for streamwise velocity profile at eight successive stations for contraction ratio CR= 2.54. It can be noticed that the maximum velocity takes place near the wall in sections 1 and 2. The inlet velocity profile to test section (section 3) is close to uniformity but the velocity profiles at sections 4, 5, and 6 are uniform in shape and approaching 2 m/s in the test section compared with the case where CR = 1.63, Fig. 7(a). Moreover, no separation and back flow in the contraction cone, as shown in Figs. 8(b, c).

Figure 9(a) shows the predictive results for streamwise velocity profile at eight successive stations for contraction ratio CR= 2.82. It is clear that the velocity is maximum near the wall in sections 1 and 2. The inlet velocity to test section (section 3) is almost in uniform profile. While the velocity profiles at sections 4, 5, and 6 are uniform with value greater than 2 m/s in the test section compared with the cases where CR = 1.63 and 2.54, as shown in Figs. (7.a, 8.a), respectively. Also, it can be seen that there is no separation and back flow in the contraction cone, as shown in Figs. 9 (b, c).

The predictive results for streamwise velocity profiles at eight successive stations for contraction ratio CR= 3.14 are shown in Fig.10(a). The maximum velocity takes place near the wall in sections 1 and 2. It can be observed that the flow seems to be irresistible for separation as shown in Figs. 10(b, c). This is attributed to the effect of contraction ratio (CR=3.14). However, the velocity profile at the inlet to test section (section 3) is nearly uniform while the velocity profiles at sections 4, 5, and 6 are uniform with value approaching 2 m/s in the test section compared with the other cases, CR = 1.63, 2.54, and 2.82.

For contraction ratio = 5.19, the flow suffers from severe separation and back flow in the contraction cone, as shown in Figs. 11(a,b,c). The effect of separation is quite clear on the value of the velocity in test section that is less than 2 m/s. However, flow separation should be avoided inside the tunnel circuit to prevent flow unsteadiness and associated noise as well as to minimize losses.

CR = 1.63
$(36^{\phi}$ to $25^{\circ})$ cm^2

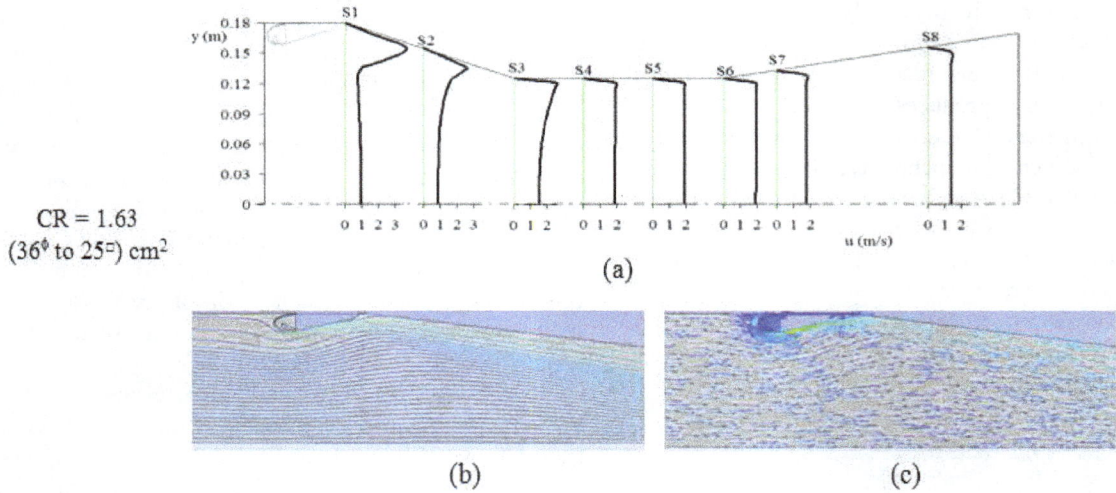

Fig 7. *(a) Velocity profiles (b) Streamlines (c) Velocity vectors for CR = 1.63.*

CR = 2.54
$(36^{\phi}$ to $20^{\circ})$ cm^2

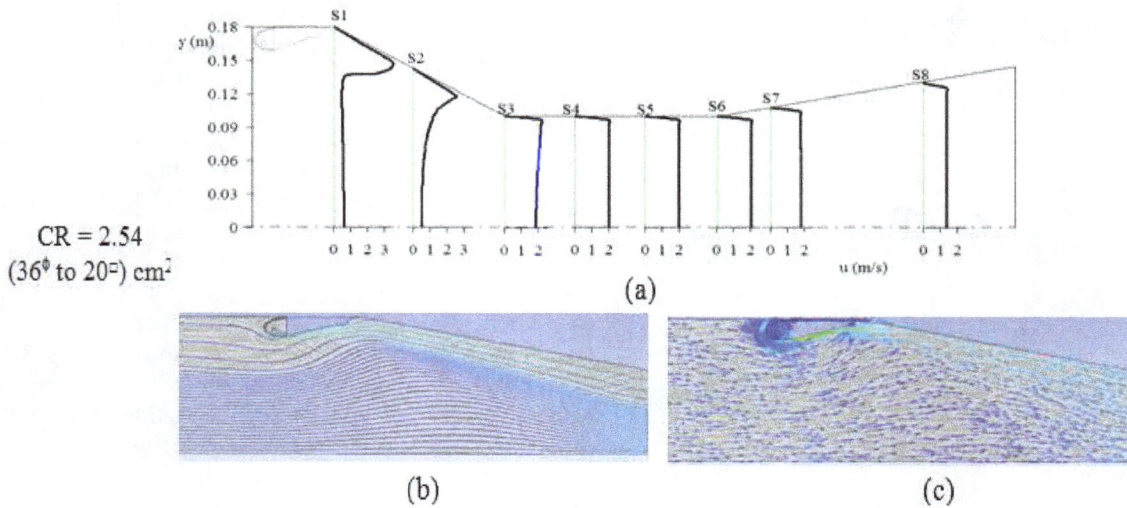

Fig 8. *(a) Velocity profiles (b) Streamlines (c) Velocity vectors for CR = 2.54.*

CR = 2.82
$(36^{\phi}$ to $19^{\circ})$ cm^2

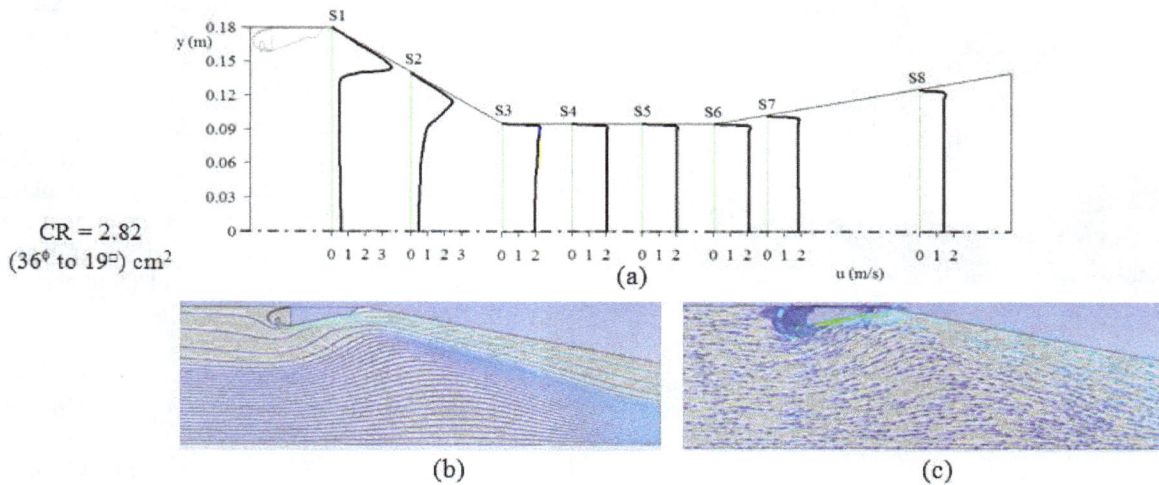

Fig 9. *(a) Velocity profiles (b) Streamlines (c) Velocity vectors for CR = 2.82.*

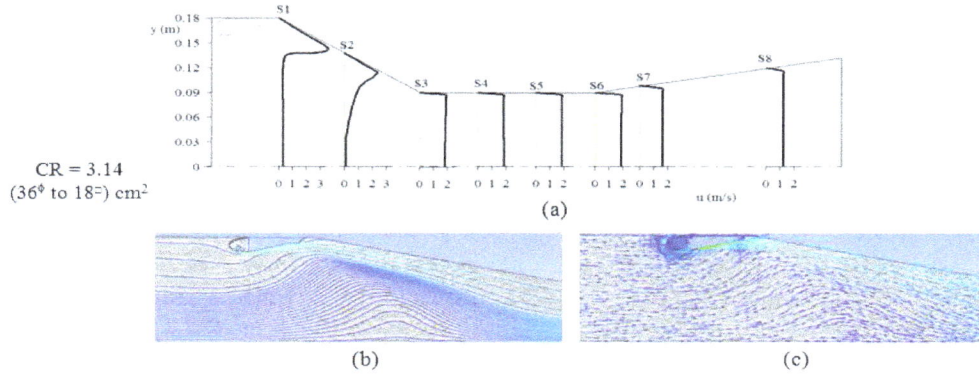

Fig 10. *(a) Velocity profiles (b) Streamlines (c) Velocity vectors for CR = 3.14.*

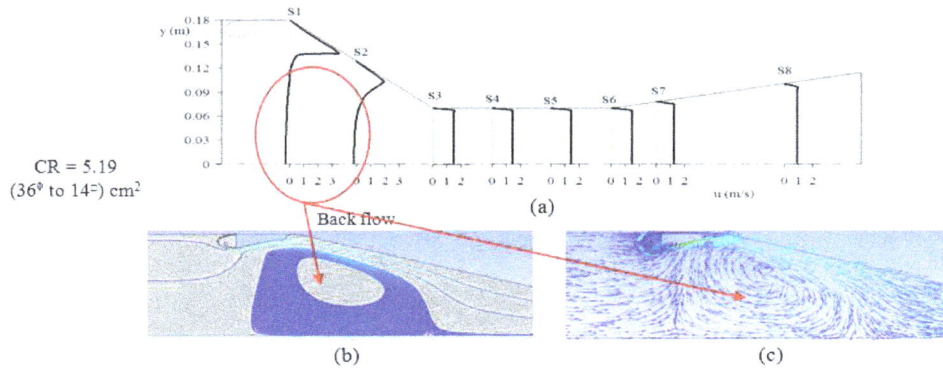

Fig 11. *(a) Velocity profiles (b) Streamlines (c) Velocity vectors for CR = 5.19.*

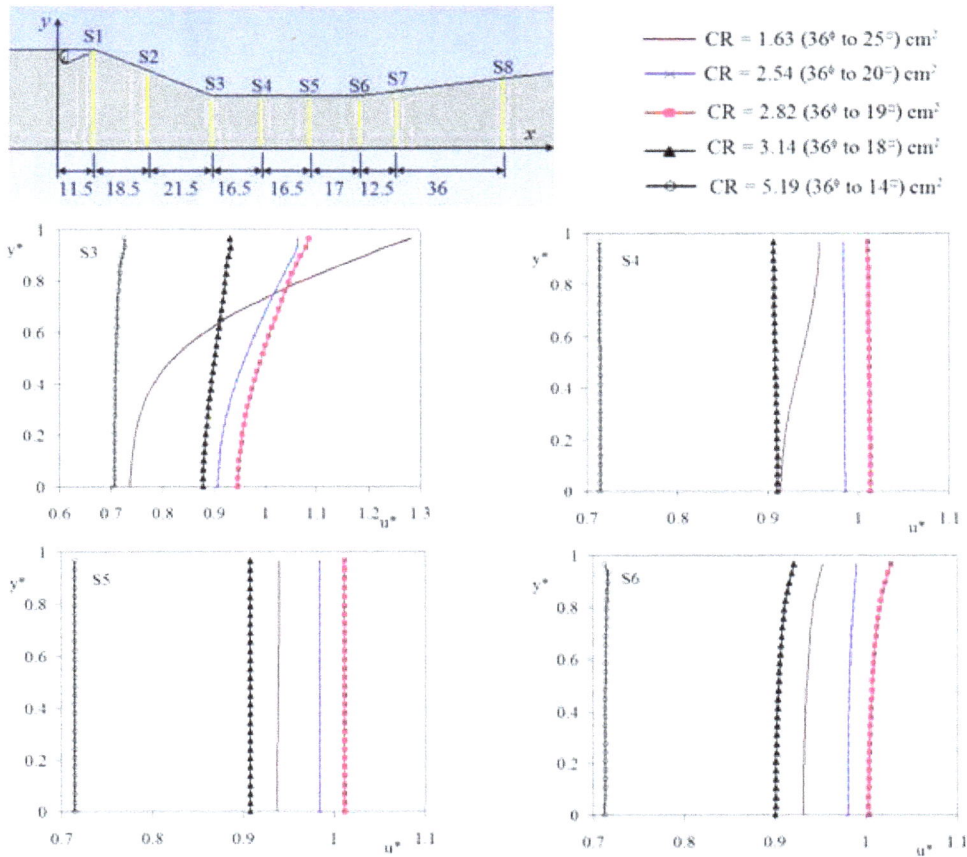

Fig 12. *Velocity profile over working section for different contraction ratios.*

The working section represents one of the important parts of the tunnel. An intensive numerical investigation is carried out in order to demonstrate the influence of contraction ratio on the velocity profile inside the test section. Figure 12 shows the predictive results for streamwise velocity profile over working section (sections 3, 4, 5, and 6) for different contraction ratios (1.63, 2.54, 2.82, 3.14, and 5.19). It can be noticed the significance of CR= 2.82 that achieves the uniformity inside the test section and high velocity value compared with the other cases of contraction ratios.

One of the most important aspects of the flow quality in a wind-tunnel is the level of turbulence intensity. Figure 13 shows the numerical results of turbulence intensity at inlet to the mini-wind tunnel for different contraction ratios. The contraction ratios (2.54, 2.82, and 3.14) demonstrated relatively reasonable estimates of turbulence intensity 0.024 %. Comparison of the numerical values of turbulence intensity, CR =2.82 appeared to have a more uniform velocity profile, no separation and back flow, and hence it was selected for manufacture.

However, the performance of the chosen mini-wind tunnel with contraction ratio (CR=2.82) still requires testing after construction, to validate the CFD simulations. The inlet contraction plays a critical role in determining the flow quality in the test section. The contraction accelerates and aligns the flow into the test section. The size and shape of the contraction dictate the final turbulence intensity levels in the test section. The contraction stretches vortex filaments, which reduces axial but intensifies lateral turbulent fluctuations. The length of the contraction should be sufficiently small to minimize boundary-layer growth and manufacturing cost but long enough to prevent large adverse pressure gradients along the wall, generated by streamline curvature, which can lead to flow separation. More comprehensive experimental data are required to validate the numerical solution. Figure 14 shows the velocity profiles for the numerical results and measurements at eight measuring sections for CR = 2.82. It can be seen that the numerical solution based on RNG turbulence model is able to represent the experimental measurements. Fair agreement between the predictive results and experimental data is achieved.

Assuming that the rate of boundary layer growth starts at the working section (0.19×0.19) m^2, both of Blasius and Von Karman Integral formulae are used to estimate the laminar boundary layer characteristics as shown in table (2). It is noticed that the wall friction drag is minimum.

Figure15 shows the boundary–layer growth based on the numerical solution (blue color) and an empirical correlation over the working section. The predicted results provide a powerful tool to demonstrate the rate of boundary–layer growth inside the working section and validate against the empirical correlations (Blasius and Karman Integral formulae).

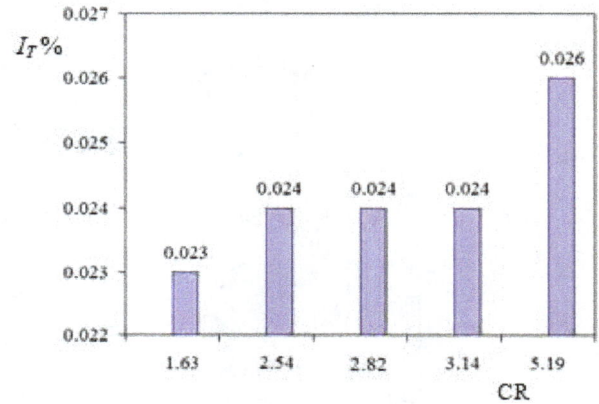

Fig 13. *Turbulent intensity for different contraction ratios.*

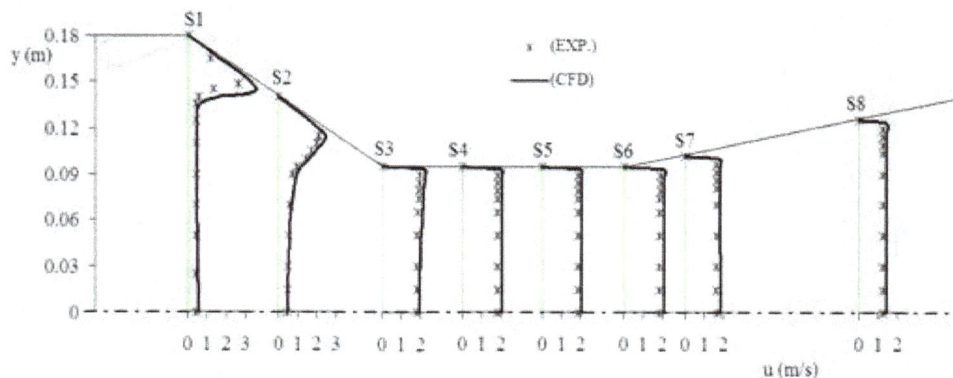

Fig 14. *Velocity profiles at eight sections for CR = 2.82.*

Table 2. *Laminar Boundary–layer characteristics (x = 0.5 m and Rex = 6.5x104).*

	Blasius	Von Karman Integral (parabolic profile) $u = \alpha y + \beta y^2$
Boundary–layer thickness δ (mm)	9.71	10.72
Displacement thickness δ^* (mm)	3.39	3.56
Momentum thickness θ (mm)	1.30	1.34
Skin–friction coefficient C_f	2.6×10^{-3}	2.86×10^{-3}
Wall–shear stress τ_w (|N/m^2)	6.39×10^{-3}	7.08×10^{-3}
Wall–friction Drag, D (N)	0.61×10^{-3}	0.67×10^{-3}

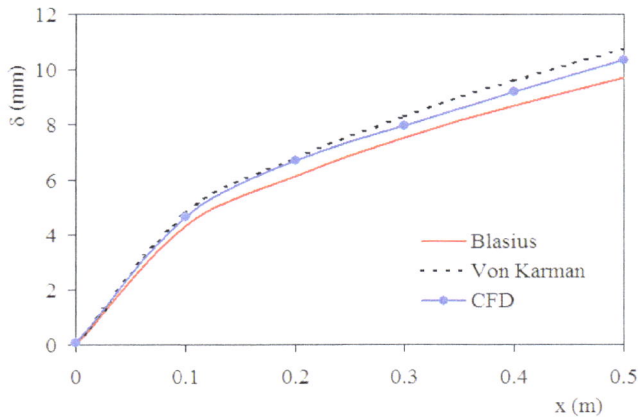

Fig 15. *Boundary–layer growth inside the working section.*

5. Conclusions

Computational treatment based on RNG turbulence model and experimental measurements are implemented to design an unconventional small–scale laboratory wind tunnel (mini–wind tunnel) based on Coanda effect suitable for laboratory model tests and basic research. Different contraction ratios are tested against uniformity in the working section and characteristics of the flow. Care must be taken to ensure no flow separation occurs in the ducts, which may also require acoustic treatment to minimize flow noise. Clearly, uniform velocity in the test section is desired as well as no separation, no back flow and minimum noise with relatively reasonable estimates of turbulence intensity. More comprehensive experimental measurements are carried out to validate the numerical solution for chosen contraction ratio. The predicted results provide a powerful tool to demonstrate the rate of boundary–layer growth inside the working section and validate against the empirical correlations with insignificant wall–friction drag. Assessment study to address the conventional wind tunnel would be considered. A significant challenge therefore exists to develop accurate and efficient design tools for flow modeling of large–scale wind tunnel in three dimensions based on Coanda effect.

Nomenclature

I_T Turbulent intensity
Re_x Local Reynolds number from the working section

u Velocity in x direction, m/s
U Mean velocity at the inlet of the working section, m/s
u^* Dimensionless velocity, $= u/U$
x Axial distance from the entrance of the working section, m
y Normal distance to wall, m
y_{max} Maximum normal distance at each section, m
y^* Non-dimensional normal distance, y/y_{max}

References

[1] Bradshaw, P., and Pankhurst, R. C., "The design of low-speed wind tunnels", Progress in aeronautical sciences 6, 1 – 69, 1964.

[2] Mehta, R.D., "Aspects of the Design and Performance of Blower Wind Tunnels Components", Ph.D. Thesis, Department of Aeronautics, Imperial College, University of London, 1978.

[3] Mehta, R.D. and Bradshaw, P., "Design Rules for Small Low Speed Wind Tunnels", Aeronaut. J., Vol. 83, No. 827, pp. 443-449, 1979.

[4] Bell, J.H., and Mehta, R.D., " Contraction Design for Small Law-Speed Wind Tunnels", National Aeronautics and space, CA 94305, 1988.

[5] Rae, W. H., and Pope, A., "Low-speed wind tunnel testing", 2nd edn. John Wiley and sons, 1984.

[6] Mansi, S., Neha, S., and Sunil, K. Y., "Review of Design and Construction of an Open Circuit Low Speed Wind Tunnel", Global J. of Researches in Eing. (A), V. XIII Issue V Version I, 2013.

[7] Björn L., and Arne V. J., "Design and Evaluation of a Low-Speed Wind-Tunnel with Expanding Corners", Technical Reports from Royal Institute of Technology, Department of Mechanics, SE-100 44 Stockholm, Sweden, 2002.

[8] Seidel, M., "Construction, design, manufacturing, and calibration of the German-Dutch wind tunnel (DNW)", Tech. Rep., Duits-Nederlandese Windtunnel (DNW), 1982.

[9] Goldstein, E.,"Wind Tunnels, Don't Count Them Out," Aerospace America, Vol. 48 4, pp. 38-43, 2010.

[10] Salam, C.A., and Ali, M.A., "Design and Fabrication of A Bench Mounted Closed Loop Wind Tunnel", MIST J. of Science and technology, Vol. 2, No. 1, 2010.

[11] Wagner, S., Bareiß, R. and Guidati, G, "Wind Turbine Noise", Springer-Verlag, Berlin, pp. 14-21, 1996.

3

On the Problem of Neutrino Mass in the Aspect of the Detected Spatial Anisotropy of Temporal Variations of Solar Neutrinos Flux

A. G. Syromyatnikov

St. Petersburg University, Department of Physics, Universitetskaya Nab., St. Petersburg, Russia

Email address:
alsyromyatnikov@mail.ru

Abstract: Mass of the muon and tau-neutrinos are calculated. Set the integer 2:5 ratio between the mass of muon neutrinos and muon ultra-light weight torsion. Found that the calculated mass of the tau-neutrino is the same as the largest with heat 3/2 T relic neutrinos. In the standard model T = 2K. Considering the temperature of the relic neutrinos as the effective gravitational temperature $T\Gamma$ = 2/9 M for some mass M, it isfollowed ratio M = $3\,m_{\nu_\tau}$, corresponding to reactions of the dissolution of the relic neutrinos by three sterile active Tau-neutrinos. Solar neutrino flux measurements data and its temporal variations in the Chlorine-Argon experiment for the period 1970-1994, examined with particular attention near dated frontiers of solar activity (FSA), which defined by means of the key criterion of 2ln2 in point a significant manifestation of the influence the Sun eccentric planets. Review shall be carried out within the framework of the standard model, taking into account the gravity. The results are presented in the form of pie charts in a consecutive temporary scan on a direct ascent of Earth's orbital position α, recalculated according to the date of the middle of each of the series of measurements. The corners of the sectors 35° and more defined for temporary length of each dimension.The amplitude of the sector is equal to the measured magnitude of neutrino flow minus the constant component 2.57 SNU. All charts show the priority direction of the solar neutrino flux detector checked by angles α = 271°, δ = 40° (-15°) axis of anisotropy of Galactic Gamma-radiation, as well as a number of less significant in transverse to the axis of the direction. Researched the relationship between solar activity, seismic activity of the Earth and neutrinos flows registered in chlorine-argon experiment for the period 1970-1994, as well as in the SAGE for the period 1970-2008years.There are correlations of solar activity with seismic activity of the Earth and the positive correlation between seismic activity of the Earth and the neutrino flow variations, reaching 95% when taking into account the peculiarities of the matching of dates of solar maximums with FSA dates 1978, 1993-5 and 2002.It turned out that all known and perhaps predicted on 2018 year the greatest average annual values of the flow of solar cosmic rays which defines corresponding dose fall exactly on the years of FSA defined on the planets eccentric.

Keywords: Solar Activity, the Seismic Activity of the Earth,
Neutrino Flux Registrated in Chlorine-Argon Experiment and Sage, Anisotropy, Correlation

1. Introduction

Currently we have only experimental lower bounds for neutrinos and practically do not have significant theoretical guidance for its difference from zero [1]. However, first, without differences of mass neutrinos from zero was impossible explanation of lack in comparison with calculations based on the standard solar model solar neutrino flow has been measuring in chlorine-argon experiment [2], which for about 24 years held in the mines in Homestake, United States, as well as gallium-germanium experiment SAGE [3] neutrino Baksan

Observatory of NRI (RAS), conducted from June 1990 to 2005 years, if the mass of the electron neutrino is anywhere from 10^{-4}-10^{-8} eV. Secondly, considerable time variation found neutrino flux, which also does not conform to the standard solar model [4]. There is a negative correlation of solar activity in Wolf numbers on the number of sunspots with seismicity of the Earth and the positive correlation between seismic activity of the Earth and variations of neutrino flow [4]. Long before that, at least in [5] 3453 outbreaks points > 2.2 f It was determined the direction of anisotropy of distribution of sunspots in the field standard Apex Sun at the second equatorial coordinate system: $\alpha = 277° \pm 5°$, $\delta = +29° \pm 5°$; in the same direction is a maximum of cometary $\alpha = 271°$, etc. In [6-7] presents the results of theoretical and experimental studies of space-time anisotropy angular distribution of Galactic Gamma-radiation in frame of the generation mechanism of cosmic rays by the method of direct conversion of extragalactic gamma-rays to the current on spin shock waves. It is shown that the angular axis anisotropy of Galactic Gamma-radiation has the following coordinates: Galactic longitude $l = 96°$, right ascension $\alpha = 271°$. There is the main maximum in this direction on a chart of the average annual release of seismic energy on among the strongest earthquakes that have occurred in the period 1990-2015 years (according to the Internet site: ceme.gsras.ru).Data of earthquakes on the CEME GS RAS base averaged sliding window method on 5 points. Then was done subtraction of geo-background. Then the resulting data were divided into $14°$ sectors. Without these operations with a data source on the chart does not receive the anisotropy. Generation of spin shock waves in the Sun is unquestionable, because the magnetic field jumps recorded since the beginning of regular observations. Their detection in the Sun means the isotropy around solar space-time according to theorem Neuter: conservation of spin angular momentum as the law of dynamics of spin shock waves means the isotropy of space-time.

The main objective of this work will be to identify and define the axes of anisotropy of flux variations of solar neutrinos.

The most interesting [1] possible explanation of anticorrelations neutrino flux with solar activity, confirmed the same detector in 1990 year [1] it was suggested by M. B. Voloshin, M. I. Vysotsky and L. B. Okhun (Russian Nuclear physics, 1986, vol. 44, pp. 440, 544, 546). According to their hypothesis, the electronic neutrino with magnetic moment 10^{-11} Bohr's magneton as a result of the influence of the solar magnetic field becomes sterile and does not interact with the nucleus of chlorine. Maximum sterilization neutrino occurs during peak solar activity (1978 year – S. A. G.), when the magnetic field of the Sun is the greatest. In theory a sterile neutrino below we follow works [8-10]. In the range 1-200 GeV mass sterile neutrinos, shared with the particles of the standard model, You can be guided by their linear relationship with the masses of the chemical elements [11-12]. In particular, this limit on the mass of the neutrino in the order of 0.3 eV $\cdot c^{-2}$, that corresponds to modern top mass measurements of electron neutrinos 2.2 eV [1]. On the other hand, their detection in the planned experiments on the LH Collider [8] can give additional information about the linear law mass particle/atomic weight.

2. Mass of Neutrinos

According to GUT (see [1] p. 35) the left light neutrinos mass is determined by the ratio of

$$m = \frac{M^2}{M_X},$$

where m is the mass of spinors-quarks or leptons, and $M_X = 10^{14}$ m_p -scale masses, GUT m_p is the mass of a proton.

This gives m = 0.28 eV $\cdot c^{-2}$ on the t-quark mass 173 GeV

$\cdot c^{-2}$. On the mechanism of "see-saw" the mass of neutrinos is proportional to the square of the fermion masses of the same family. It is expected (see. [1] p. 35) that

$$m_{\nu_e} : m_{\nu_\mu} : m_{\nu_\tau} = m_e^2 : m_\mu^2 : m_\tau^2.$$

Fixing M as the mass of muon 105.66 MeV $\cdot c^{-2}$ [13] on the formula for the mass of the muon neutrinos get value 2.5874 $m_T = (5/2 + 0.035) m_T$, where $m_T = 4.70 \cdot 10^{-8}$ eV- axial torsion weight [14], freezing in heavy electrons spin density distribution. This is an integer value with a precision 3.5% complete (the law of conservation of energy in the center of mass) in the reactions of $\gamma / \nu_\mu \bar{\nu}_\mu / 5TS$ of muon neutrino pair production by photon with followed the release of the five lightest axial torsions TS. Of course, in case of a non-zero neutrino pulse more likely could create from one torsion.

Fixing M as the tauon mass 1777 GeV $\cdot c^{-2}$ [13], is similar to the formula for the mass of tauon neutrinos we get the value of 0.000301 eV $\cdot c^{-2}$. The estimated mass M of the tau neutrino is the same as the largest with heat energy 3 /2 T of relic neutrinos with a spread of 14%. In the standard model T around 2K. However, considering the temperature of the relic neutrinos as the effective gravitational temperature $T_\Gamma = 2/9$ M for some M, mass of m = 3/2 T_Γ ratio turns out M = 3 m, corresponding reactions of the dissolution of the relic neutrinos by three sterile active Tau-neutrinos [8]. On the theory [8] in this energy range can only be one sterile neutrinos. A life time of sterile neutrinos, such a mass of not less than the lifetime of the proton. On the other hand this value [1] evaluation of the upper limit of the mass of the tau neutrino 10 MeV reaches 13 billion years. When considering the effective gravitational temperature commonly used generalized principle of equivalence: the system of graviting fermions, which fit this definition is equivalent to the quantum state oscillator in the coordinate system, moving with some acceleration. When the effective temperature of the oscillator is proportional to acceleration. When the effective temperature of the oscillator is proportional to acceleration. Thus, we see that mass of relic neutrinos, tau-sterile neutrinos generated as dark matter [8-10], in some ways generated by accelerating expanding universe. More accurate staging of this connection is given in [14].

To conduct the experiment on finding heavy axial torsion with weight 3 TeV at the LH Collider of interest reaction $\gamma / Z / \bar{N} N / TS$. From kinematics, it follows that the mass of the sterile neutrino N must be in the order of 1.5 Tev. Conversely, when the mass of sterile neutrinos at least 1.5 Tev mass heavy axial torsion shall be not less than 3 TeV. Therefore, accounting of loops of sterile neutrinos can lead to substantial growth of the axial torsion weight.

3. Correlation of Neutrino Fluxes from Seismicity of the Earth

In the experiments (see [15] reference [5]), conducted in 2009-2012 years simultaneously in Moscow at the Institute of terrestrial magnetism, ionosphere and radio propagation RAS (TMIRAS) and at the point of integrated observations of the Kamchatka branch of geophysical survey of RAS near Petropavlovsk-Kamchatsky were registered neutron flows associated with earthquake with a magnitude M 8.8 in Chile February 27, 2010 year, volcanic eruption in Iceland, March-April 2010 years, an earthquake with a magnitude M 9 March 11, 2011 in Japan, the earthquake with a magnitude M 8.6 in Indonesia April 11, 2012. On complex for registration of neutrons used gas-discharge counters filled with gas helium-3.These devices detect thermal neutrons, the energy which is 0.025 eV and fast neutrons with energy greater than 2 MeV. In these events increase the particle flows began to occur a few months before these events. When this happened as a continuous monotonic increase in the flow of particles and individual short-term increase with amplitude on the minute data from several thousand, hundreds of thousands of per cent. The duration of these spikes ranging from several minutes to several hours [15]. These and other data displayed in Figure 1.

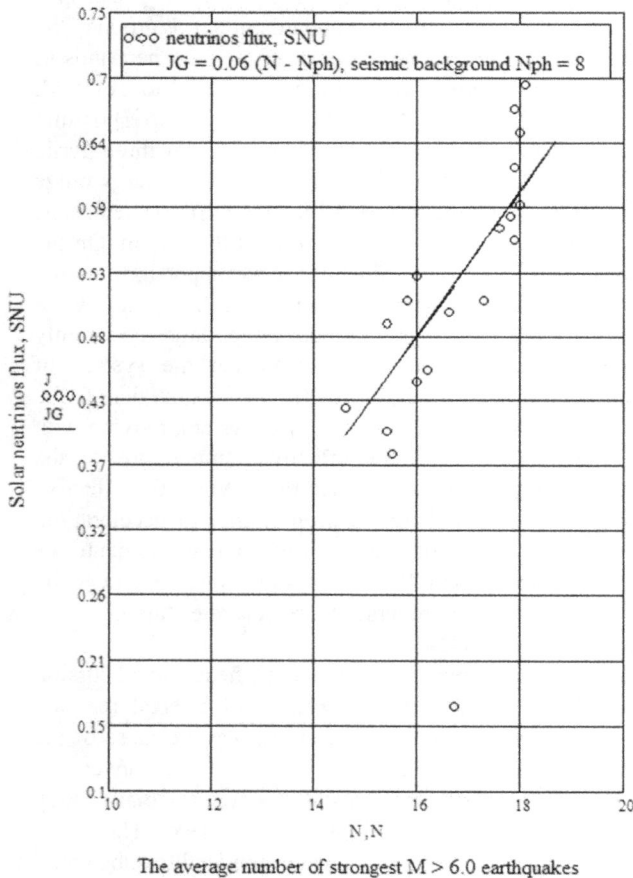

Figure 1. The flow of solar neutrinos for the 1970-1994 year s depending on the release of seismic energy on average among the strongest root from 400 km of earthquakes ([15], fig. 3, p. 8). Point at the bottom is dated 1979 year of solar maximum, which fell on the frontier of solar activity (FSA) 1978 year [16]. Neutrino highs everywhere for 45 years of observations fall on the dates of the FSA. Here, at the time of solar maximum, the negative correlation is replaced by a positive. Therefore, this point falls from JG linear dependence graph.

Shown in Fig. 1 data is approximated by a linear dependence between the flow of solar neutrinos and the

release of seismic energy on the average annual number of the strongest deep from 400 km earthquakes [15].Point at the bottom is dated by the 1979 year of the Frontier of solar activity (FSA) 1978 [16]. Neutrino maximums mainly occur in periods of recession of solar activity (see, e.g., [16]), as well as 45 years observing everywhere for fall on dates FSA. Here, however, at the time of solar maximum, the negative correlation is replaced by a positive. Therefore, this point falls out of graphics JG of the linear dependence. As a result, after removal of this sample point correlation coefficient of the neutrino fluxes from seismicity of the Earth grows dramatically from 73% to 95%.Standard deviation of data from linear dependence JG 0.03 ± 0.05.Relative variation is within 7%.

These data demonstrate a deep solar-terrestrial relationships. In the author's treatment by the method of [17, 18] stochastic seismic processes occur with 22.3 years period the solar cycle and inside the loop reaches a peak dates FSA, caused by the orbital motion of the Sun due to the planetary eccentric. Telling in this regard, parade of planets from the date FSA 1986 etc., when all the planets line up so that the center of mass of the solar system (CMSS) goes beyond the solar disk. The interval between two FSA inside the solar-terrestrial 22-year cycle for 1000 years is an average of 8 years. The offset maximum dates the strongest earthquakes regarding the dates for FSA 200 years does not exceed 2 days. On the other hand CMSS output beyond the solar disk on the date the FSA1978, 1986, 1993-4, 2002 and may 2018 years. In theory [14, 15] at the time of the achievement of the FSA has an imbalance of the gravyelectroweak protective potential, which is proportional to Newtonian gravitation potential, inside (220 MeV) and out of the Sun. This happens, as shown above, mostly away from solar maximum to the number of Wolf, so the formation of sunspots and the output of the neutrinos are going mainly in antiphase.So perhaps sunspots processes generally stem from some other, more energy processes.

Experiments on the LH Collider reached a threshold energy 3 TeV differences between gravitational and inertial masses of GUT scales. Physical effects of this estimated range [14] on the protons from the Sun at the scale of the solar spots size of 20-50th. km 700 km deep photosphere is sufficient to ensure the increased solar activity for billions of years.

Anyway, identified correlations neutrino fluxes from deep earthquakes approach [15] can be generalized to the entire seismic land. The results are shown below in Fig. 2.

Shown in Fig. 2 data as well as in Fig. 1 are approximated by the linear dependence between the flow of solar neutrinos and the release of seismic energy on average among the strongest earthquakes m > 6.6 according to the earthquake catalog GS RAS CEME [19] for 1990-2008 years in averaged over five points. The standard deviation of the data from the linear dependence JG on neutrino- 0.06 ± 9.8 SNU, among the strongest earthquakes 0.06 ± 2.8.Relative variance for neutrinos is within 10%. This is somewhat more than for deep earthquakes, but takes into account all the points.

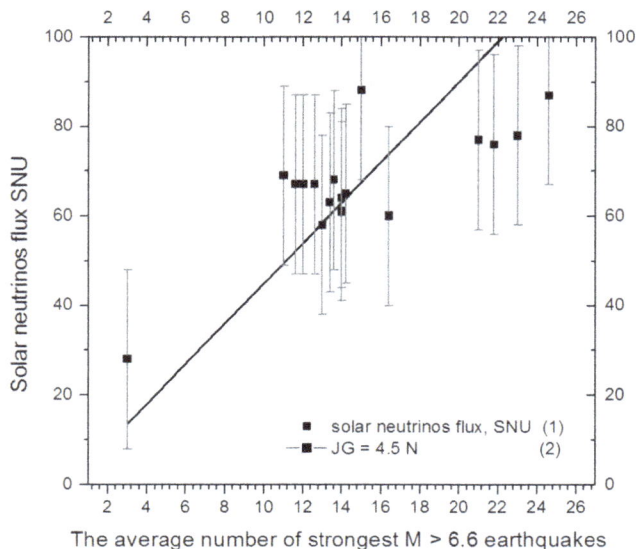

Figure 2. The flow of solar neutrinos for 1990-2008 years according to ([3], fig. 2, p. 4) averaged over three points depending on the release of seismic energy on average among the strongest M > 6.6 earthquakes catalogue CEME GS RAS [19] averaged over five points.

4. Correlation of Solar Activity on the Annual Course of the Earthquakes

Frontiers of solar activity above under paragraph 2 determined by the time the center of mass of the solar system beyond the solar disk (according to the results of the calculations of the main Astronomical Observatory RAS officer G. Ya. Vasilyeva, A. A. Shpital'naya and others, the movement of the Sun relative to the center of mass of the solar system (see, e.g., [20]). The period between the two closest Frontiers is within 22 year solar cycle, with maximums of solar activity may not match, very variable and the average is about 8 years old. Near the Frontiers of the solar system as a whole is destabilized (it is a correction of its Galactic orbit) -disturbed balance between the protective gravielectroweak energy potential (PGEWP) [16] of the center of mass of the solar system and Galactic Centre on the one hand, and synergistic gravitational partially screened potential Universe [14, 16]. This is accompanied by powerful emissions of substances and radiation of the Sun, as well as bursts of seismic activity of the Earth, the following followed by transition through Frontiers delayed by 2 days on average for the period of regular observations from beginning of the 16th century.

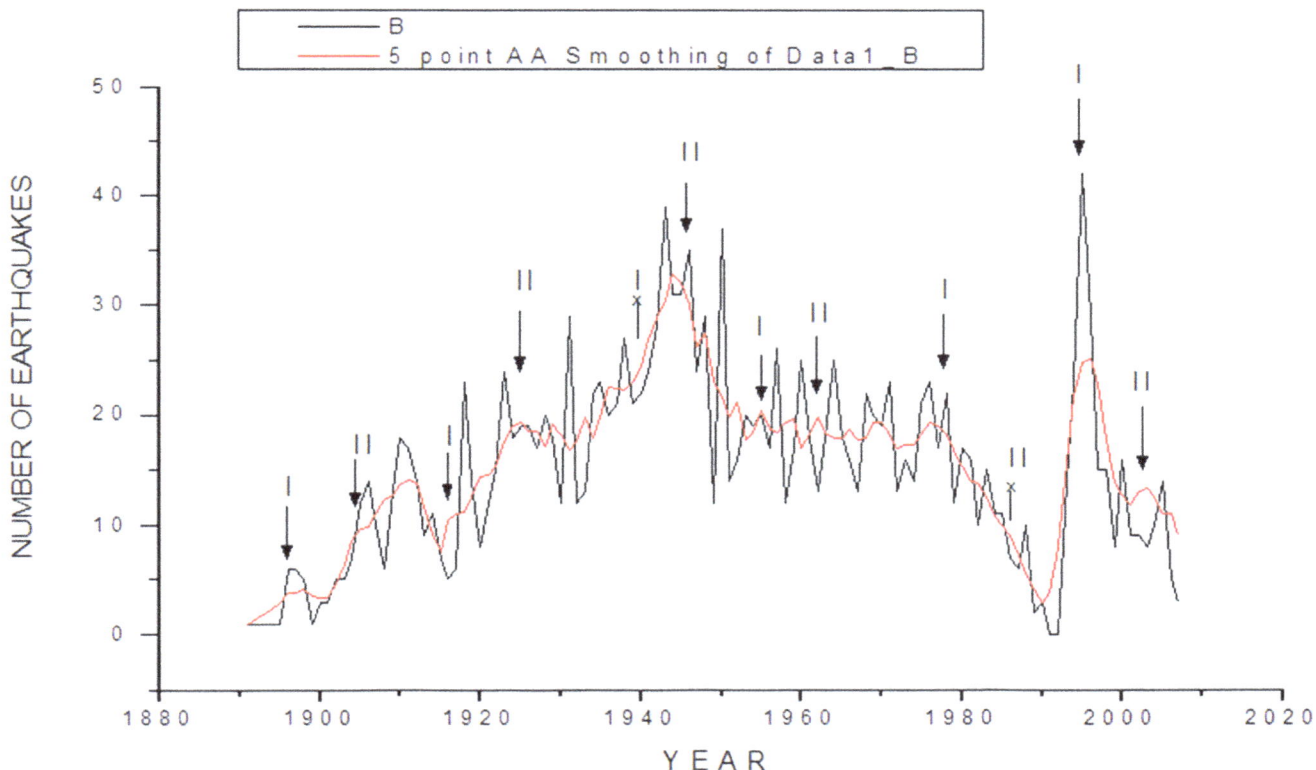

Figure 3. The annual course of the strongest earthquakes on the frontiers of solar activity (FSA). The date of the first (I) and second (II) FSA marked by arrows. Shows the smoothed curve by the sliding window method with averaging on 5 points (the smoothed curve by 5 points).Two frontiers 1940 and 1986, inconspicuous amid rapid growth/decline of seismicity, indicated by an arrow with the icon "x".

Here is a comparison of solar output at Frontiers of solar activity (FSA) in Wolf numbers on the angular coordinate α directional effects, registered in the ground experiment [5] (see, e.g., [7], fig. 3, pie charts semitones), with the annual course of the strongest earthquakes. The difference is that everything happens in the dense solar plasma order the water density. Blast on the Sun 1946 year [5], took place in a context where directed impact clamping the solar fiber length about solar diameter to the surface of the Sun, on model calculations occurred a powerful plasma response in the opposite direction to action of the clamping force so, what's with the delay because of the rotation through the floor of the turn over that led to the explosive release of the solar flash. Locally, in the face of advances in friction electrization breed layers it is possible to generate acoustic waves through electro acoustic effect or piezoelectric effect anisotropy (underground lightning).In the crystal lattice in addition to acoustic sound branch phonons excited and optical branch. Output optical phonons is similar to the output of the visible sunlight. Therefore, in the context of the planetary impact of possible mapping between the release of seismic energy and output of solar radiation in the days of Wolf numbers. In the model, similar to the mechanism of the formation of Galactic Gamma-radiation [7]: directional effects on the Sun accompanied by the development of the opposite directional response of the graviting solar plasma, experiencing as a whole moving in the speed of the Sun direction around the center of mass of the solar system (SSCM) [20]. The influence of the planets is a response to this motion of the Sun. Following this the angular coordinate (number of weeks) initiates impact rebounding in, opposite Sun speed according to the date of the FSA.

With an asterisk in the table 1 included corners defined by the date of the FSA for the possible cumulative effect (at an angle in the transverse direction). As you can see, made assumption leads to good results.

From Fig. 4 shows that the frontiers of solar activity (SAR) average annual course of earthquakes with 7% spread follows the release of solar energy in Wolf numbers as on fig. 1.

5. Correlation on the Output of the Neutrino

Table 1. *Comparative analysis of solar energy output on Frontiers of solar activity (FSA) and seismic energy on the annual course of the strongest earthquakes.*

Date Of The FSA	No. a week	Solar activity W/155	The annual course of the strongest earthquakes with geo-background subtraction, N – 19	Deviation 37W/155 – (N-19)
1962	13	0.48	17	1
1986	19	0.12	0	4
1956*	28	0.69	23	2.5
1946	37	0.54	18	2
1986*	39	0.12	3	1
1962*	51	0.42	10	5.5
1978	33/34	0.66	30/15	-6
1986	14	0.12	1	3
1978*	21	0.69	26	-0.5
1994	2	0.23	8	0.5
		The standard deviation for the number of Wolf ± 13		1.4 ± 3.1

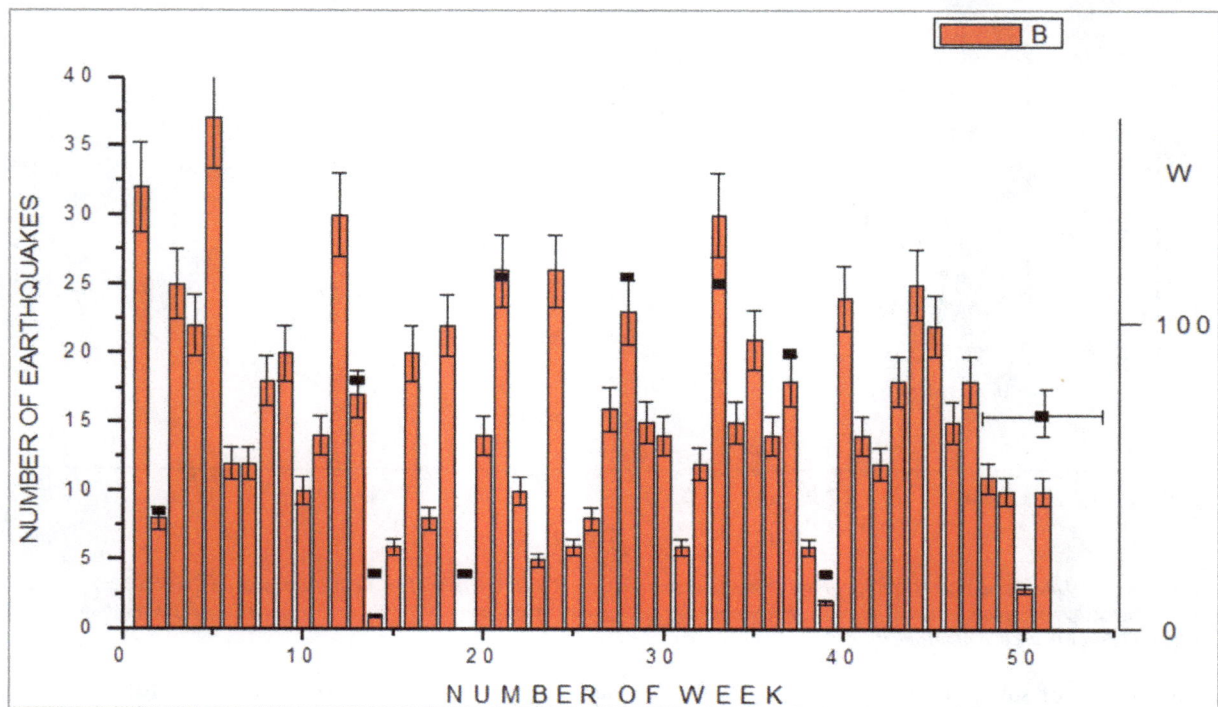

Figure 4. *The output of solar activity in Wolf numbers W at dates FSA table. 1 (squares) and on average go for the 1730 strongest earthquakes of M> 6.6 minus geo-background19 /week.*

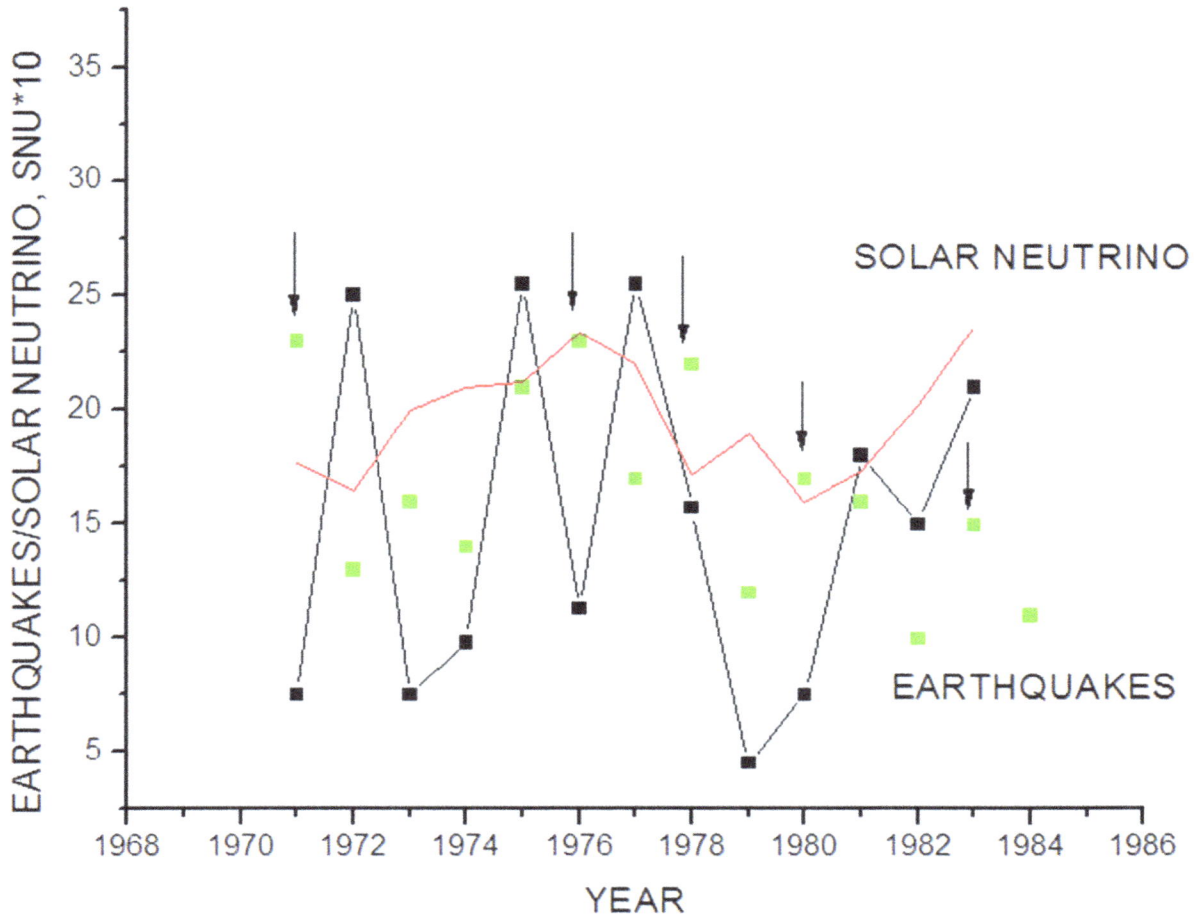

Figure 5. The average annual flux of solar neutrinos the Sun Neutrino Units (SNU) at a scale of 10:1 (solid curve). Shows the smoothed curve by 5 points. The point is the number of the strongest earthquakes per year.

In Fig. 5 shows the dependence of the average annual flow of solar neutrinos ([1], Fig. 5.22, p. 186).The average value of the neutrino flow is 2 solar neutrino units that 3 times smaller than predicted by the standard solar model. Squares points shows the progress of the strongest earthquakes on the A. A. Shpital'naya database, GAO RAS, between frontiers of solar activity 1978 and 1986 years. You can see that the earthquakes repeat the neutrino oscillation. Neutrino minimum 1979 year - precise measurement errors chlorine detector on the release speed of argon -37. This shows the sensitivity of the release of neutrinos by the time the center of mass of the solar system (CMSS) outside of the solar disk on the frontier of the solar activity. Just as dramatically in 1979 year falls seismic activity.

The most interesting [1] possible explanation of neutrino flow anticorrelation with solar activity, confirmed the same detector in 1990 year [1], was suggested that M. B. Voloshin, M. I. Vysotsky and L. B. Okhun (Russian Nuclear Physics, 1986, vol. 44, pp. 440, 544, 546).According to their hypothesis, the electronic neutrino with magnetic moment 10^{-11} Bohr's magneton as a result of the influence of the solar magnetic field becomes sterile and does not interact with the nucleus of chlorine. Maximum sterilization neutrino occurs

during peak solar activity (1978 year– S. A. G.) when the magnetic field of the Sun is the greatest.

6. Time Variation of Solar Neutrino Flux in Chlorine-Argon Experiment [2]

Constant component $J_\nu = 2.56$ SNU = 0.51 ^{37}Ar atom/day will be regarded as a constant background which will be deducted from the signal detector to determine the unknown factor of temporal variations of the flux of neutrinos. In our analysis of such deductions will be made for each individual measurement cycle. The results of these operations will be presented in the form of pie charts in a sequential scan of the coordinate α of the Earth the second equatorial coordinate system, the recalculated according to the date of the middle period, each individual cycle measurements. Angle 35° sector and more defined the temporary duration of the measurement. Value of the radius of the sector-its length is equal to the measured magnitude of neutrino flow minus constant component 0.51 ^{37}Ar atom/day.

1)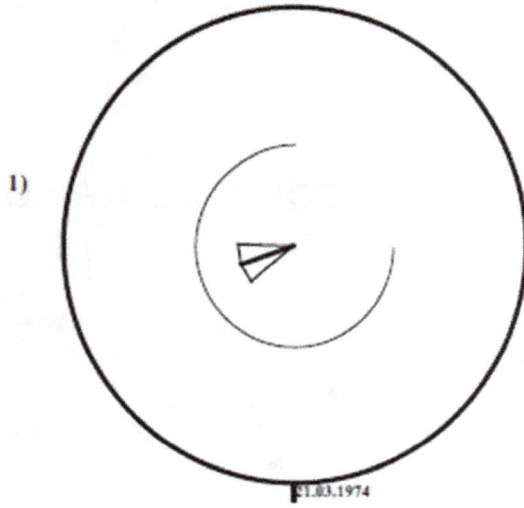

21.03.1974

2)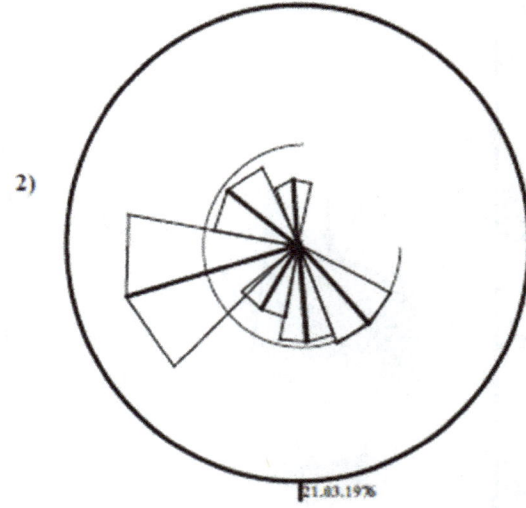

21.03.1976

3)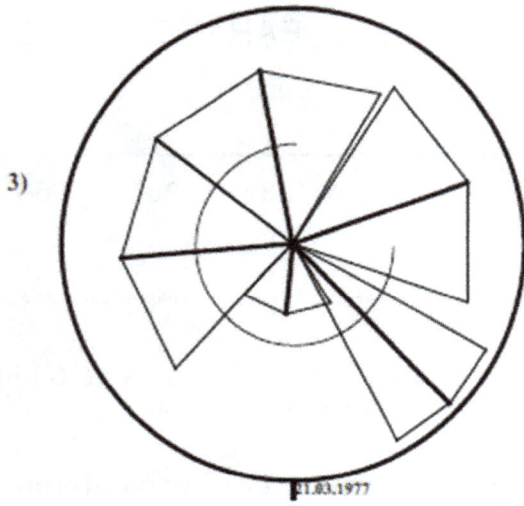

21.03.1977

4)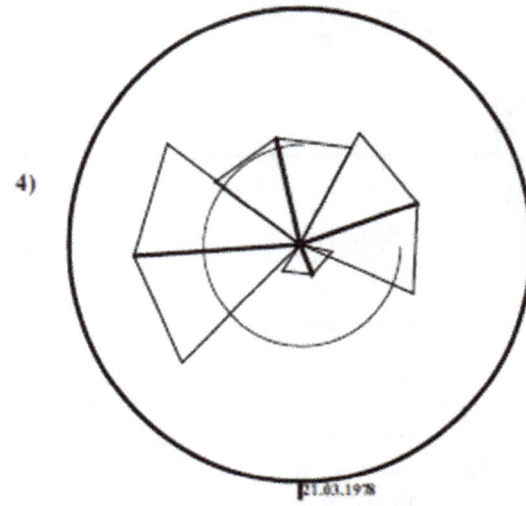

21.03.1978

5)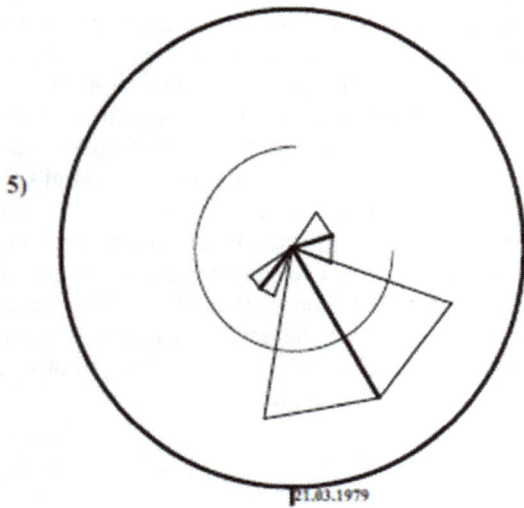

21.03.1979

6)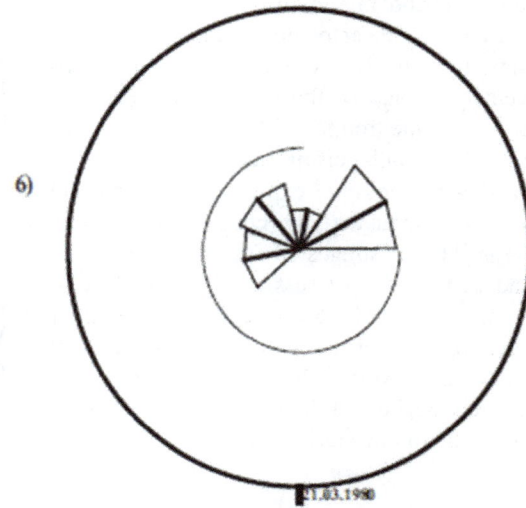

21.03.1980

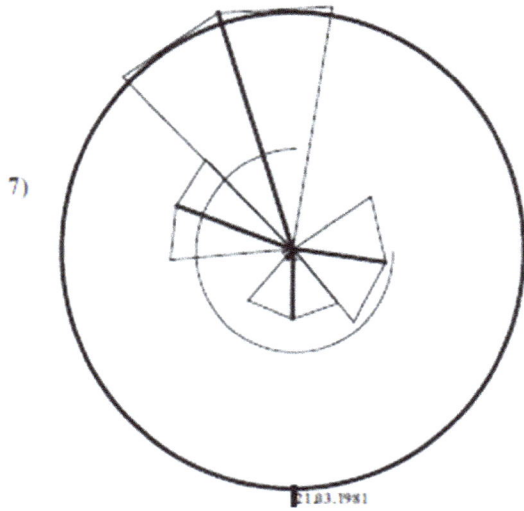

7)

21.03.1981

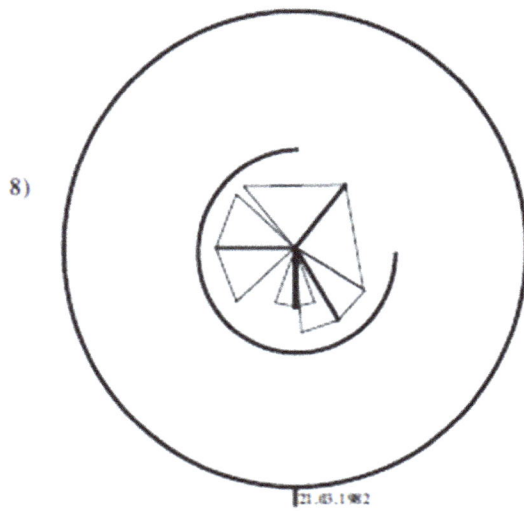

8)

21.03.1982

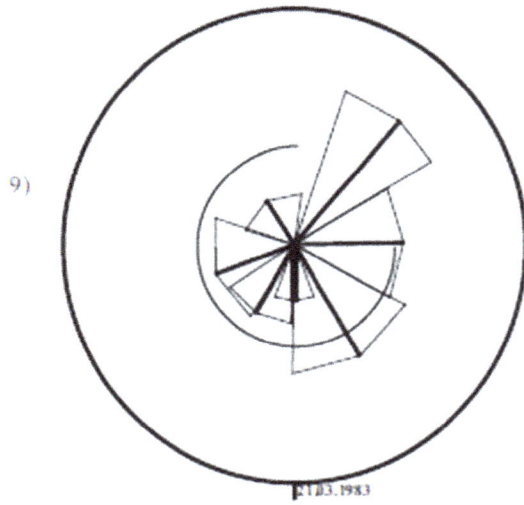

9)

21.03.1983

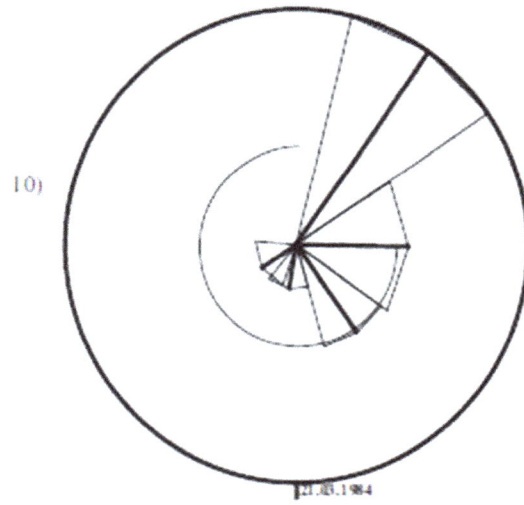

10)

21.03.1984

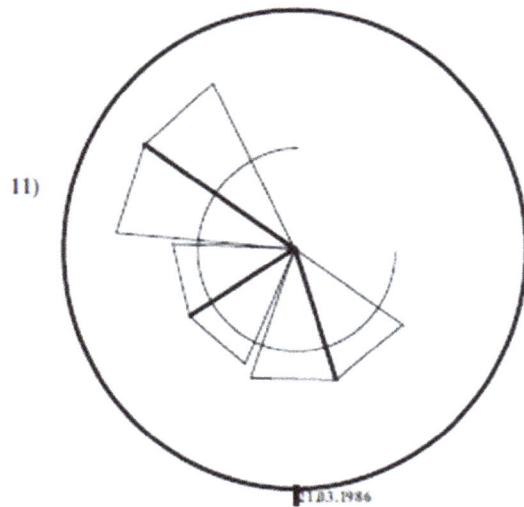

11)

21.03.1986

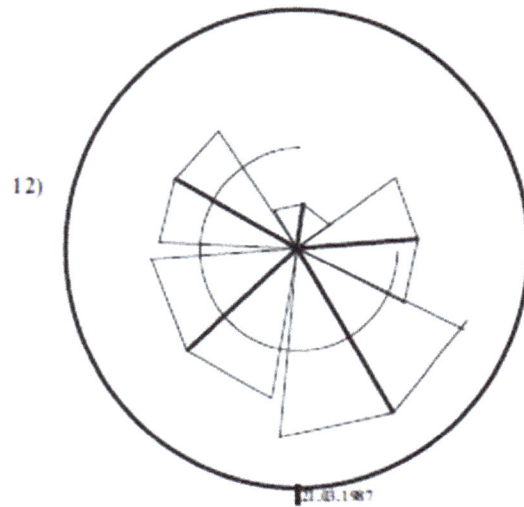

12)

21.03.1987

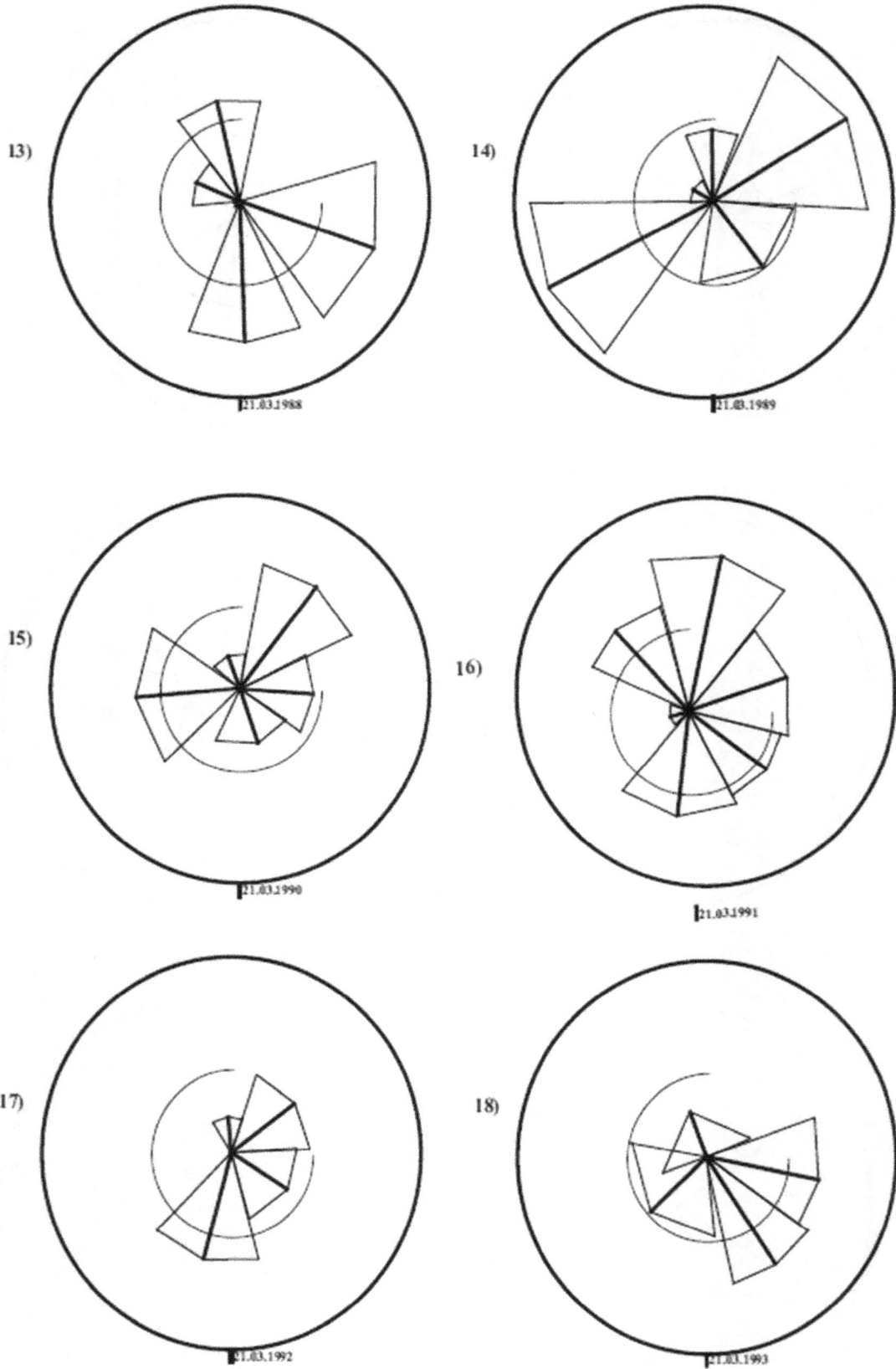

Figure 6. *Pie charts flow variations of solar neutrinos for the period 1990-1994 years. Constant component of 0.51 ^{37}Ar Atom/day marked by an open circle.*

1)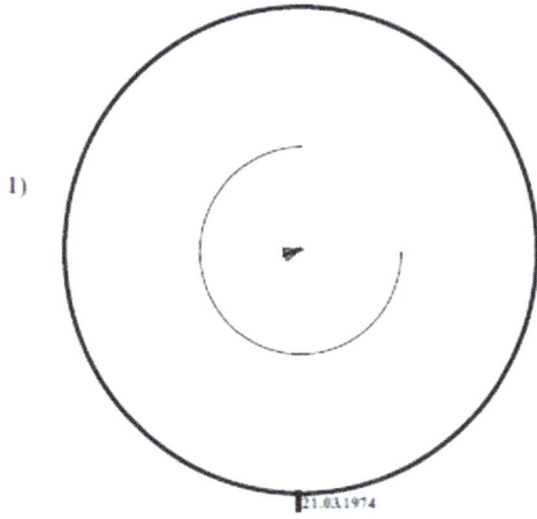

21.03.1974

2)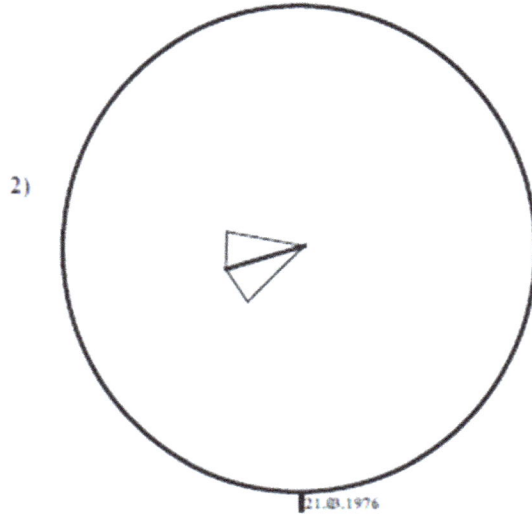

21.03.1976

3)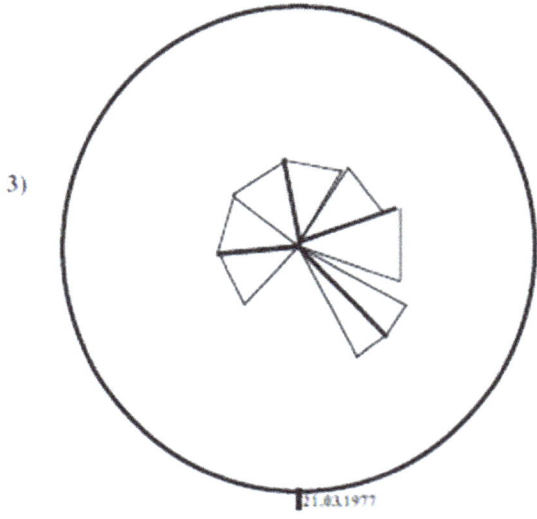

21.03.1977

4)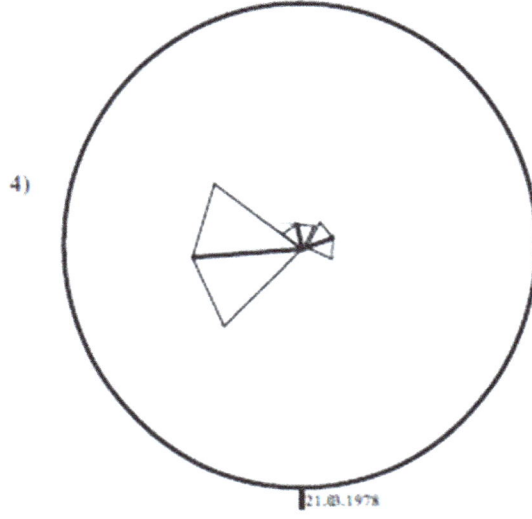

21.03.1978

5)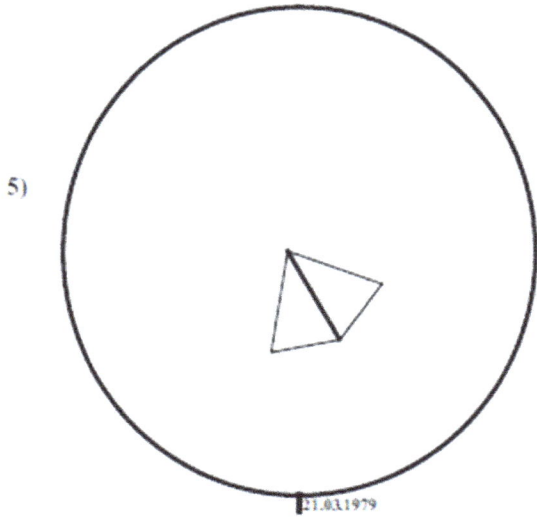

21.03.1979

6)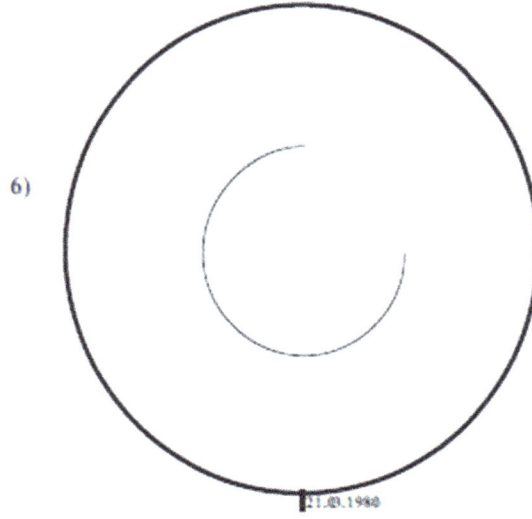

21.03.1980

7)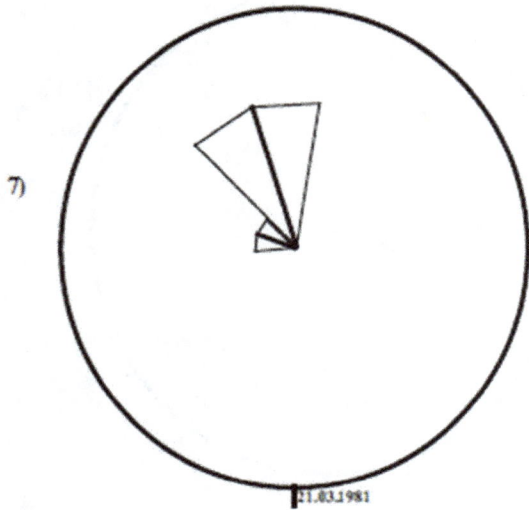

8)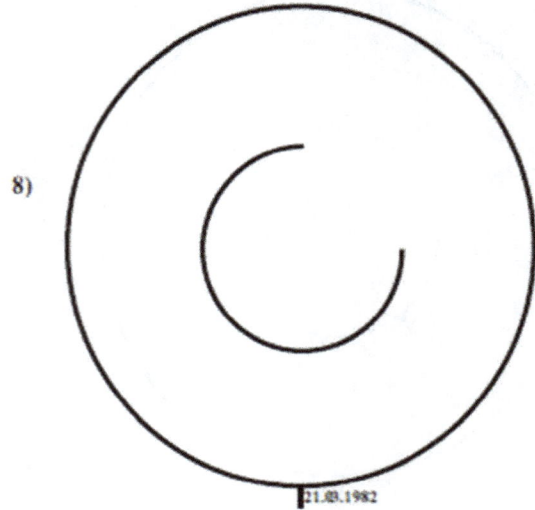

9)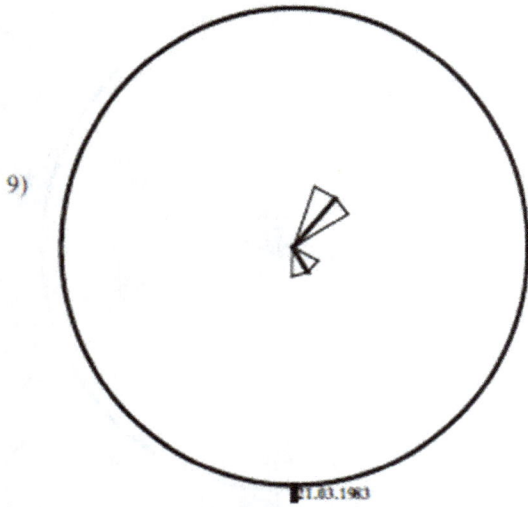

10)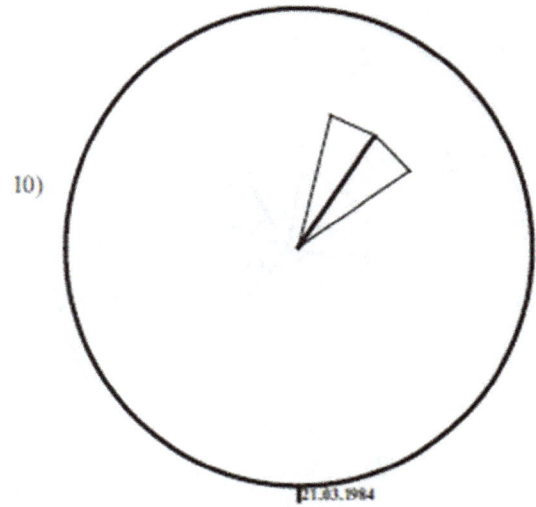

11)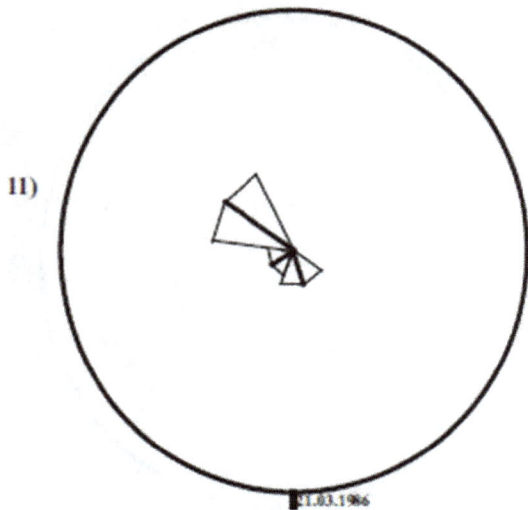

12)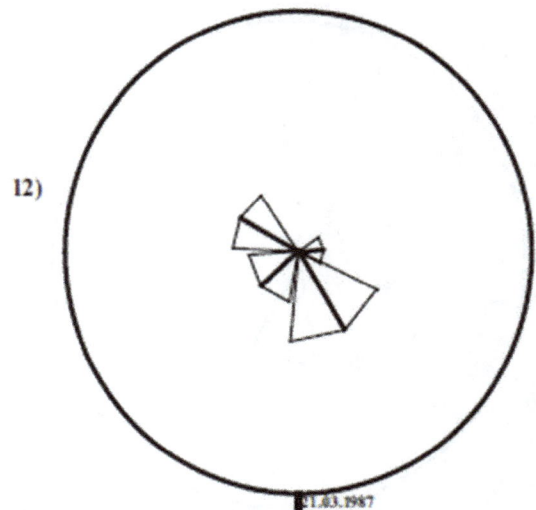

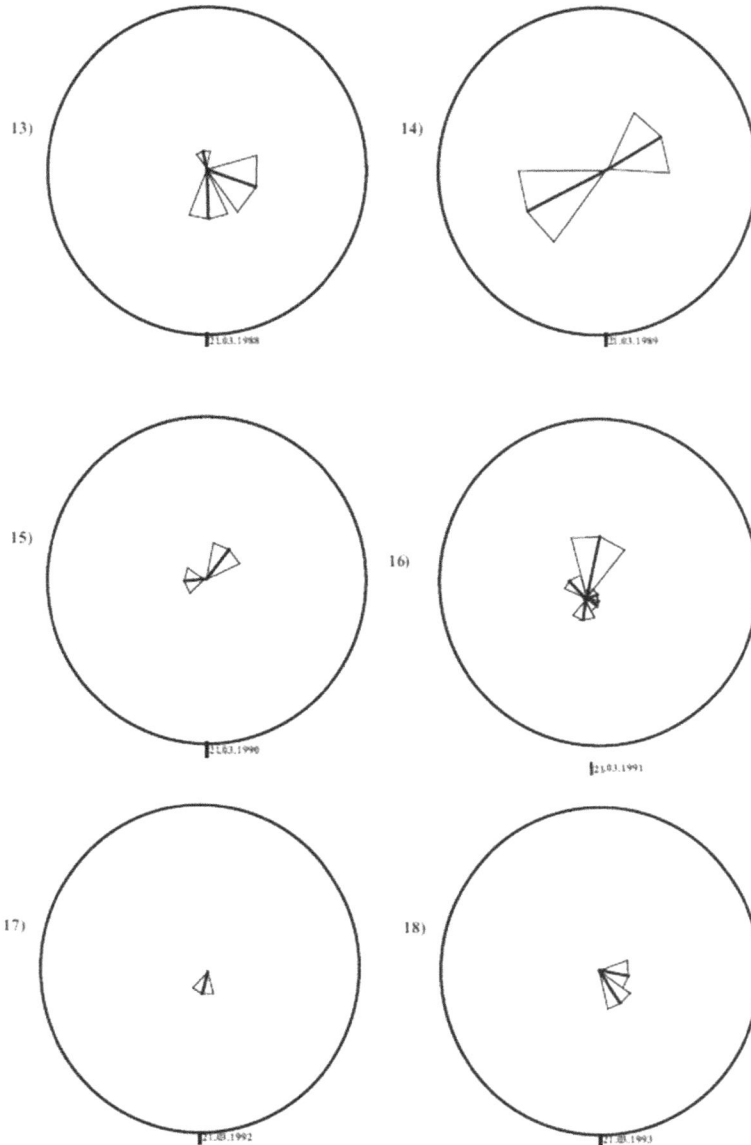

Figure 7. *Pie charts of solar neutrino flux variations during the period 1990-1994 years, minus the constant component 0.51 ^{37}Ar atom/day (open circle).*

It should b noted that although and notes the quasi two-year cycle of variations of the solar neutrino flux [21], we will not produce averaging of biennium, won't do it is because of the informative sharp variations. Separately for each characteristic angle α (310°, 271°, 210° and ± 20°) estimated average value of scan period measurements. It turned out that just such sectors 11-13. The average period of solar neutrino flux measurements amounted to 73° ± 6°, where the value of 6°, scatter to the boundaries of the intervals specified split duration 33, 45, 60 days, etc. small enough to consider all angles except pie the biggest order 100 ° (but all these great angles are also included in the review).

The resulting 14 pie charts (from 18) see fig. 6 and 7 are characterized by sharp anisotropy for 4 selected sector directions (centered sector); 4 sector chart after deduction of the constant component of the Chlorine-Argon experiment attributed to noise on the anisotropy. 2 of these 4 are far from

FSA 1978, 1986 and 1994.Hence, outside dates FSA therearethe predominantly an isotropic neutrino flux and the whole effect of anisotropy from the remaining 14 charts focus on the dates of the FSA. This conclusion is reinforced by the similar diagram anisotropy strongest earthquakes 1990-2015, with also local maxima distribution release of seismic energy Fig. 3 on the average annual number of earthquakes, exactly localized on dates Frontiers of solar activity.

Time variation of the flow of solar neutrinos is not covered in the standard solar model. On the other hand here, we can see that these are the same effects of spatial anisotropy, characterized by the same angle α = 271° identified on Galactic Gamma-radiation [6-7], on the chart ground experiment [22] of anisotropy detection cosmological vector A$_G$, the same effect which is described in the model of a quantum oscillator on the string [14] within the extended standard model, taking into account the gravity with a nonzero torsion. Thus, the data flow time variation of solar neutrinos could well be included in the synthesis of the standard Solar model. You can take it that the selected axis of anisotropy in accordance with [6-7] set on the isotropy of space and time.

7. Time Variation of the Flow of Solar Neutrino Experiment SAGE [3]

A permanent component of the neutrino flow 81.4 SNU will consider as a constant background, which will be deducted from the signal detector to determine the unknown factor of temporal variations of the flux of neutrinos. In our analysis of such deductions will be made for each individual measurement cycle. The results of these operations will be presented in the form of pie charts in a sequential scan of the coordinate α of the Earth the second equatorial coordinate system, the recalculated according to the date of the middle period, each individual cycle measurements. Angle 28° sector

and more defined the temporary duration of the measurement. Value of the radius of the sector-its length is equal to the measured magnitude of neutrino flow minus constant component 81.4 SNU. It should be noted that although and notes the quasi two-year cycle of variations of the solar neutrino flux [21], we will not produce averaging of biennium, won't do it is because of the informative sharp variations.

The resulting pie charts (from 18) see fig. 8 and 9 are characterized by sharp anisotropy for 4 selected sector directions (centered sector). Just as in chlorine-argon experiment outside dates FSA there are the predominantly an isotropic neutrino flux and the whole effect of anisotropy focus on the dates of the FSA. This conclusion is reinforced by the similar diagram anisotropy strongest earthquakes 1990-2015, with also local maxima distribution release of seismic energy Fig. 3 on the average annual number of earthquakes, exactly localized on dates Frontiers of solar activity.

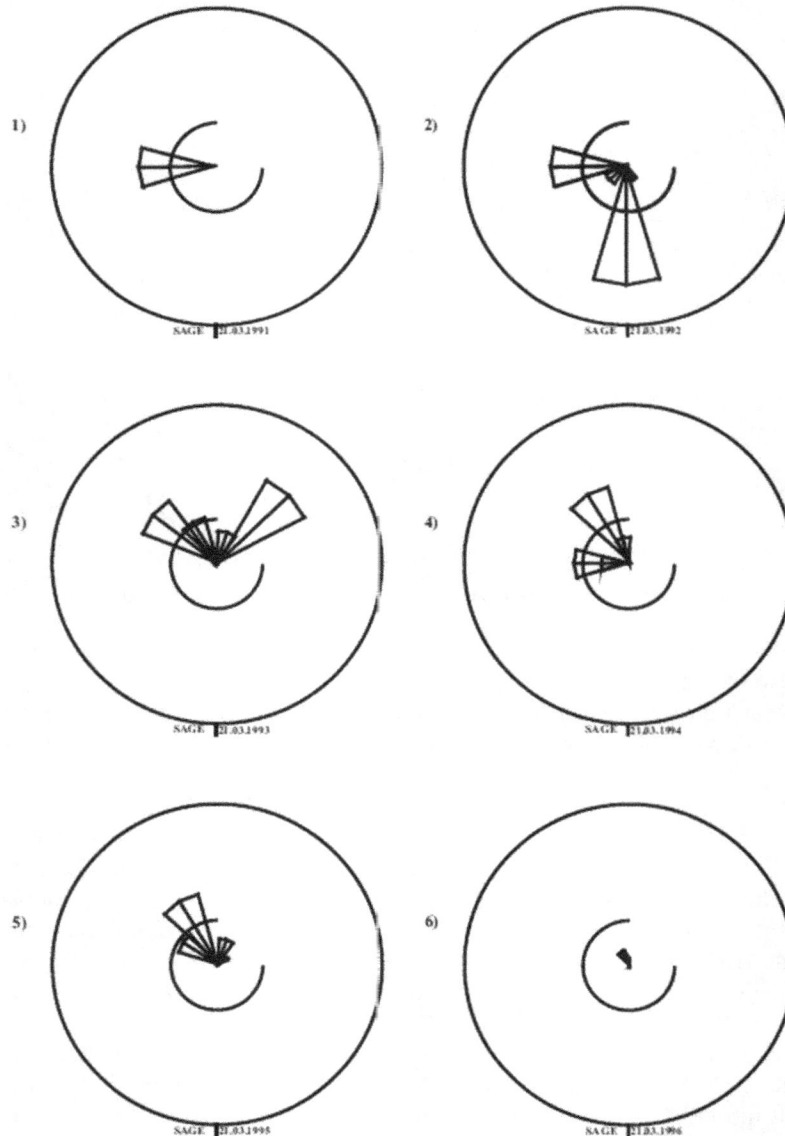

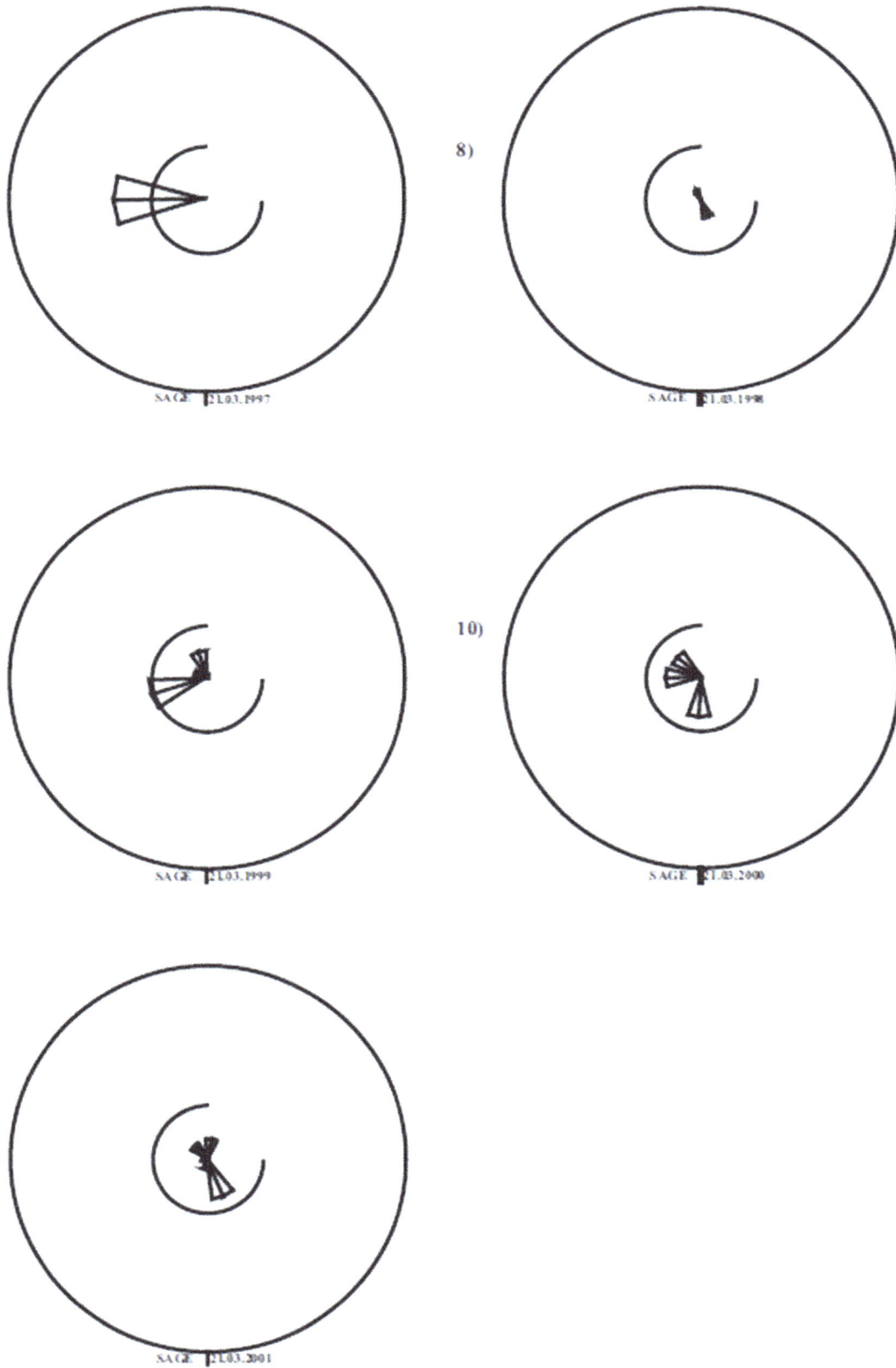

Figure 8. *Pie charts flow variations of solar neutrino experiment SAGE for the period 1990-2005 years minus constant component 81.4 SNU (open circle).*

1)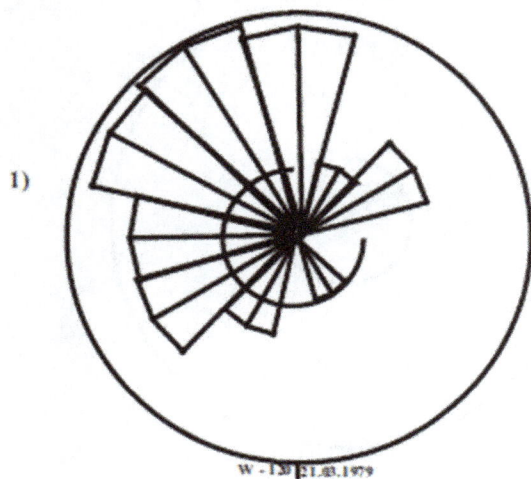
W - 120 21.03.1979

2)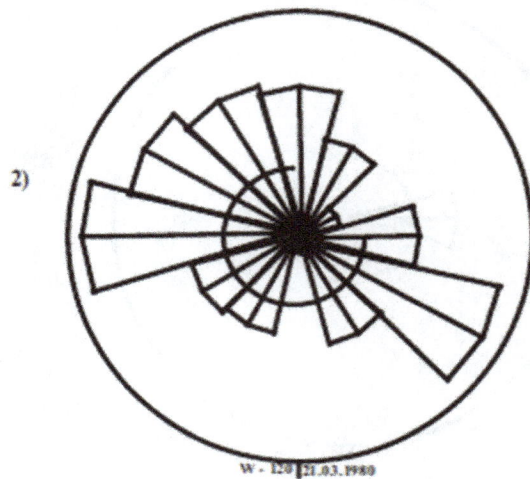
W - 120 21.03.1980

3)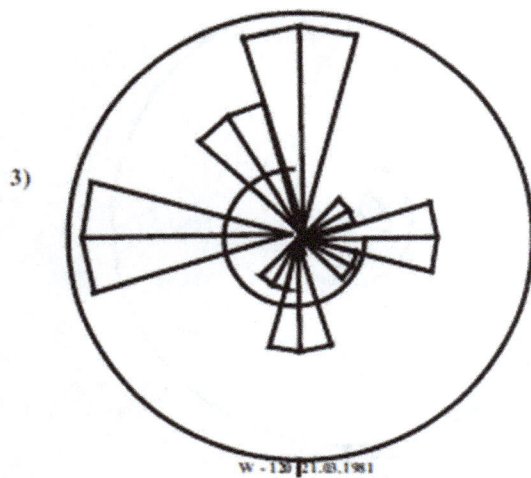
W - 120 21.03.1981

4)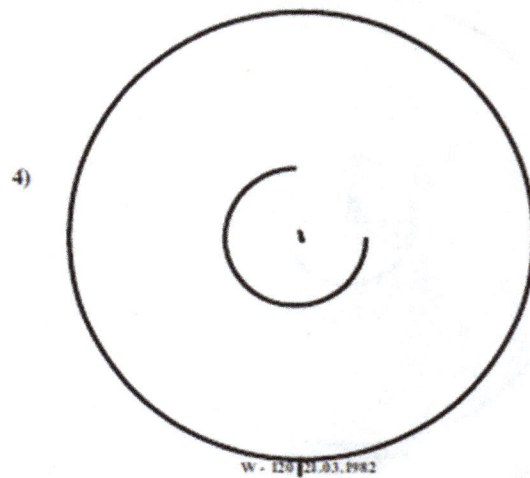
W - 120 21.03.1982

5)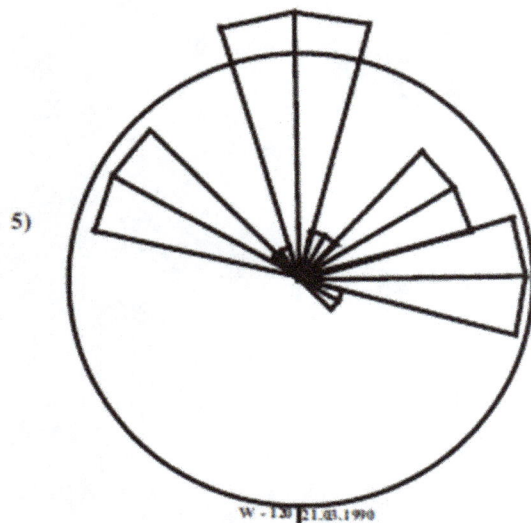
W - 120 21.03.1990

6)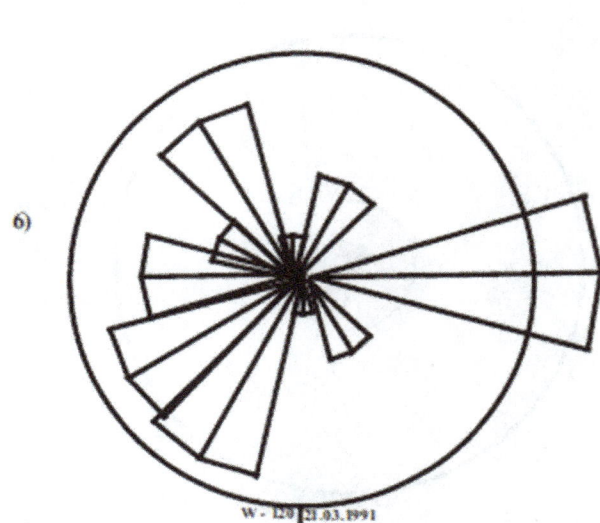
W - 120 21.03.1991

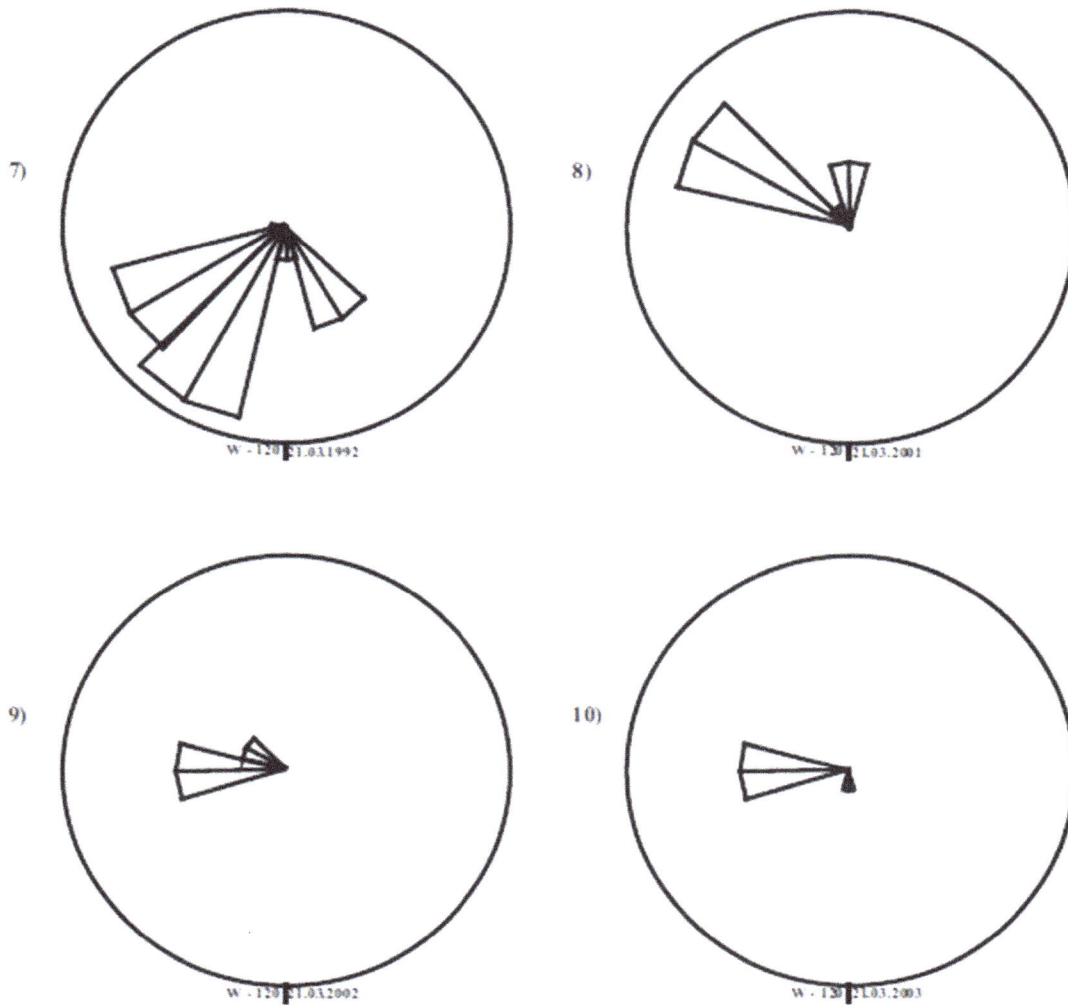

Figure 9. Pie charts the variations in solar activity for the period 1979-2003 years.

For comparison the fig. 9 pie charts are temporal variations in solar activity [3] by number Wolf W during the period 1970-2005 years.

8. Discussion of Results

As a result of processing rice pie charts in Fig. 7, measuring in Chlorine-Argon Homestake experiment [2] and time variation of the solar neutrino flux were distributed on the following directions:

series A of 12 diagram units-
$\alpha_A = 271° \pm 8°, \delta = 23°$;
series B of 16 diagram units-
$\alpha_B = 310° \pm 22°, \delta = 23°$;
series C of 5 diagram units-
$A_C = 208° \pm 7°, \delta = 23°$;
series D of 7 diagram units-
$A_D = 21° \pm 23°, \delta = 23°$.

All these areas are within the dispersion by the time of the cycle experiments 17°-48°.

Pie directions of temporary variations of solar neutrino flux,

dedicated to Fig. 8 in SAGE experiment [3]:
series A
$\alpha_A = 273° \pm 12°, \delta = 23°$;
series B
$\alpha_B = 312° \pm 15°, \delta = 23°$;
series C
$A_C = 210° \pm 0°, \delta = 23°$;
series E
$A_E = 232° \pm 7.5°, \delta = 23°$.

All these areas are also within the dispersion by the time of the cycle experiments 15°. In addition, you can take it confirmed overlapping the main directions of anisotropy of 271° and 310° in limits permissible under 2σ.

Pie charts of temporal variations of the solar neutrino flux and the seismic energy of the Earth in Fig. 3 discover the apparent similarity in periods of FSA 1978 the EARTH and Neutrino 1977 Fig. 7; 1979 the EARTH and 1985-1986 the EARTH and 1986 Fig. 7; 1993 the EARTH and 1994 Fig. 8; 1995 the EARTH and 1995 Fig. 8; 1996 the EARTH and the 1996-97 Fig. 8; 1999 the EARTH and 1999 Fig. 8; 2000 the EARTH and 2000 Fig. 8, etc. Interestingly, in some cases, the neutrino pie chart repeats the pie chart release seismic energy of the Earth only through 1 year, and a year earlier-you can only see the beginning in neutrino 2-year cycle. It cannot serve as a justification for those solar neutrinos could be the source of earthquakes. Maybe that own time of neutrinos on strongly Non equilibrium closed torsion string, moving at a speed very little-10^{-13} [14] (relativistic factor 10^{13}) is different from the speed of light in vacuum, in the strongest grade of slowly- in 10^{13}so, the entire process of generating neutrinos for 1 microseconds, the entire process of generating neutrinos for 1 microsecond in the laboratory coordinate system (on Earth) will last exactly 1 year. All of these processes in the unity of the solar-terrestrial relationship [15] are taking place simultaneously.

In the second Equatorial coordinate system the plane of the ecliptic, to which the orbital velocity vector belongs everywhere land, has an angle $\delta = 23$, ° 5. Declination δ radiation from the Sun during the year varies from 23, ° 5 in late December to -23, ° 5 in late June, taking the value zero during the vernal and autumnal Equinox. Summary on the decline the axes of anisotropy of flux of solar neutrinos defined in section 7, are listed below in the table 2.

Table 2. Coordinates α, δ the axes of anisotropy of temporal variations solar neutrino flow minus the constant component 2.57 SNU in Davis 1970-1994 defined in section 7, in the second Equatorial coordinate system in comparison with coordinates of spatial earthquakes anisotropy axes over the 1990-2015 on chart [7], Fig. 3 and ground experiment [22].

No	Solar neutrinos, 1970-1994		Ground experiment [22]		Earthquake 1990-2015
	α	δ	α	δ	α
1	271° ± 8°	23,°5	293° ± 20°	36° ± 10°	250°-263° 278°-293°
			Early experiments		
			283°	26°	
			293°	36°	
			303°	46°	
2	310° ± 22°	< 23°	310° ± 20°	40	307° -322°
3	21° ± 23°	~0°	0° -20°		0° -15°
4	210° ± 7°	0°	200° ± 10°		210° -225°
5	232° ± 8°	12°	230° -240°		225° -240°
6	162° ± 25°	-12°	170° -190°		150° -165°

9. Correlation of Solar Cosmic Rays from Seismicity of the Earth

Similar data for solar variations may be obtained by variations in dose RAD detector, measured inside the spacecraft during flight to Mars [23].RAD started working 9.12.2011-14.07.2012.The measured dose averaged 458-461 μGy/day. Dose from solar Proton events-23-25.01, 7-17.03 at 65-70 day flight and 140-e 17.05.2012. are equal respectively to 4, 19.5 and 1.2 mSv, total 25 mSv. This is 5% of the total dose of 466 mSv. In [23] calculated the coefficient of anticorrelation between doses with a solar neutron monitor potential modulation control feature A + B/Dl^2 in Volts (D1-the number of counts per minute, A and B are constants): 0.77-0.80.

In [24] was presented a series of reconstructed monthly value of the modulation potential for the period from 1936

through December 2009. The modulation potential parameterizes the energy spectrum of Galactic Cosmic Rays (GCR) near Earth with accuracy sufficient for practical application. The presented series is a composite of three parts. The most reliable reconstruction, which based on data from the world network of neutron monitors, covers the period since 1964 and is characterized by the mean 68% significance level uncertainties of 26 mV. Reconstruction for the period between 1951 and 1964 is based on a few mountain neutron monitors of other type. It is characterized by large uncertainties (formal error is 44 mV). Reconstruction before 1951 is based on ionization chambers. It is characterized by large uncertainties of about 140 mV. The comparison with date of measurements of flux of CR in the stratosphere since 1957 shows agreement since 1964, within 10% in the overall level and with only a few discrepancies which fall upon period of enhanced rate of solar activity minima.

In [23] according to the year 2012 made recalculation of the average annual solar modulation potential on the neutron monitor up to the year 2020, and on it, using the well-known linear relationship with registered radiation dose, it is obtained that the peak dose mSv/day accounted for at the Frontier of Solar Activity (FSA) date 2002 ± 2.7, as well as on table 3 and the fig. 10 the peaks of the dose for all other FSA dates 1994, 1986, 1978, 1955, 1946 and 1940 without exception. All this peaks correspond to modulation potential minima [23].

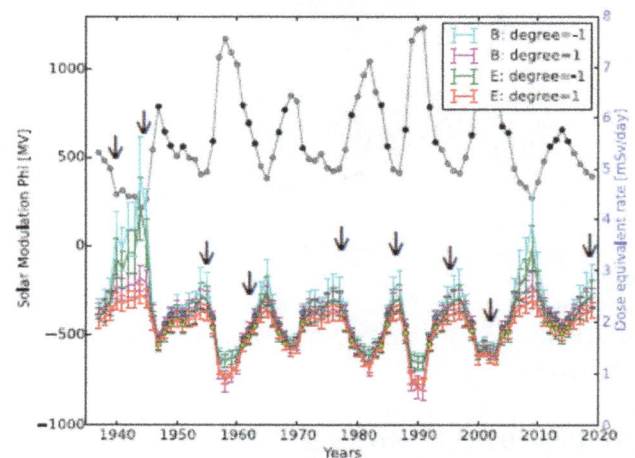

Figure 10. Annual average of reconstructed modulation potential Φ since 1937 (gray line and dots with units on the left axis) and estimated annual dose equivalent rate derived from different models (with units on the right axis) ([23], Fig. 4). Black, yellow, and blue dots represents the data within the range of modulation Φ between 550 and 810 MV which the RAD cruise measurements covered [23]. Arrows show the year of Frontiers of the solar activity.

Such as in according to ([24], table 3) only at FSA 1940 was exact minimum of the modulation potential and since 1940 there are not FSA years when was exact minimum of the modulation potential. Of course, years of exact minima of the modulation potential were within the 2.7 years interval near FSA years. So that in whole with above caveats we have amazing match between FSA years of the most numbers of strongest earthquakes per year on fig. 3, 5, when there we stated maximum of the solar neutrino production, in another side and years of the annual average dose maxima in transit to Mars on table 3.

Table 3. The annual average dose in transit to Mars calculated on the annual average of reconstructed solar potential neutron monitor ([23], fig. 4, p.4).

Year	Dose, mSv/day	Year of the Frontier of Solar Activity (FSA)
1940	3.2 maximum	1940
1943	3.4 maximum	
1946	4.4 maximum	1946
1948	2	
1955	2.5 maximum	1955
1962	1.5	1962
1964	2.5 maximum	
1978	2.4 maximum	1978
1986	2.5 maximum	1986
1995	2.2 maximum	1994
2002	2	2002
2004	2.7 maximum	
2018	2.5 maximum	2018

This match with dose maximum take place and for the FSA 2018 year. However, it is difficult to say how much dose extrapolation [23] is justified in this case. Whereas the date of the FSA calculated per thousand years [16].

10. Conclusion

1. There is a difference between 1978 year solar maximum on output of neutrinos from other solar maximums.
2. Difference of point 1 occurs at the moment of coincidence of the year solar maximum with the year of Frontier of solar activity (FSA).
3. From points 1-2, it follows that the solar neutrino yield significantly associated with FSA dates.
4. In whole with above caveats we have amazing match between FSA years of the most numbers of strongest earthquakes per year on fig. 3 and 5, when there we stated maximum of the solar neutrino production, in another side and years of the annual average dose maxima in transit to Mars on table 3.
5. Date of the FSA is determined by the Sun eccentric planets.
6. Proportionality of solar activity to the exit seismic energy of the Earth on the average annual number of strongest earthquakes in oscillatory response model of dense solar plasma by date all FSA before year 2000 coincided on 7% spread with the schedule release linear dependence of depth seismic energy [15] from the solar neutrino flux temporal variations on a continuous series of chlorine-argon experiment in 1970 to 1974.

Mass of the muon and tau-neutrinos are calculated. Set the integer 2:5 ratio between the mass of muon neutrinos and muon ultra-light weight torsion. Found that the calculated mass of the tau-neutrino is the same as the largest with heat 3/2 T relic neutrinos. In the standard model T = 2 K. Considering the temperature of the relic neutrinos as the effective gravitational temperature $T_\Gamma = 2/9$ M for some masses M, it is

followed ratio $M = 3\, m_{\nu_\tau}$, corresponding to reactions of the dissolution of the relic neutrinos by three sterile active Tau-neutrinos. A life time of sterile neutrinos such a mass is not less than the life time of the proton. On the other hand this value on the numerical score of the upper limit of the mass of the tau neutrino 10 MeV reaches 13 billion years.

Solar neutrino flux measurements data and its temporal variations in the Chlorine-Argon experiment for the period 1970-1994, examined with particular attention near dated frontiers of solar activity (FSA), which defined by means of the key criterion of 2ln2 in point a significant manifestation of the influence the Sun eccentric planets. Review shall be carried out within the framework of the standard model, taking into account the gravity. The results are presented in the form of pie charts in a consecutive temporary scan on a direct ascent of Earth's orbital position α, recalculated according to the date of the middle of each of the series of measurements. The corners of the sectors 35° and more defined for temporary length of each dimension. The amplitude of the sector is equal to the measured magnitude of neutrino flow minus the constant component 2.57 SNU in chlorine-argon experiment and 81.4 SNU in experiment SAGE. All charts show the priority direction of the solar neutrino flux detector checked by angle $\alpha = 271°$, $\delta = 23°$ (40° -15°) near axis of anisotropy of Galactic Gamma-radiation, as well as a number of less significant in transverse to the axis of the direction within the spread over time of cycle 15° -58° experiments. Researched the relationship between solar activity, seismic activity of the Earth and neutrinos flows registered in chlorine-argon experiment for the period 1970-1994, as well as in the SAGE for the period 1970-2008 years. There is a correlation of solar activity with seismic activity of the Earth and the positive correlation between seismic activity of the Earth and the neutrino flow variations, reaching 95% when taking into account the peculiarities of the matching of dates of solar maximums with FSA dates 1978, 1993-5 and 2002. It turned out that all known and perhaps predicted on 2018 year the greatest average annual values of the flow of solar cosmic rays fall exactly on the years of FSA defined on the motion of the Sun around the center of mass of the solar system because of the influence of the planets eccentric.

References

[1] BOEHM F., VOGEL P., Physics of massive neutrinos (Cambridge University Press, Cambridge, 1987).

[2] CLEVELAND Bruce T., DAILY T., DAVIS R. etc., Measurement of the solar electron neutrino flux with the Homestake chlorine detector. *Astrophysical journal*, 496: 505-526, 1998 March 20.

[3] The SAGE Collaboration. arXiv: 0901.2200v3 [nucl-ex] 10 Aug 2009.

[4] BELOV S. V., SHESTOPALOV I. P., KHARINE. P., On the Interrelations between the Earth Endogenous Activity and Solar and Geomagnetic Activity. *DAN [Academy of Science reports]*. Vol. 428. No. 1. 104-108 (2009).

[5] BAUROV Yu. A., EFIMOV A. A., SHPITALNAYA A. A., *Fizicheskaya misl Rossii,* No. 1, 1-9 (1997).

[6] *SYROMYATNIKOVA. G., LXV International Conference "Nucleus 2015".* New horizons in nuclear physics, nuclear engineering, femto- and nanotechnologies. Book of abstracts. June 29 – July 3, 2015, Saint-Petersburg, Russia. P. 173.

[7] SYROMYATNIKOV A. G., When Anisotropy of Vacuum Set The Space Isotropy. International Journal of High Energy Physics. Special Issue: Breaking of Space Symmetry in the Masses Spectrum Problem. 2015(to be published).

[8] GORBUNOV D. S., Sterile neutrinos and their roles in particles physics and cosmology, *Uspekhi Fizicheskikh Nauk* 184 (5) 545 ± 554 (2014).

[9] PILAFTSISA, UNDERWOOD T. E., J *Phys. Rev.* D 72 113001 (2005); hep-ph/0506107

[10] ASAKA N., SHAPOSHNIKOV M., *Phys. Lett.* B 620 17 (2005); hep-ph/0505013.

[11] SYROMYATNIKOV A. G. On Similarity between All-Known Elementary Particles and Resonances Mass Spectrum and Nuclear Atomic Weight. *Univ. J. Phys. and Appl.* 2(2)76-79. DOI: 10.13189/ujpa.2014.02023.

[12] SYROMYATNIKOV A. G., Comparison Analogy Between Properties of Hypernucleus and Supernucleus with Properties of the Elementary Particles and Resonances Electroproduction by Spin Shock Waves. International Journal of High Energy Physics. Special Issue: Breaking of Space Symmetry in the Masses Spectrum Problem. 2015 (to be published).

[13] NAKAMURA K. et al. [Particle Date Groups], *J PG* 37, 075021 (2010) and 2011.

[14] SYROMYATNIKOV A. G., On some feature of possible torsion effects on observables at hadron colliders, *Int. J. Geom. Meth. Mod. Phys* (2015) DOI: 10.1142/S0219887815500802

[15] SHESTOPALOV I. P., KUZHEVSKY B. M., KHARIN E. P., Correlations flows neutrinos with seismicity of the Earth. The hypothesis on elocation neutrinos in a period of deep strong earthquakes, *Inzhenernaya fizika.* No 1. 2014. P. 4-12.

[16] SYROMYATNIKOV A. G. Physical effects in Conformal Gauge Theory of Gravitation (LAP Lambert Academic PublishingGmbH & Co. KG, Saarbrucken, Germany, 2012).

[17] SYROMYATNIKOV A. G., On connection between a Solar activity and Solar system dynamics. – *Works of III International Congress "Weak and Hyperweak FIELDS and RADIATIONS in Biology and Medicine",* Saint-Petersburg, 01-04.07.2003, p. 90.

[18] SYROMYATNIKOV A. G., ZAKOLDAEV Yu. A., Depth distribution of rocks under the Geospace Universal X-structure of the Earth crust and upper mantle. (Petropolis, Saint-Petersburg, 2011).

[19] Web-site: http://www.ceme.gsras.ru.

[20] VASILIEVA G. F., NESTEROV M. M., CHERNIKCH Yu. V., The generation process in the Sun's magnetic field when you change the dynamic parameters of the solar system, *A series "Problems of research of the Universe".* Issue 25. Part II. Saint-Petersburg, Russia, 2002). P. 303-320.

[21] BAKAL Dzh., Nejtrinnaja astrofizika [Neutrino astrophysics]. Transl. from English. M.: Mir [Moscow: Publishing House "Peace"], 1993. 624 p.

[22] BAUROV Yu. A., *J. Mod. Phys.* Vol. 3. 2012. P. 1744.

[23] GUO G., ZEITLIN C., WIMMER-SCHWEINGRUBER R. F. et al.: Variations of dose observed by MSL/RAD in transit to Mars: arXiv: 1503.0663v1 [physics.space-ph] 23 May 2015.

[24] ISOSKIN I. G., BAZILEVSKAYAG. A., and KOVALTSOV G. A. (2011), Solar modulation parameter for cosmic rays since 1936 reconstructed from ground-based neutron monitors and ionization chambers, *J. Geophys. Res.,* 116, A02104, doi: 10.1029/2010JA016105.

4

On Optimal Trajectory in Space Flight

Sergey Orlov

Petrozavodsk State University, Petrozavodsk, Russia

Email address:

ion@sampo.ru

Abstract: Are investigated a trajectory of new type in distant, space flights unlike usual trajectories of direct flight to heavenly object (Moon) it is supposed to use asymmetry of a gravitational field and to carry out flight bypassing the most power gravitational impact on the spacecraft. It leads to economy of power for 20-30%.

Keywords: The Theory of Vortex Gravitation, Space Flights with an Optimum Trajectory

1. Introduction

Now in a science there is a statement that force of gravitation is created by any body. According to this law force of gravitation decreases under the law of a return square of distance from this body equally in all directions.

In the theory of vortex gravitation [1] proposed study, the force of gravity - a force push, which is caused by the decrease in pressure in the space medium called ether. In turn pressure reduction in heavenly points is caused by vortex rotation of ether round these points, according to hydro aerodynamics laws. Speed of orbital rotation of streams of each whirlwind is inversely proportional to distance from square of center of this whirlwind.

As the whirlwind rotates in one plane, and the law of dependence of speed, pressure and force of gravitation from square of distance to the center of rotation of ether, operates too only in one plane of rotation of ether.

2. Model of the Origin of the Universal Gravitation Force

In this section, a model of appearance of the gravitation attraction force is considered from the viewpoint of aerodynamics. Namely, the two-dimensional model (Fig. 1) is considered on the basis of the following initial postulates. These postulates will be expanded and defined more exactly below.

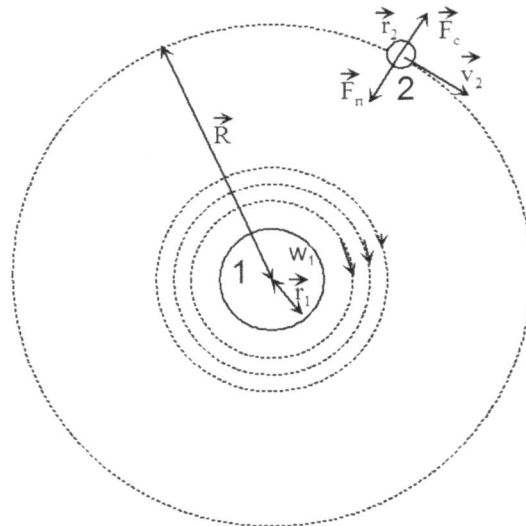

Fig. 1. Two-dimensional model of gravitational interaction of two bodies. The forces are shown acting on body 2: F_c – the centrifugal force, F_n – the force of attraction of body 2 from body 1; v_2 – linear velocity of body 2 at the orbit, R – the radius of the orbit, r_1 – the radius of body 1, r_2 – the radius of body 2, w_1 – angular velocity of ether rotation at the surface of body 1, and m_2 are the mass of body 2.

1. There exists an ether vortex around any physical object.

2. The ether motion in the vortex has laminar nature and obeys the laws of hydro- or aero-dynamics; the ether viscosity is low.

3. The pressure gradient, arising during the vortex motion of the ether gas, is the reason for an attractive force from

body 1 to body 2 (see Fig. 1).

4. The direction of the force $F_\text{п}$ does not depend on the direction of the ether angular velocity, which is necessary for the attractive force between the bodies, irrespective of their relative position. This implies the absence of the Magnus force – the force of interaction between the two vortexes which appears in the classical aerodynamics. Such an assumption can take place at a weak interaction between the two ether flows, as if they would move one through another, not affecting mutual motion.

5. The appearing attraction force must describe the experimentally obtained law of gravity:

$$F_\text{п} = G \cdot \frac{m_1 \cdot m_2}{r^2} \qquad (1)$$

where m_1, m_2 are the masses of bodies 1 and 2, respectively, $G=6.672 \cdot 10^{-11}$ N·m²/kg² – the gravitation constant, and r – the distance between the bodies.

Next we consider the appearance of the attraction force in more detail and derive a formula describing it. As was said above, a pressure gradient arises as the result of the vortex motion. Let's find the radial distribution of the pressure and the ether velocity. For this purpose, we write the Navier-Stokes equation for the motion of a viscous liquid (gas).

$$\rho \left[\frac{\partial}{\partial t} + \vec{v} \cdot \text{grad} \right] \vec{v} = \vec{F} - \text{grad } P + \eta \Delta \vec{v} \qquad (2)$$

where ρ is the ether density, $\vec{v}$ and P are, respectively, its velocity and pressure, and η - the ether viscosity. In cylindrical coordinates, taking into account the radial symmetry $v_r = v_z = 0$, $v_\varphi = v(r)$, $P = P(r)$, the equation can be written as the system:

$$\begin{cases} -\dfrac{v(r)^2}{r} = -\dfrac{1}{\rho}\dfrac{dP}{dr} \\[2mm] \eta \cdot \left(\dfrac{\partial^2 v(r)}{\partial r^2} + \dfrac{\partial v(r)}{r \partial r} - \dfrac{v(r)}{r^2} \right) = 0 \end{cases} \qquad (3)$$

In case of a compressible substance (ether), there will be a function $\rho = f(P)$ (instead of ρ).

From the first equation of system (3), one can find $P(r)$ provided that the dependence $v(r)$ is known. The latter, in turn, should be found from the second equation of that same system (one of the solution of which is the function $v(r) \sim 1/r$). At zero viscosity, the system permits any dependence $v(r)$ [2].

The force affecting the body can be estimated from the formula

$$\vec{F}_\text{п} = - V \bullet \text{grad} P(r) \qquad (4)$$

where V is the volume of body 2.

In cylindrical coordinates the modulus of $\vec{F}_\text{п}$ is

$$F_\text{п} = V \cdot \frac{\partial P}{\partial r} \qquad (5)$$

Then, comparing equations (3) and (5), for the incompressible ether (ρ=const) we find that

$$F_\text{п} = V \cdot \rho \cdot \frac{v(r)^2}{r} \qquad (6)$$

For the correspondence of the ether rotation to the planet motion law (according to Kepler 3-rd law) in one cosmic (e.g., Solar) system, $v(r)$ must obey the dependence $v(r) \sim \dfrac{1}{\sqrt{r}}$, and not the $v(r) \sim \dfrac{1}{r}$.

Taking into account the edge condition $v(r_1) = w_1 \cdot r_1$,

$$v(r) = \frac{w_1 \cdot r_1^{\frac{3}{2}}}{\sqrt{r}} \qquad (7)$$

Thus

$$F_\text{п} = V \cdot \rho \cdot \frac{w_1^2 \cdot r_1^3}{r^2} \qquad (8)$$

Here we make one more supposition (№ 6) – Ether penetrates through all the space, including the physical bodies. The volume V in formula (8) is an effective volume, i.e. the volume of elementary particles, which the body is composed of. All the bodies are composed of electrons, protons, and neutrons. The radius of an electron is much smaller than that of a proton and neutron. The radii of the latter are approximately equal to each other, $r_n \sim 1.2 \cdot 10^{-15}$ m. The same is true as to the masses: $m_n \sim 1.67 \cdot 10^{-27}$ kg (r_n and m_n are the radius and the mass of a nucleon). Therefore, the volume in formula (8) is:

$$V = \frac{m_2}{m_n} \cdot \frac{4\pi}{3} \cdot r_n^3 \qquad (9)$$

Taking into account the formula (9), Eq. (8) can be rewritten as

$$F_\text{п} = \frac{4 \cdot \pi \cdot r_n^3 \cdot \rho}{3 \cdot m_n} \cdot \frac{w_1^2 \cdot r_1^3 \cdot m_2}{r^2} \qquad (10)$$

From the obtained formula for vortex gravitation, it is obvious that, in the existing Newton's law of gravitation, instead of the reason of gravity (the gradient of pressure), the consequence of that (i.e. the mass) is used.

3. Cause of the Ellipsoidal Shape of Orbits

It is known that the planets circulate around the Sun by an ellipse with a small eccentricity.

This fact is accounted for from the viewpoint of vortex gravitation; moreover, it serves as a convincing proof of the existence of this gravitation with its discoid plane-symmetrical configuration (Section 2).

The cause of the planets orbit "compression" is the inclination of these orbits to the sun torsion plane. This statement is based on the following.

As is known, the planes of orbital motion of all the planets are situated with small deviations one from another. Consequently, planet orbit planes have inclinations to the plane of the sun gravitation torsion, where the highest gravitation force for this orbit acts, and the planets should intersect the sun torsion in two points during their orbital motion. As will be shown below, these intersection points coincide with the centers of perihelion and aphelion.

In the aphelion and perihelion, the sun gravitation force acts onto the planets with the highest magnitude at this orbit, and hence the orbit possesses a maximum curvature. At going out (deviation) from the sun torsion plane, the gravitation forces decrease and the planet trajectory "unbends" (Fig. 2). As such the cycle of the gravitation force and motion trajectory change repeats for each planet and for each turn around the Sun. The more the planet circulation trajectory is deviated from the central sun torsion plane, the higher is the degree of the gravitation force decrease in these regions, and hence the higher is the degree of "straightening" or "compression" of the orbit. Due to a permanent cyclic change of these forces, the orbit becomes ellipsoidal.

At significant inclinations and high speeds, the orbit of a satellite (meteorite, comet) have a hyperbola or parabola trajectory, and, correspondingly, the celestial object, once turning around the Sun, abandons the sun gravitation torsion field forever.

Determining of the sun torsion direction

On the basis of the stated above, it is obvious that the orbit trajectory eccentricity value of any planet depends on the value of inclination of this orbit to the sun torsion. Therefore, a reverse relation takes place, i.e. the lower the orbit eccentricity, the lower the inclination of the planet orbital plane to the sun torsion plane.

Since the Venus orbit has the least eccentricity, for preliminary calculations, it is permissibly to accept the following property of the sun torsion:

- the direction of the sun gravitation torsion in the World coordinates coincides with the Venus orbital plane direction to a highest degree.

Therefore, all the inclinations and latitudes of any astronomical point can be determined with regard to the orbital plane of Venus with a small correction up to 0.5 degree.

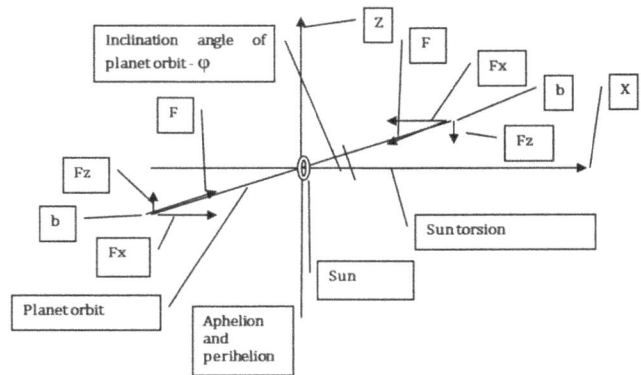

Fig. 2. *Kinematical scheme of orbital motion.*

Let's consider the planet circulation in more detail with the Mercury motion as an example, in accordance with its heliocentric coordinates of 1993 [3].

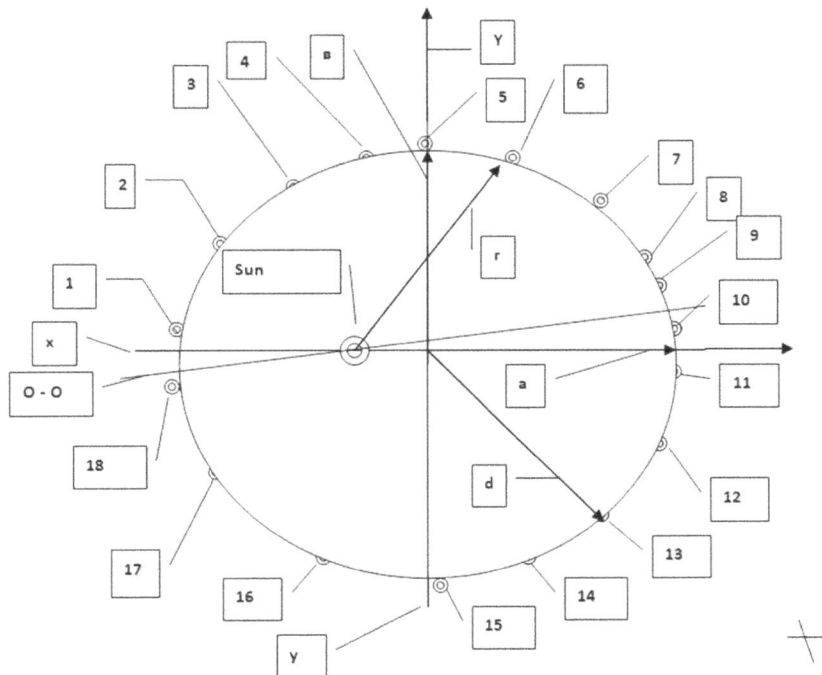

Fig. 3. *Orbit of Mercury.*

In Table 1 and Fig's 2 and 3, the following denotations are used:

Z – the torsion rotation axis

№ - numbers of the Mercury orbit points according to the astronomy calendar of 1993

Д – the heliocentric longitude J2000.0 of these points

r – the radius-vector, in A. U.

d – the distance from the ellipse center to the point under study, in million km.

V – the orbital speed, in km/s

R – the curvature radius, in million km, $R = a^2 b^2 / d^3$ where a, в are the major and minor axes

Fc – the centrifugal forces

Fg – the gravitation forces.

The values of centrifugal and gravitation forces are in portions of the planet mass.

O – O – the apse line coinciding with the line of intersection of the Mercury orbital plane with sun torsion plane. The center of Mercury perihelion has the longitude of (85.83 – 8.18) degree regarding to the point №1 in 1993.

On the basis of comparison of latitudes from the astronomy calendar, it has been established that the Mercury traverses the Sun (Venus) torsion in the aphelion and perihelion. The same is true for the other planets. Therefore, at these parts of the planet motion, the orbit curvatures are maximum and equal to each other, and the gravitation forces correspond to their classical values, i.e. they are inversely proportional to square of the distance to the Sun or are equal to the centrifugal forces.

Thus, the exact position of the sun torsion by its latitude relatively to the ecliptic is indicated by the latitudes of the aphelion and perihelion centers of each planet and by the apse line directions of these planets.

Comparing the astronomy point latitude values with the ratio of the gravitation and centrifugal forces in these space points, one can find that the more the planet orbit is inclined to the sun torsion, the higher is the difference between the Newtonian calculated gravitation forces and the actual centrifugal forces in those same points.

We consider two points of the Mercury orbit (№ 9 и № 10 in Fig. 3).

Table 1. Mercury orbit parameters.

№	Д, degree	r	d	V	R	Fc	Fg
9	250,04	0,4657	58,11	38,41	55,46	26,60	27,32
10	263,78	0,4659	58,03	38,96	55,69	27.26	27,29

The distance between point 10 and the Sun is 0.4659 x 150 = 69.885 million km.

For point 9 it is - 0.4657 x 150 = 69.855 million km.

The distance of point 10 from the Sun is 1.0004 times longer than that for point 9. Therefore, in point 10, the Newtonian sun gravitation forces have to be 1.001 times less than those in point 9 (see Table 1). In reality, according to the calculation, the value of centrifugal forces in point 10 are

1.025 times higher as compared to point 9 which is associated with a larger orbit curvature in this point (see Table 1). Since the planet circulation centrifugal forces are reactive and always equal to the gravitation forces, it is follows from the above-said that, in this region of Mercury motion trajectory, the classical gravitation law is not fulfilled.

On the basis of the vortex gravitation model with a plane-symmetrical configuration, this paradox has a physical-mathematical ground.

The Newton's world attraction law or formula 10 in Section 2 can describe the action of the gravitation forces only in the plane of the gravitation torsion.

The above-presented calculation of the centrifugal forces appearing at the planet motion in the aphelion shows that the inertial circulation of the planets along an ellipsoidal trajectory in a central-symmetrical gravitation field is impossible in accordance with the classical ideas.

It should be noted that the planet orbit perihelion revolving round the Sun is also accounted for by a permanent change of the force magnitudes acting upon the planets.

4. Calculation of Gravitation in Three-Dimensional Model

The change of the dynamical properties of the planets at their inclination, discussed in Section 3, gives a possibility to obtain a formula describing the change of gravitation forces in the three-dimensional model.

Comparing the orbit compression coefficients for all the planets with cosine of the angle of inclination of these orbits to the sun torsion, one finds that these values are directly proportional to each other:

$$b/a \sim Cos\ \varphi \qquad (11)$$

Proofs of equation (11)

axis X – the direction of the parent torsion central plane.

axis Z – the rotation axis of the parent torsion.

φ – the inclination angle of the satellite (planet) orbit torsion.

OB – the curvature radius of the torsion-satellite revolving at the coincidence of the satellite-torsion motion trajectory with the parent torsion rotation plane, i.e. at the perihelion or aphelion, or at the apex of the orbit major semiaxis:

$$OB = b^2 / a \qquad (12)$$

OD1 - the curvature radius of the torsion-satellite revolving when it moves in a region possessing the inclination of angle φ from the parent torsion central plane, i.e. at the apex of the orbit minor semiaxis:

$$OD1 = a^2 / b \qquad (13)$$

We prove that the equation cos φ = b/a is fulfilled at equalities (12) and (13)

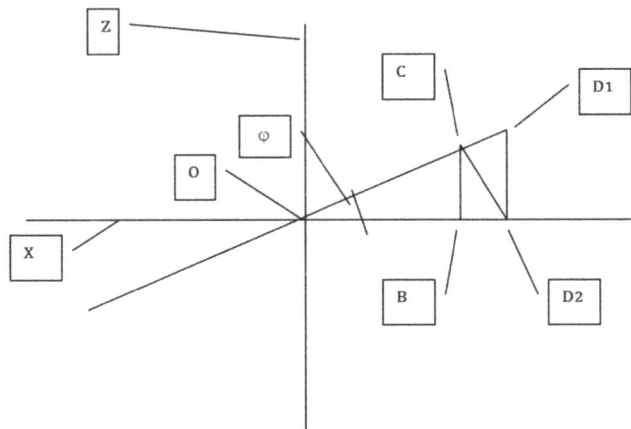

Fig. 4. Plane projections of minor and major orbital semiaxes.

Proof:

First we draw a segment OB on the axis X (fig. 4) coinciding with the apses line. This segment is to be equal to the curvature radius in the major semiaxis apex and is directed along the sun torsion central plane or the apses line.

Let's now draw a line from the center O with the angle φ; the direction of this line has to coincide with the minor semiaxis apex.

Since, from the problem condition, $\cos \varphi = b/a = OB/OC$, then:

$$OC = OB\, a/b = (b^2/a)\,(a/b) = b$$

Let's drop a perpendicular from point C on axis X, as the angle OCD2 is right:

$$OC/OD2 = \cos \varphi = b/a,\ \text{whence}$$

$$OD2 = OC\, a/b = b\,(a/b) = a,$$

And finally we drop a perpendicular from point D2 on line OC, as the angle D1D2O is right:

$$OD2 / OD1 = \cos \varphi = b/a,\ \text{whence}\ OD1 = OD2\,(a/b) = (a^2/b)$$

Therefore, equations (12) and (13) are fulfilled provided that $\cos \varphi = b/a$. That is, the cosine of the planet orbit inclination angle in the minor semiaxis apex to the sun torsion plane is equal to the compression coefficient of this orbit.

Note 1. The inclination φ of an orbital point does not coincide with the angle of inclination of this point indicated in astronomy calendars, because, according to the astronomy rules, all the coordinates in the Solar system are measured heliocentrically and from the ecliptic plane.

Since the centrifugal forces are reactive and always equal to the sun attraction forces, these centrifugal forces may be considered as experimental or etalon values for the estimation of the accuracy and correctness of the results of gravitation forces calculations. Therefore, the change of the value of the planet centrifugal forces at a change of their coordinates is always equal to the change of the value of the gravitation force acting onto this planet.

Determining of the three-dimensional gravitation coefficient Kg.

Let's write the formulas to determine the orbit (ellipse) curvature radius:

- in the major semiaxis apex or in perihelion and aphelion:

$$R\text{кр}.a = b^2 / a \qquad (14)$$

- in the minor semiaxis apex:

$$R\text{кр}.в = a^2 / b \qquad (15)$$

On the basis of the 2nd Kepler law, the planets change the orbital velocity (V) as a function of the distance to the Sun (R), in the limits of their orbits, in the following proportion:

$$Va \sim 1/\,Ra \quad Vb \sim 1/\,Rb \qquad (16)$$

where

Va – the orbital speed in the perihelion (aphelion), i.e. in the apex of the planet orbit major semiaxis,

Vb – the orbital speed in the apex of the planet orbit minor semiaxis

Ra – the distance from the Sun to the aphelion (perihelion).

Rb – the distance from the Sun to minor semiaxis apex.

The centrifugal force is determined from the formula:

$$Fc = m\,V^2 / R\text{кр} \qquad (17)$$

Substituting (14) – (16) into (17):

$$Fca = m\,Va^2 / R\text{кр}.a \sim m\,a\,/Ra^2\, b^2 \qquad (18)$$

$$Fcb = m\,Vb^2 / R\text{кр}.b \sim m\,b\,/ Rb^2\, a^2 \qquad (19)$$

Since the gravitation forces in the aphelion and perihelion Fa correspond to their classical values or to the centrifugal forces, then, to determine a deviation of the gravitation forces in the torsion periphery (in the minor semiaxis apex – point b), it is necessary to determine the analogous deviation of the values of the centrifugal forces as compared to those same forces in the perihelion. For this purpose, we divide formula (19) by formula (18):

$$Fcb / Fca = [b^3 / a^3]\,[Ra^2 / Rb^2]$$

Here the relative value Ra^2 / Rb^2, in accordance with formula 10 in Section 2 or with the Newton formula, determines the gravitation force change as a function of the change of the distance from the torsion center to the points under consideration.

According to the expression (11), the value b/a equals to the cosine of the inclination angle in the considered point. Hence, this value determines the change of the gravitation forces as a function of the inclination of the considered point to the sun torsion.

Therefore, one can write:

$$b^3/a^3 = \cos^3 \varphi = Kg \qquad (20)$$

The gravitation forces in any point of the cosmic space are

determined by the formula:

$$Fv = Fg\,Cos^3\,\varphi, \qquad (21)$$

where

Fg – the gravitation force in the two-dimensional model (formula 10 in Section 2 or Newton equation)

Fv – the gravitation force in the three-dimensional model

Consequently, using the gravitation coefficient Kg, one can determine the gravitation forces in any point distant from the center of a cosmic torsion.

Formula (21) shows that, when moving away from the gravitation torsion plane, parallel to the torsion axis, the gravitation force decrease inversely as the cube of the distance - $1/s^3$

5. Calculation of Vortex Gravitation

In general, the force of gravity can be calculated by the formula 21 -

$$Fgv = Fgn\,Cos^3\,\varphi$$

where

Fgn - the force of gravity in the two-dimensional model (Eq. 10 in [1], which corresponds to the empirical formula for the law of universal gravitation Newton)

Fgv - the force of gravity in a three-dimensional vortex model.

φ - the angle between the straight line connecting the center of the torsion from this point, and the plane gravitational torsion.

The location of the plane of cosmic torsion can determine the coordinates of celestial bodies - satellites of the torsion.

In the solar system, the heliocentric latitude, the gravitational torsion coincide with latitude of the center of the perihelion and aphelion of the orbits of all the planets.

The earth latitude, the gravitational torsion coincide with latitudes of apogee and perigee of the orbit of the moon.

Thus, the coordinates of the gravitational torsion, we can determine the coordinates of the plane in which the gravitational force decreases at the lower, that is inversely proportional to the square of the distance from the center of torsion. As Earth is in the center earht gravitational torsion, at removal from it at distant space flights it is necessary to move to detour earth torsion, instead of on a direct trajectory, as in case of flight on the Moon.

The following shows the calculation of the physical work required to make the spacecraft during flight to the moon in two different routes.

Let's consider a problem of comparing the works expended on getting over the gravitation attraction forces (F) by a body, when traveling from point A to point C (see Fig. 5) by the paths AC and ABC at two different F(r, φ) dependences. The OAS line – a face projection gravitational torsion of Earth.

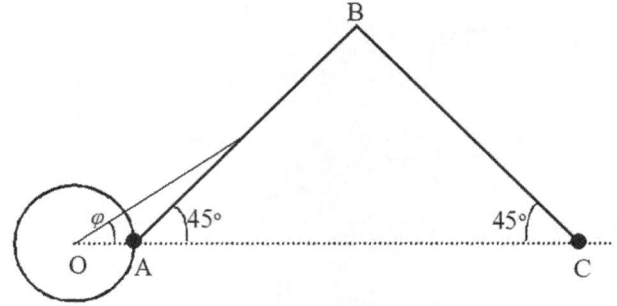

Fig. 5. *Scheme of space flight.*

O – centre of Earth
A – start of flight
C – Moon (finish)
AC – projection of gravitational flat

In the first case, F is independent of φ and obeys the Newton law

$$F(r) = G\frac{m_1 \cdot m_2}{r^2} \qquad (22)$$

where m_1 and m_2 are the masses of bodies, G – the gravitation constant, and r – the distance between the bodies.

In the second case, F depends on φ in accordance with formula (21)

$$F(r,\phi) = G\frac{m_1 \cdot m_2}{r^2}\cdot\cos^3(\phi) \qquad (23)$$

where φ is the angle between axis OC and the position radius-vector of the replaced body.

As is known, the work equals to the path integral

$$A = \int \vec{F}\cdot d\vec{r} \qquad (24)$$

Let A_{AC} be the work expended at the transference AC for the case of the dependence (22). We determine the works A_{AB} and A_{BC}. For A'_{AC} being the work expended at the transference AC for the case of the dependence (23) we determine, respectively, the works A'_{AB} and A'_{BC}.

Now we write the integral (24) for each case

$$A_{AC} = \int_{r_1}^{r_2} G\cdot\frac{m_1 \cdot m_2}{r^2}\,dr \qquad (25)$$

$$A_{AB} = \int_{0}^{\phi_{BOC}} G\cdot\frac{m_1\cdot m_2\cdot(\cos(\phi)-\sin(\phi))\cdot\cos\left(\dfrac{3\cdot\pi}{4}+\phi\right)}{r_1\cdot\sin\left(\dfrac{\pi}{4}-\phi\right)}\,d\phi \quad (26)$$

$$A_{BC} = \int_{0}^{\phi_{BOC}} G\cdot\frac{m_1\cdot m_2\cdot(\cos(\phi)+\sin(\phi))\cdot\cos\left(\dfrac{\pi}{4}+\phi\right)}{r_2\cdot\cos\left(\dfrac{\pi}{4}-\phi\right)}\,d\phi \quad (27)$$

$$A'_{AC} = A_{AC} \qquad (28)$$

$$A'_{AB} = \int_0^{\phi_{BOC}} G \cdot \frac{m_1 \cdot m_2 \cdot (\cos(\phi) - \sin(\phi)) \cdot \cos(\frac{3 \cdot \pi}{4} + \phi) \cdot \cos^3(\phi)}{r_1 \cdot \sin(\frac{\pi}{4} - \phi)} d\phi \tag{29}$$

$$A'_{BC} = \int_0^{\phi_{BOC}} G \cdot \frac{m_1 \cdot m_2 \cdot (\cos(\phi) + \sin(\phi)) \cdot \cos(\frac{\pi}{4} + \phi) \cdot \cos^3(\phi)}{r_2 \cdot \cos(\frac{\pi}{4} - \phi)} d\phi \tag{30}$$

where r_1 – the distance OA, r_2 – OC, and ϕ_{BOC} – the angle BOC.

Formula (28) is valid because, in this direction, the forces (22) and (23) are equal to each other.

Calculating the integrals (28-30) numerically for the case of moonflight ($r_1 = 6400 \cdot 10^3$ m, $r_2 = 40000000$m, $m_2 = 6 \cdot 10^{24}$ kg, $m_1 = 1$ kg), one obtains $A_{AC} = 6.1554643 \cdot 10^7$ J, $A_{AB} = 6.1140242 \cdot 10^7$ J, $A_{BC} = 4.1440045 \cdot 10^4$ J, $A'_{AB} = 4.5279719 \cdot 10^7$ J, $A'_{BC} = 3.5727542 \cdot 10^5$ J.

One can see that $A_{AC} = A_{AB} + A_{BC}$, which just must be the case for the Newtonian forces when the work does not depend on the transference path from point A to point C.

In the case of the law (23), the work on the path ABC equals to $A'_{ABC} = A'_{AB} + A'_{BC} = 4.5636994 \cdot 10^7$ J. This is less than the work $A'_{AC} = A_{AC} = 6.1554643 \cdot 10^7$ J.

The ratio (decrease) of the works is $s = A'_{ABC} / A_{AC} = 0.7414062$. The value of s depends on the distances r_1 and r_2 and on the transference path.

Thus, the transference by the path ABC in the case of the law (23) is more energetically preferable than that directly by the path AC.

6. Conclusion

The above calculation shows that the moonflight with a detour of the Earth torsion should decrease the fuel consumption on 25%.

At present, most interplanetary cosmic apparatus get accelerations which can not be explained on the basis of cosmic calculations in the relativity theory of Einstein. Particularly, deviations have been found for the apparatus of «Galileo», «Rosetta» and «Cassini». The suggested model of vortex gravitation (formula 21) shows that, if the trajectory of the satellite flight does not coincide with the Sun gravitation torsion plane, then one should take into account the value of gravitation coefficient in the calculation of solar gravity acting onto the satellites. This coefficient ($\cos^3 \phi$) reduces the value of solar gravity, which gives a certain acceleration to cosmic satellites and results in a deviation of the motion trajectory.

References

[1] S. Orlov. *Foundation of vortex gravitation, cosmology and cosmogony*. Global journal of science Frontier research. Physic and Space Science Volume 12 issue 1 Version 1.0 January 2012 https://globaljournals.org/GJSFR_Volume12/3-Foundation-of-vortex-gravitation-cosmology.pdf

[2] L V Kiknadze, Yu G Mamaladze. *Classical hydrodynamics for physicists-experimentalists*. Tbilisi University Press. Tbilisi, Georgia. 1979. Page 136.

[3] A P Gulyaev. *Astronomy calendar*. Cosmosinform. Moscow, Russia. 1993. Page 285.

5

Proposed simple electro-mechanical automotive speed control system

Ahmed Farouk AbdelGawad[1, *], **Talal Saleh Mandourah**[2]

[1]Professor of Computational Fluid Mechanics, Mech. Eng. Dept., Umm Al-Qura Univ., Makkah, Saudi Arabia
[2]Mech. Eng. Dept., Umm Al-Qura Univ., Makkah, Saudi Arabia

Email address:

afaroukg@yahoo.com (A. F. AbdelGawad)

Abstract: Millions of people are killed or seriously injured on the roads due to terrified accidents every year. Most of these accidents are attributed to the over-speeding of the road vehicles. Thus, the road speed limiter (*RSL*) is a very important technique to reduce the possibility of road accidents. An interesting idea to control the speed of the vehicle is to apply electronic control of the air-supply that enters the vehicle carburetor according to road transmitters that are connected and operated either by local network or satellite. In the present paper, a control system was designed and implemented. It is consisted of a control mechanism and an electronic circuit to control the air-inlet to the carburetor according to pre-set programming based on the vehicle speed. Although, it is a challenging job to design and implement modifications to existing systems, the present speed control system was successfully implemented and tested. The present proposed mechanism is simple, inexpensive and suitable to be implemented in developing countries where a big number of cars still work using the traditional carburetor mechanism.

Keywords: Road Speed Limiter, Control Mechanism, Electronic Circuit, Car Carburetor

1. Introduction

Every six seconds, someone is killed or seriously injured on the world's roads due to horrible accidents. With 1.3 million road deaths each year, this is a global epidemic comparable to Malaria or Tuberculosis. And like those killer diseases, road crashes prey on the young, the poor and the vulnerable. Yet by comparison to other global killers, road injury is utterly neglected [1]. The main reason for these accidents is the high-speed driving. So, the question is how to enforce the moving vehicle to follow the speed limit. This speed limit changes according to the location, topography, weather, cultural standards, etc.

An interesting idea to control the speed of the vehicle is to apply electronic control of the air-supply that enters the vehicle carburetor. An integrated control circuit may be designed and implemented to control the air-inlet to the carburetor according to road transmitters and operated either by road network or satellite. So, the road is to be divided according to the speed limit of the vehicle running on it as desired. This may occur by using a transmission device that is put on the beginning of the road. This device sends signal to the receiver device that is inserted in the car. Then, the receiver takes this signal and translates it to a suitable mechanism to fix a new limit for the speed of the vehicle. The transmitter can be replaced by a satellite receiver.

The main idea of this research is to control the speed of the vehicle during its running along the high-speed roads by the control of the inlet air-flow that enters the vehicle carburetor. An integrated control circuit was designed and implemented. This circuit controls the air-inlet to the carburetor according to pre-set programming based on the vehicle speed. A mechanism was added to control the carburetor-opening using an electrical motor according to the required signal, which is received from the control circuit. The electrical motor rotates a pulley to take a new position according to the control signal and limits the movement of the carburetor throttle arm.

2. Previous Investigations

The problem of automotive speed control was considered by many investigators and researchers. Most of the research work concentrated on the impact of the speed limitation on the reduction of accidents and causalities. Few scientific publications are available that concern the technical aspects of the speed control system. The following paragraphs give a quick summary of the important previous publications.

Some of the publications handled the technical perspectives of the automotive speed control systems in the recent few years.

Paine [2] presented a classification of speed limitation devices showing the advantages and disadvantages of each type. He found that properly designed *ISA* (Intelligent Speed Assistance) systems can be highly effective in encouraging motorists to obey speed limits and should be encouraged by governments. Pérez *et al*. [3] presented a new Infrastructure to Vehicles (I2*V*) communication and control system for intelligent speed control, which is based upon Radio Frequency Identification (*RFID*) technology for identification of traffic signals on the road. Their results suggested that an automatic intelligent speed control system can be used to prevent any unexpected traffic circumstances and improve the safety of the occupants of the vehicle. Kameswari *et al*. [4] presented a design to control the speed of the automobiles at remote places for fixed time. Their proposed model used a microcontroller unit that receives the pedal position and then transfers appropriate signal to the Electronic Control Unit (*ECU*) that in turn controls the automobiles' throttle position. They stated that their theoretical study needs further extension to consider more than one vehicle.

In the early seventies of the past century, Ford Motor Company assigned patents concerning maximum vehicle speed limiter for a vehicle that has a pedal connected to a carburetor throttle valve through a linkage means [5, 6, 7]. The patents were based on mechanical and/or electromagnetic-circuit systems. Other inventors produced patents for engine maximum speed limiter with operator control. The control may be carried out by setting the upper limit of engine speed and then not be exceeded once a key-operated switch has been activated to an "off" position [8]. Other method of control can be applied by a programmable device that interfaces with the vehicle and identifies the operator who is allowed to set the maximum speed limit [9, 10].

Some researchers discussed the impact of speed limit on the reduction of crashes and pedestrian fatalities in city of Zurich [11], South Australia [12], USA [13] and India [14]. Comte [15] described a driving simulator experiment using The University of Leeds Advanced Driving Simulator to test two speed control systems against an advisory system and a baseline control (no system). His results indicated that there are safety benefits of control systems including a reduction of maximum speed, speed variance and inappropriate speed at hazardous locations.

Safety impacts of speed limiter installations on commercial vehicles; trucks and buses were investigated by some authors.

The experience of national speed-limit legislation [16] and the assessment of the safety efficiency of speed limiters [17] were reported for Australia, Europe and North America. Very recently, Hanowski *et al*. [18] objected to identify the impacts of implementing road speed limiters (*RSL*) in commercial vehicle fleet operations. Their study included data from 20 truck fleets, approximately 138,000 trucks, and analyzed more than 15,000 crashes. Their findings showed strong positive benefits for *RSLs* and that the cost of the technology is negligible and would not be expected to be cost-prohibitive for fleets/owners.

The acceptance of intelligent speed adaptation by car drivers was also concerned by some researchers.

Field trials with in-car speed limiter were reported by Mäkinen and Várhelyi [19, 20]. Their investigations covered three European countries, the Netherlands, Spain and Sweden representing different regions and driving cultures. They concluded that the majority of the drivers accepted the speed limiter as a driver operated system. Also, Duynstee and Katteler [21] discussed the public acceptance of a trial involving Intelligent Speed Adaptation (*ISA*) for passenger cars in an urban area in the Netherlands for a one-year period. Their test results showed substantial public support for *ISA*. Abraham [22] analyzed speed data from Ontario highways using the standards of the Institute of Transportation Engineers and proved that the speed limits on these highways should be increased. He also stated that in order to implement the new speed limits, a public education campaign is recommended that would educate road users to reserve the left lane for passing, and to restrict the left lane to mature drivers who have several years of road experience.

3. Road Speed Limiter (*RSL*)

A Road Speed Limiter (*RSL*) means a device whose primary function is to control the fuel feed to the engine, in order to limit the vehicle speed to a preset value, Fig. 1.

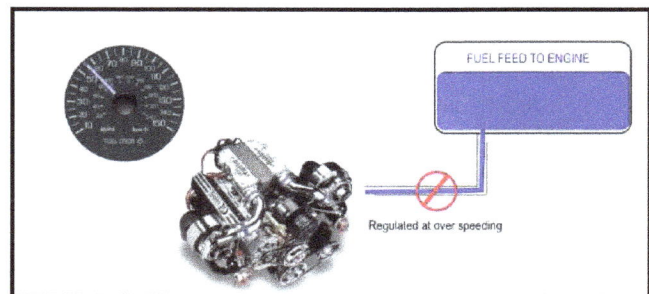

Fig 1. Description of the function of the road speed limiter (RSL) [23].

3.1. Types of Road Speed Limiter (RSL)

There are many types of speed limiters. These types may be classified according to either the technique of applying the speed control or the functionality of the speed limiter.

3.1.1. Types of Road Speed Limiter (RSL) Based on the Control Technique
(1) Accelerator control
(2) It is also known as the "Cable Type", Fig. 2. The idea of this type is adopted in the present investigation.
(3) Direct fuel control (solenoid valve type)

(4) Electronic pedal control (for electronic accelerators)

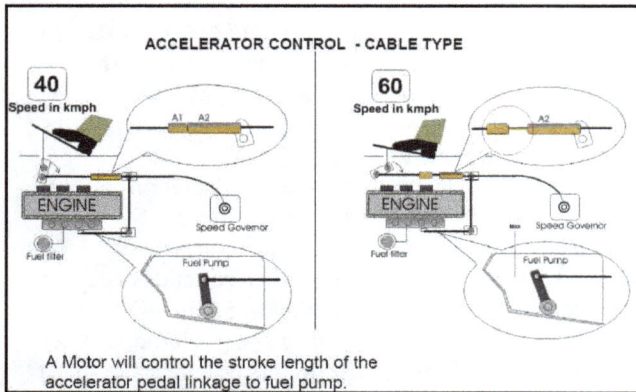

Fig 2. Description of the accelerator control (cable type) [23].

3.1.2. Types of Road Speed Limiter (RSL) Based on the Functionality [22]

(1) Top-speed limiting

It prevents the vehicle from exceeding a set speed. Most modern vehicle engine-management systems have a top speed setting but it is usually well in-excess of maximum national speed limits and could not be regarded as a safety device.

(2) Speed alarm set by the driver

It alerts the driver if a selected speed is exceeded. Some vehicles have this feature.

(3) Speed limiter set by the driver

It prevents the vehicle from exceeding the selected speed, except for temporary over-ride situations (*e.g.*, "kick-down" of accelerator pedal). A few vehicle models have this feature (*e.g.*, Renault Megane). These are also known as "Adjustable Speed Limitation Function (*ASLF*)".

(4) Intelligent speed alarm

The system "knows" the speed limit of the current section of road and direction of travel and alerts the driver if that speed is exceeded by an audible alarm, a visual signal or a vibrating throttle pedal or a combination of these.

(5) Intelligent speed limiter

The system "knows" the speed limit of the current section of road and direction of travel and prevents the vehicle from

being accelerated beyond this speed.

3.2. Limitations of Manually-Set Road Speed Limiter (RSL)

Systems that require the driver to manually set the speed have several limitations, namely:

(1) They assume that the driver knows the speed limit or can decide a "safe" speed. Actually, in both situations the driver can be in serious error.

(2) The task of setting the speed is tedious and may be distracting.

(3) In practice, these voluntary systems are unlikely to be used on a regular basis.

3.3. Examples of Automatically-Set Road Speed Limiter (RSL)

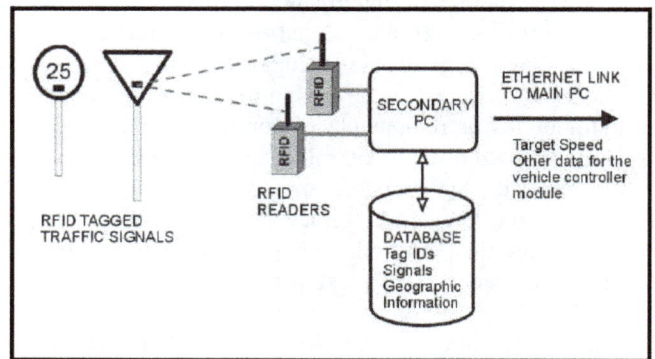

Fig 3. Operational block diagram of the RFID subsystem [3].

(1) First example [3]

It is an intelligent speed controller that is based upon Radio Frequency Identification (*RFID*) technology for identification of the traffic signals on the road. The operation of the *RFID* subsystem onboard the vehicle is described with the block diagram of Fig. 3.

The proposed architecture of Ref. [3] for cruise control is shown in Fig. 4. It comprises two parts: placement of *RFID* sensors (tags) in the road's traffic signals, and the on-board systems in the vehicle.

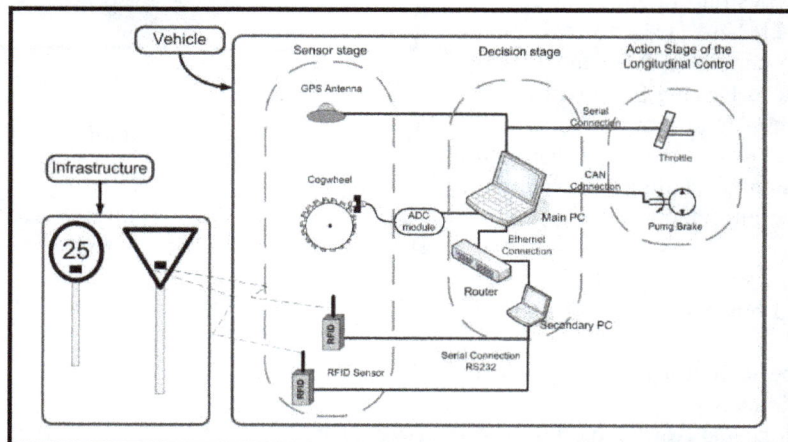

Fig 4. Control scheme onboard the vehicle and its interaction with the infrastructure [3].

(2) Second example [4]

A microcontroller unit was used to receive the pedal position from the corresponding sensor, Fig. 5. Then, the microcontroller unit transfers appropriate signal to the Electronic Control Unit (*ECU*) that in turn controls the automobiles' throttle position, Fig. 6. The microcontroller unit can also interface with a wireless module that is capable of detecting any other transceiver through radio frequency (*RF*) signals.

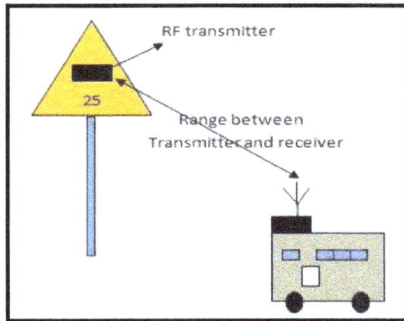

Fig 5. Traffic-signal-posts equipped with RF transmitter (left side) and automobile equipped with RF receiver (right side) [4].

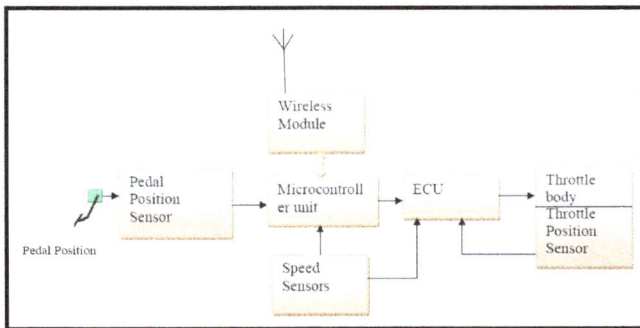

Fig 6. Hard scheme of automobile control system [4].

4. Present Proposed Speed Limiter Model

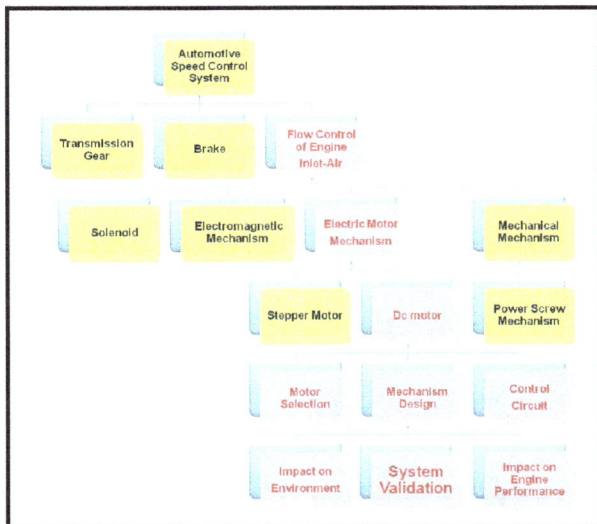

Fig 7. Main ideas of automotive speed control systems.

Based on the two examples of RSL that are shown in Sec. 3.3, we adopted the general idea to propose a much simpler and less-expensive RSL. Before carrying out the present speed limiter, we had to survey the available techniques for speed control and choose the promising technique from our point of view. Main ideas about automotive speed control systems are shown in Fig. 7. The blocks in "red" are adopted in the present speed limiter model.

4.1. Engine Specifications of the Present Work

The main specifications of the engine that was used in the present investigation are listed in table (1).

Figure 8 illustrates the main components of the present investigated engine.

Table 1. Main specifications of the present engine.

Item	Specifications
Type	Four-stroke, Two valves per cylinder
Bore × stroke	85.00 mm × 70.00 mm
Bore / stroke ratio	1.21
Displacement	1588 cm3 (96.90 in3)
Compression ratio	9
Fuel system	One air carburetor
Max. output	55.9 kW at 5200 rpm
Max. torque	115 Nm at 3800 rpm
Coolant	Water

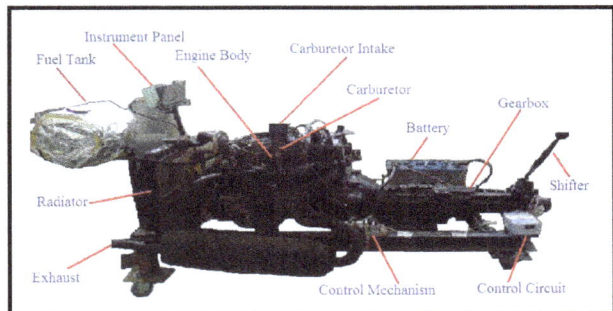

Fig 8. Main components of the present investigated engine.

4.2. Accelerator System

4.2.1. Carburetor

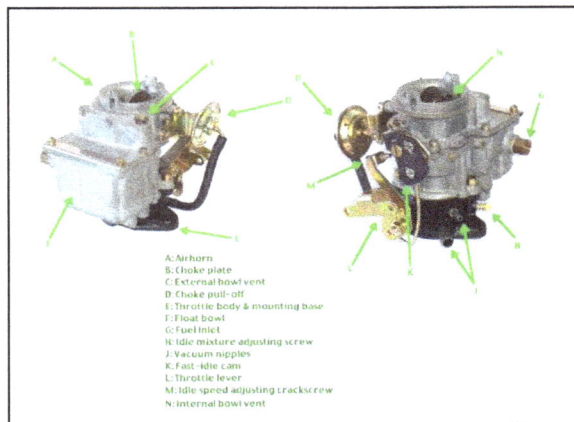

Fig 9. Main parts of the carburetor [24].

A carburetor is basically a device for mixing air and fuel in the correct amounts for efficient combustion. It is a network of passages and related parts that help control the air-fuel ratio under specific engine-operating conditions. Figure 9 shows the main parts of a conventional carburetor. The carburetor bolts to the engine intake manifold. The air cleaner (filter) fits over the top of the carburetor to trap dust and dirt.

4.2.2. Throttle Arm

Figure 10 demonstrates the throttle arm that controls the opening of the carburetor of the present engine. The accelerator steel wire transmits the mechanical signal of the accelerator pedal to the throttle arm. The present speed limiter model is based on resisting the response (rotation) of the throttle arm to the signal of the accelerator pedal. This resistance is applied by a suitable mechanism that obeys the signal of an electronic circuit according to the permissible speed limit.

Fig 10. Throttle arm of the carburetor of the present engine.

4.3. Speed Calibration

As the engine was not really operating a vehicle, the testing of the validation of the speed limiter was based on the rotational speed of the engine (*RPM*). Thus, there was a real need to carry out a calibration procedure to relate the rotational speed of the engine to the readings of the speedometer, which reads the corresponding speed of the vehicle (*km/h*). Figure 11 shows the used arrangement of the speedometer and the reader of the engine rotational speed (*RPM*).

The calibration was carried out in a special workshop with a suitable facility using a real car.

Fig 11. Arrangement of speedometer and reader of the engine rotational speed.

Table (2) illustrates the results of the calibration process. These results were used for the adjustment of the control mechanism as well as the programming of the control circuit.

To ease the control process, eight values of the vehicle speed (*km/h*) were selected and the corresponding values of the engine rotational speed (*RPM*) were recorded. The values that appear in table (2) are average values.

Table 2. Results of the calibration process

Stage No.	Engine speed (*RPM*)	Vehicle speed (*Km/h*)
1	500	20
2	1000	30
3	1500	40
4	2000	55
5	2500	70
6	3000	80
7	3500	105
8	4000	110

4.4. Force Measurement

For proper design and application of the control mechanism, the force required to resist the pulling of the accelerator cable must be estimated at every stage of table (2). Thus, the vehicle speed can be limited by preventing excess opening of the carburetor. As shown in Fig. 12, a spring balance was hooked to the throttle arm to measure the resistance force (F_r). These values of resistance force (F_r) were used to design the control mechanism and program the control circuit.

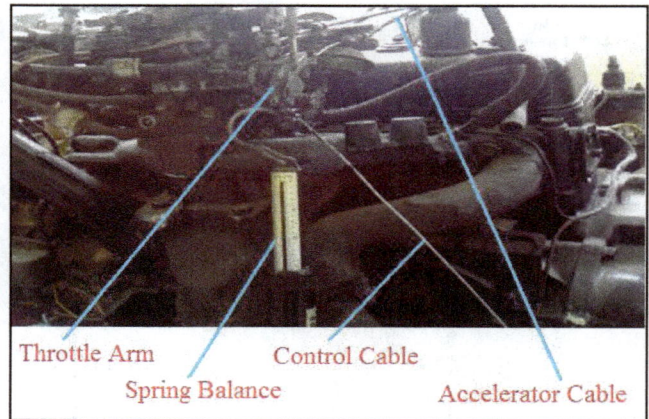

Fig 12. Measurement of the control force using a spring balance.

4.5. Control Mechanism

Fig 13a. First view.

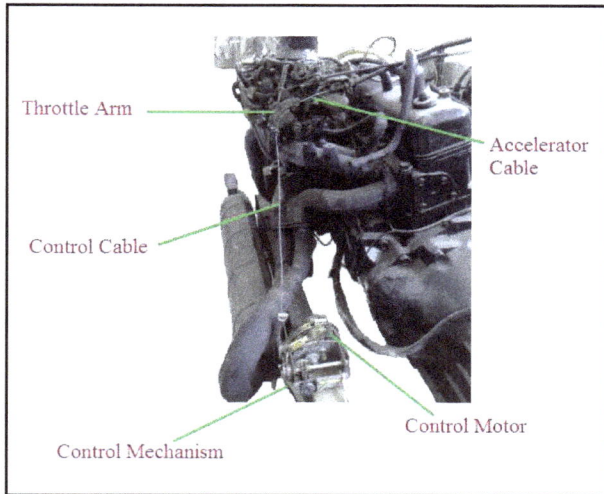

Fig 13b. *Second view.*

Fig 13. *Two pictures of the control mechanism in its place.*

The job of the control mechanism is to produce the necessary force to limit the movement of the throttle arm of the carburetor.

Figure 13 shows two pictures of the control mechanism in its place relative to the other components of the engine.

4.5.1 Main Components of the Control Mechanism

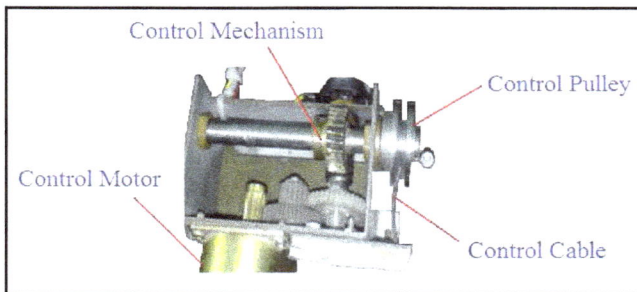

Fig 14. *Main components of the control mechanism.*

Figure 14 shows the main components of the control mechanism. These components can be listed as:

(1) Control motor

The function of the control DC motor is to produce the necessary torque to resist the movement of the throttle arm that obeys the accelerator cable. Figure 15 shows the DC motor that was used in the present work.

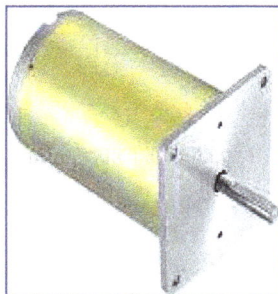

Fig 15. *Picture of the used DC motor.*

This *DC* motor was selected to be suitable for the torque requirements and compatible with the control circuit. The motor is connected to the pulley through the gear train, Fig. 14.

(2) Wire-pulley assembly

The wire-pulley assembly is used to transfer the limiting torque of the motor to the throttle arm. The steel wire is welded to the pulley. The dimensions of the pulley were chosen to be suitable to the overall response of the control mechanism. Figure 16 shows a drawing of the pulley which was fabricated from steel.

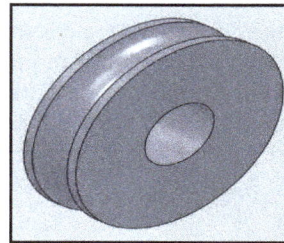

Fig 16. *A drawing of the steel pulley.*

(3) Gear train

A gear train was designed and fabricated to transfer the torque and movement of the *DC* motor to the shaft of the pulley, Fig. 14. The gear train contains a combination of spur and worm gears. The overall gear ratio of the train was considered in the programming of the control circuit. Figure 17 shows a drawing of the gear train. Some of the gears were fabricated from metallic materials and others from plastic materials.

Fig 17. *A drawing of the gear train.*

4.5.2. Design Steps

The design process of the present control mechanism can be summarized in the following steps:

(1) Considering the maximum value of the force (F_r) that is required to limit the pulling movement of the accelerator wire.
(2) Determination of the necessary pulley torque (T_p) based on the maximum limiting force (F_r) and the diameter of the pulley (d_p).
(3) Calculating the overall gear reduction-ratio based on the required pulley torque (T_p) and the maximum torque of

the available *DC* motors. Also, the ratio between the rotational movements of the shaft of the pulley and the shaft of the motor.

(4) Finding the required motor torque (T_m) based on the above parameters.

(5) Final step is the motor selection from the available *DC* motors. Suitability to the control circuit is an essential parameter in motor selection.

4.6. Present Control Circuit

An electronic circuit, Fig. 18, was designed and assembled to control the operation of the *DC* motor. Thus, the necessary torque and rotation of the *DC* motor are performed that result in resisting the movement of the accelerator cable. The final result is the speed limiting of the vehicle at different speed stages, table (2). Figure 19 shows a block diagram of the control circuit.

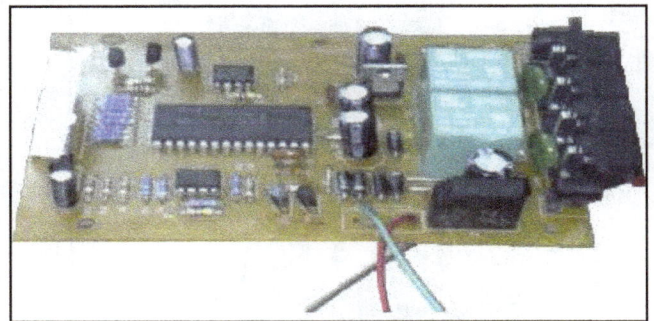

Fig 18. Picture of the present control circuit.

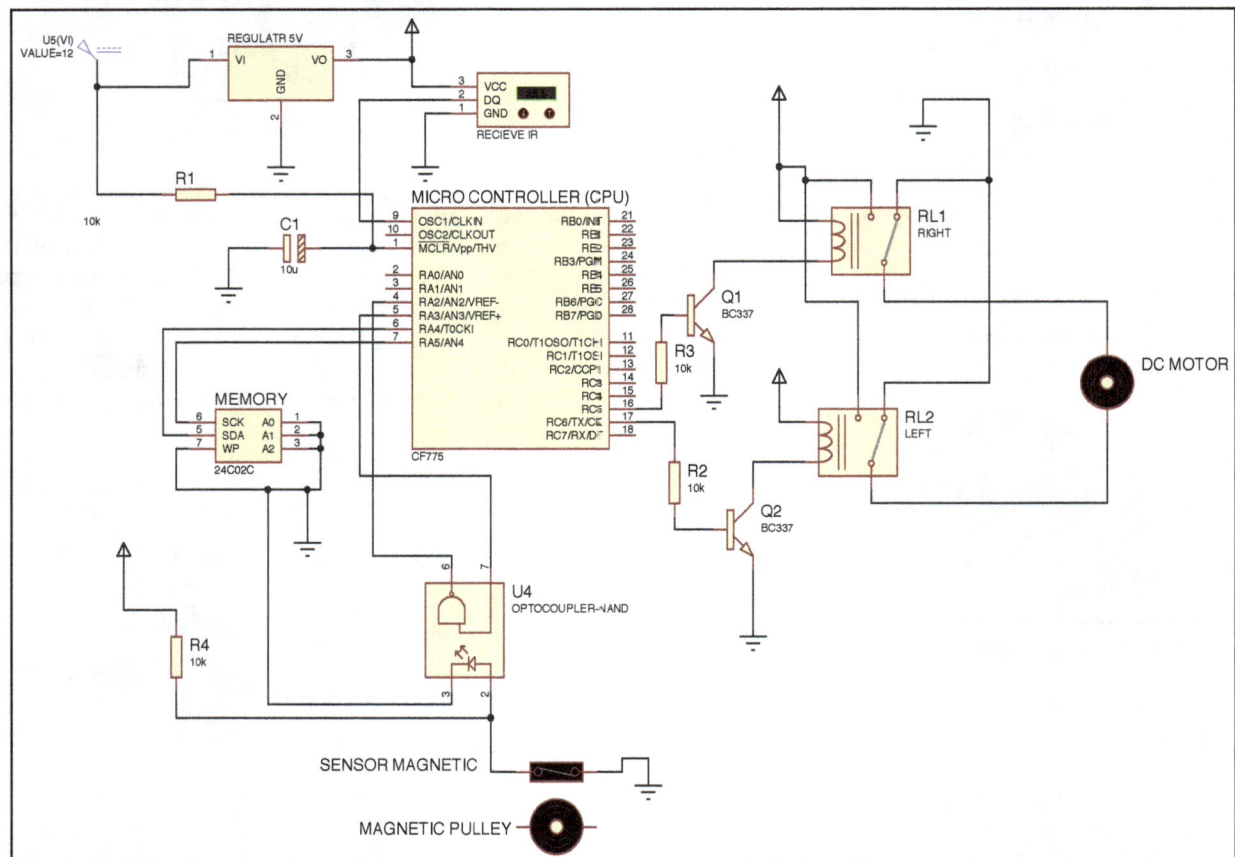

Fig 19. Block diagram of the present control circuit.

4.6.1. Components of the Present Control Circuit

The main components of the present control circuit, Fig. 18, can be listed as:

1. *Remote-control unit:* It sends the signal to the circuit receiver using infrared waves (*IR*). The signal informs the control circuit the stage of vehicle speed as shown in table (2).

2. *Receiver:* It receives the signal from the remote-control unit.

3. *Screen:* It shows the data of the operating conditions.

4. *Amplifiers:* They amplify the signal that comes from the remote-control unit and convert the microcontroller command signals into the power necessary to energize the *DC* motor windings.

5. *Microcontroller:* It is the most important component of the circuit. The microcontroller is a microprocessor capable of generating steps pulses and direction signals for the driver (amplifier). It is programmed to control the action of the *DC* motor to perform the necessary operation to limit the vehicle speed at each stage of table

(2). The microcontroller was programmed by the commercial software "MIKROBASIC" [25].

6. *Two relays:* The relay is a switch that is used to control a circuit by a low-power signal. Thus, the two relays are used to control the operation of the electronic circuit.

7. *Transistors:* A transistor is a semiconductor device that is used to amplify and switch the electronic signals and electrical power of the control circuit.

8. *Memory:* It is used to save vehicle speed at each stage of table (2).

9. *Regulator:* It keeps the voltage of the circuit at $5V$.

10. *Battery (12V):* It provides the DC current to the control circuit.

4.6.2. Programming of the Microcontroller

The microcontroller was programmed by the commercial software "MIKROBASIC, *V.5.0*" [25]. This software was developed as a simple programming tool of microcontrollers. Its language is similar and has characteristics of many BASIC implementations with an editing environment to allow the user to create, test and deploy BASIC programs. MIKROBASIC allows math and string operations. Also, a series of commands to control the program flow and *I/O* operations are implemented in its interpreter engine. Generally, the software disk contains two binaries: "based.exe" and "mbasic.exe" and a set of example programs. The binary "based.exe" is the editor environment and can be used to edit and run MIKROBASIC programs. The binary "mbasic.exe" is the actual interpreter engine and it is necessary to run the developed programs. The software engine is automatically called by the editor environment. Such functionality makes very easy to develop and test applications inside the editor environment.

4.6.3. Operation Steps of the Speed Control System

The automotive speed control system can be operated by carrying out these steps:

1. Switching on the electronic circuit.

2. Pressing one of the buttons of the remote-control unit. Every button represents specific vehicle velocity for the engine, table (2).

3. The signal is received to the electronic circuit by the receiver and amplified by the amplifier.

4. The microcontroller compares the previous value of speed that is stored in the circuit memory with the new value that was received from the remote-control unit.

5. As a result of this comparison, the microcontroller gives the correct order to the *DC* motor to rotate by the proper amount. This order is given based on the pre-programming of the microcontroller.

6. The motion is transferred to the shaft of the pulley through the gear train.

7. The pulley rotates to take a new position to fix the speed of the car. The rotation may be either clockwise or anti-clockwise to increase or decrease the resistance (F_r) to the accelerator cable at the throttle arm.

8. The microcontroller checks the new position and makes necessary adjustments until reaching the correct pre-set

position.

4.6.4. Important Remarks on the Present Control System

There are some important remarks that have to be recorded concerning the present control system, namely:

(1) The present speed control system was successfully implemented and tested. Careful observation of the operation of the control system was paid. The control system succeeded in limiting the engine speed according to the pre-set vehicle-speed stages of table (2).

(2) The signal of the remote-control unit that is used in the present model is a symbol of the external control signal to the vehicle speed control system. The infra-red (*IR*) signal is used for the demonstration of the idea of the speed control system. In real-life applications the control signal can be received from the control transmitter on the road itself or directly from satellites. This control signal is most likely to be a radio frequency (*RF*) signal.

(3) A stepper motor is a good choice whenever controlled movement is required. It can be used with advantages in applications where controlling of rotation speed, speed and position is needed. However, we found that the stepper motor torque is weak to overcome the force of the accelerator pedal and this torque decreases with increasing the speed of the stepper motor. So, we rejected the idea of using a stepper motor in the present control system.

(4) Experiments were carried out to make sure that there was no effect of the present speed control system on the engine performance. Experiments revealed that there is no change on either the air/fuel ratio or the percentages of the exhaust components.

(5) It should be noted that the present test of the speed control system was carried out without actually loading the engine. This situation is due to the lack of suitable facility and resources to carry out such a test. Thus, we strongly recommend testing the present speed control system in real-life operating conditions. We think that the only difference is a new programming of the microcontroller to adjust for the loading conditions.

5. Conclusions

Based on the previous explanations and discussions, the following concluding points can be stated:

1. The present speed control system was successfully implemented and tested. The control system succeeded in limiting the engine speed according to the pre-set vehicle speed limits. Thus, the objective of the investigation was achieved.

2. Generally, the present mechanism has no noticeable bad effect on the engine performance. There is no change on either the air/fuel ratio or the percentages of the exhaust components.

3. The present proposed mechanism is simple, inexpensive

and can be added to the engine with minimum modifications. It is suitable to be implemented in developing countries where a big number of cars still work using the conventional carburetor mechanism.

4. It is a challenging job to design and implement modifications to existing systems.

Recommendations for Future Work

Based on the present work, the following points can be recommended for future investigations:

1. Using the radio frequency (*RF*) signal as the input signal to the control circuit.
2. Using satellite signal, based on the Global Positioning System (*GPS*), as input signal to the control circuit.
3. Development a new control circuit to be suitable to modern cars that have a fuel-injection arrangement.
4. Utilizing other control techniques and hardware such as neural networks, fuzzy logic, fuzzy-neural, programmable logic controller (*PLC*), *etc*.

Acknowledgements

The authors would like to acknowledge Eng. M. Gameaa, Eng. M. Ghorab and their colleagues for helping in the assembly, programming, and testing of the control circuit as well as the manufacturing of the control mechanism.

Nomenclature

d_p	Diameter of the pulley
T_m	Motor torque
T_p	Pulley torque
F_r	Resistance force

Abbreviations

ASLF	Adjustable Speed Limitation Function
DC	Direct Current
ECU	Electronic Control Unit
RPM	Engine rotational speed
GPS	Global Positioning System
I2V	Infrastructure to Vehicles
IR	Infrared Waves
ISA	Intelligent Speed Adaptation
PLC	Programmable Logic Controller
RF	Radio Frequency
RFID	Radio Frequency Identification
RSL	Road Speed Limiter

References

[1]   http://www.makeroadssafe.org/about/Pages/Issues.aspx: March 2013

[2]   M. Paine, "Devices to Assist Drivers to Comply with Speed Limits", Vehicle Design and Research Pty Limited, Australian Business No. 63 003 980 809, January 2009, mpaineATtpg.com.au

[3]   J. Pérez, F. Seco, V. Milanés, A. Jiménez, J. C. Díaz, and T. de Pedro, " An RFID-Based Intelligent Vehicle Speed Controller Using Active Traffic Signals", Sensors-Open Access Journal, ISSN 1424-8220, Vol. 10, pp. 5872-5887, 2010.

[4]   U. J. Kameswari, M. Satwik, A. Lokesh, and G. V. Reddy, "A Design Model for Automatic Vehicle Speed Controller", International Journal of Computer Applications (0975 – 8887), Volume 35, No.9, pp. 19-24, December 2011.

[5]   B. G. Radin, O. Park, and L. J. Vanderberg, "Maximum Vehicle Speed Limiter", United States Patent Office, No. 3,520,380, July 14, 1970. Assignee: "Ford Motor Company, dearborn, Michigan, USA".

[6]   L. F. Mieras, "Maximum Engine Speed Limiter ", United States Patent, No. 3,563,219, Feb. 16, 1971. Assignee: "Ford Motor Company, dearborn, Michigan, USA".

[7]   Z. J. Jania, and L. J. Vanderberg, "Maximum Vehicle Speed Limiter", United States Patent, No. 3,708,031, Jan. 2, 1973. Assignee: "Ford Motor Company, dearborn, Michigan, USA".

[8]   W. A. Snell, and A. R. Fillman, "Engine Maximum Speed Limiter", United States Patent, No. 5,549,089, Aug. 27, 1996. Assignee:"Textron Inc., Providence, R. I".

[9]   R. Fiske, A. Surabian, and K. Weigold, "Motor Vehicle Operator Identification and Maximum Speed Limiter", United States Patent, No. 7,757,803 B2, July 20, 2010.

[10] R. Fiske, A. Surabian, and K. Weigold, "Motor Vehicle Operator Identification and Maximum Speed Limiter", United States Patent, No. 7,959,177 B2, June 14, 2011. Assignee: "Kar Enterprises, LLC, Shrewsbury, MA, USA".

[11] F. H. Walz, M. Hoefliger, and W. Fehlmann, "Speed Limit Reduction from 60 to 50 Km/h and Pedestrian Injuries", Twenty-Seventh Stapp Car Crash Conference Proceedings (P-134) with International Research Committee on Biokinetics of Impacts (IRCOBI), San Diego, California, October 17-19, 1983.

[12] A. J. McLean, R. W. G. Anderson, M. J. B. Farmer, B. H. Lee, and C. G. Brooks, "Vehicle Travel Speeds and The Incidence of Fatal Pedestrian Collisions-Vol. I", NHMRC Road Accident Research Unit, The University of Adelaide, South Australia, for the Federal Office of Road Safety, Report No. CR146, October 1994.

[13] S. Ferguson, "Relation of Speed and Speed Limits to Crashes", National Forum on Speeding, Washington, D.C., USA, June 15, 2005.

[14] N. V. Malyshkina, and F. Mannering, "Effect of Increases in Speed Limits on Severities of Injuries in Accidents", Journal of the Transportation Research Board, No. 2083, pp. 122-127, 2008.

[15] S. Comte, "Evaluation of In-Car Speed Limiters: Simulator Study", Master Project, , Project Funded by he European Commission under the Transport RTD Program of the 4th Framework Program, Contract No. Ro-96-Sc.202, May 1998.

[16] "Learning from Others: An International Study on Heavy Truck Speed Limiters", Prepared for Transport Canada on behalf of the Council of Deputy Ministers Responsible for Transportation and Highway Safety, Canada, March 2008.

[17] "Safety Impacts of Speed Limiter Device Installations on Commercial Trucks and Buses", Commercial Truck and Bus Safety Synthesis Program, CTBSSP Synthesis 16, Sponsored by the Federal Motor Carrier Safety Administration, Transportation Research Board, Washington, D.C., 2008, www.trb.org

[18] R. J. Hanowski, G. Bergoffen, J. S. Hickman, F. Guo, D. Murray, R. Bishop, S. Johnson, and M. Camden, "Research on the Safety Impacts of Speed Limiter Device Installations on Commercial Motor Vehicles: Phase II", U.S. Department of Transportation, Federal Motor Carrier Safety Administration, Report No. FMCSA-RRR-12-006, March 2012.

[19] T. Mäkinen, and A. Várhelyi, "Field Trials with In-Car Speed Limiter", 9th International Conference Road Safety in Europe, Bergisch Gladbach, Germany, September 21-23, 1998.

[20] A. Várhelyi, and T. Mäkinen, "The Effects of In-Car Speed Limiters: Field Studies", Pergamon-Transport Research Part C, pp. 191-211, 2001.

[21] L. Duynstee, and H. Katteler, "Acceptance of Intelligent Speed Adaptation in Passenger Cars by Car Drivers", Proceedings of the 7th World Congress on Intelligent Transport Systems (ITS), Torino, Italy, 6-9 November 2000.

[22] J. M. Abraham, "Analysis of Highway Speed Limits", Department of Civil Engineering, Faculty of Applied Science and Engineering, University of Toronto, Italy, December 2001.

[23] "Speed management: A road safety manual for decision-makers and practitioners", 2008, http://www.autograde.ae/SPEED%20LIMITER_GLOBAL.pdf : March 2013

[24] http://en.wikipedia.org/wiki/File:CarbNomenclature.jpg: March 2013

[25] http://www.mikroe.com/mikrobasic/: March 2013

6

Tele-operated robotic arm and hand with intuitive control and haptic feedback

Monica Dascalu[1, 2], Mihail Stefan Teodorescu[1], Anca Plavitu[2, 3], Lucian Milea[1], Eduard Franti[2, 4], Dan Coroama[1], Doina Moraru[1]

[1]Faculty of Electronics and Telecommunications, Politehnica University of Bucharest, Bucharest, Romania
[2]Center for New Electronic Architecture, Research Institute for Artificial Intelligence, Bucharest, Romania
[3]Faculty - Exact Sciences and Engineering, Hyperion University, Bucharest, Romania
[4]National Institute for Research and Development in Microtechnologies, Bucharest, Romania

Email address:

monicad@artsoc.ro (M. Dascalu), mihait@artsoc.ro (M. Teodorescu), ancap@artsoc.ro (A. Plavitu), edif@artsoc.ro (E. Franti), lucian@artsoc.ro (L. Milea), dcoroama@artsoc.ro (D. Coroama), doinam@artsoc.ro (D. Moraru)

Abstract: The paper presents a robotic arm having as end effector an anthropomorphic hand and its control system. The robotic arm and hand are controlled using a Complex Interactive Control Glove (CICG) and operator joint sensors. The robotic hand imitates the finger and joint movements of the human operator. The anthropomorphic hand sends pressure feedback from a pressure sensor array mounted at the robotic hand's fingers and palm to the human operator wearing a Complex Interactive Control Glove that comprises haptic actuators. The pressure exerted by the robotic hand on various objects is perceived as vibrations on the corresponding hand area of the human operator. The robotic arm adjusts its position in correlation with the human operator's arm, placing the end effector at the right position, corresponding to the operator's hand. Data for the movement of the robotic arm are collected from the movements of the human operator by means of three joint sensors placed on the shoulder, elbow and hand wrist. Targeted applications of the tele-operated robotic arm and hand with intuitive control and haptic feedback include all situations where a human-like operation is needed in a hazardous or remote environment: space environment, operations executed in toxic atmosphere, working in high-radiation level environments, marine applications. In such cases, the robotic hand and arm that are executing the same movements as the human operator can replace the actual human operator. This will control the robotic arm form a safe, possibly remote, environment, and will be able to process the haptic feedback of the systems.

Keywords: Robotic Arm, Robotic Anthropomorphic Hand, Haptic Feedback, Complex Interactive Control Glove, Hazardous Environments

1. Introduction

In the past years, robotic arms and robotic hands have gained more and more attention due to their diversified applications in the industrial field. Significant advances were also made in medical domain, military applications, marine and space exploration, and even entertainment and home applications. Robotic arms usually use as end effector a task-specific designed gripper. Robotic arms are nowadays available in a variety of realizations, from industrial types, with ranges, degrees of freedom, speed and force by far exceeding the ones of a human arm, to anthropomorphic robotic arms that partially or nearly fully reproduce the capabilities of a human arm.

Some basic characteristics that define the performance of a robotic arm are: number of limbs and joints, degrees of freedom, force, speed, accuracy, repeatability, performances of the control system [1]. A robotic arm has attached an end effector that is specifically designed to interact with the environment and its exact nature depends on the application of the robot. Usually the end effector consists of a gripper or a tool and in some cases of an anthropomorphic hand. When the end effector is a tool, it serves various purposes, such as spot welding in an assembly, spray painting where uniformity of painting is necessary and for other purposes where the working conditions are dangerous for human beings. Surgical robots have end effectors that are specifically manufactured. In this case the robot arm can be used only for that particular purpose, any other operation requiring a tool change, if possible.

When referring to robotic prehension there are four general categories of robot grippers.

- Impactive – claw-like grippers
- Ingressive – needle type gripper
- Astrictive – suction gripper
- Contigutive – require glues, freezing or other types of adhesion.

The most known impactive grippers consist of two, three or even five fingers. They can have different degrees of freedom (DOF) and can be used in different environments exerting various movement types and forces. A particular case occurs when the gripper is an anthropomorphic hand, which is useful or even imperative when the robotic harm has to perform operations with human-like dexterity, replacing a human in hazardous environments, hard-to-reach places or wherever a life-threatening or risky situation appears. One of the key issues in the field is the control system, mainly the human-robot interaction [2] [3].

Difficulties an challenges encountered in designing robotic anthropomorphic hands comes first from the decomposition of continuous natural movement of the hand in distinct components that can be imitated by the artificial hand. Other typical limitations occurs in reproducing/implementing the feedback data from a human hand in an artificial one. Tactile, temperature, position and force feedback would be very useful in controlling the hand and taking the right decisions, but even very modern sensors cannot cover the whole range of natural sensory feedback.

Different methods are proposed in scientific literature in order to improve the feedback of a robotic hand [4]. A number of spots for tactile feedback can be applied to the robotic hand and the force feedback can be read from the power absorbed by the driving motors. Many other methods to gain sensory feedback are used, considering also accuracy and cost. The computing power needed to process all the incoming data and run the implemented software algorithms is also to be considered.

Another important challenge regards the control method of the anthropomorphic hand. The dynamics of the human hand can only be partially modeled and dexterity implies brain-like computing power, by far not available nowadays. Automatized procedures for limited fixed tasks or a human operator for unforeseen operations are the best choice for controlling the robot hand and arm. Preprogrammed sets of movements can be stored in a computer memory and put in use for the appropriate situations. If a human operator is implied, that remotely controls the robotic hand attached to a robotic arm, several control methods have been researched and applied [5] [6] [7] [8].

The traditional method is controlling the robotic limbs with a joystick-type controller and/or a keyboard. This is also the most difficult method requiring a specific training for the operator since the movements of the operator and the desired movements of the robot differ significantly. Consequently, no force or pressure feedback can be given directly to the operator and usually few or no warning signals are used.

Another much more intuitive method is to have a duplicate robotic arm that is physically moved by the operator. All the movements of the joints are sensed and reproduced in the operating remote arm, as accurate as possible and in real time. This method implies mostly the control of a robotic arm and a rough end effector, but it can also offer force feedback to the operator, if constructed accordingly.

The operator driven arm and hand can also take the shape of an exoskeleton covering the human operators shoulder, arm and even fingers, which could produce maximum accuracy in reproducing the operator's movements and also some force feedback, but it implies a mechanical and electrical complexity difficult to deal with.

Other recently reported control methods are position and motion detectors for the operators' arm or even mind controlled robots, but research is only at the beginning [9] [10]. The sensors used to track movement and position can be accelerometers, Hall Effect sensors, potentiometers, tension sensors or others.

For teleoperated robots, haptic feedback from the robotic hand implies using tactile sensors that reproduce the sense of touch, sending the corresponding signals to the human operator or control system and again translating the signals in visual, audible or preferable tactile stimuli sensed by the operator [4]. The haptic actuator provides tactile sensing by means of vibration to the human operators hand. The vibrations can be produced by different actuator types.

The simpler and cost effective option is the Eccentric Rotating Mass vibration motor, or ERM. Linear resonant actuators (LRA) are widely used in haptic feedback applications because of their low response times and feasibility.

The paper presents a robotic arm having as end effector an anthropomorphic hand controlled using a Complex Interactive Control Glove (CICG) and operator joint sensors. The robotic hand imitates the finger and joint movements of the human operator and sends feedback. The pressure exerted by the robotic hand on various objects is perceived as vibrations on the corresponding hand area of the human operator. The robotic arm adjusts its position in correlation with the human operator's arm, placing the end effector at the right position, corresponding to the operator's hand.

The architecture of the system, including control system and the robotic arm, put together for the first time different ideas from our own research and scientific literature. The novelty of the system consists mainly in control of the robotic arm through imitation of the operator's arm (see fig.1), based on flex sensors and sustained by complex processing algorithms. The structure is capable to execute fluently sequences of complex high precision movements that imply combinations of simultaneous movements of the shoulder, the elbow and the wrist. The computing algorithms that process the signals from the sensors and generate the commands for the robotic arm and hand are also new and original – they are only intuitively explained in this paper.

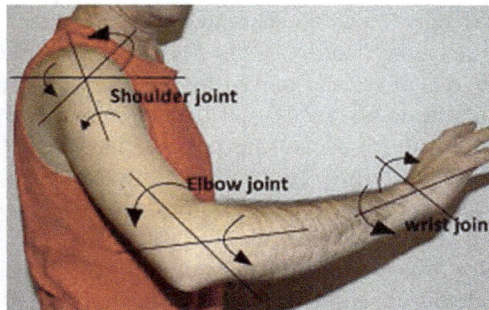

Fig. 1. Movement types of the human arm and hand

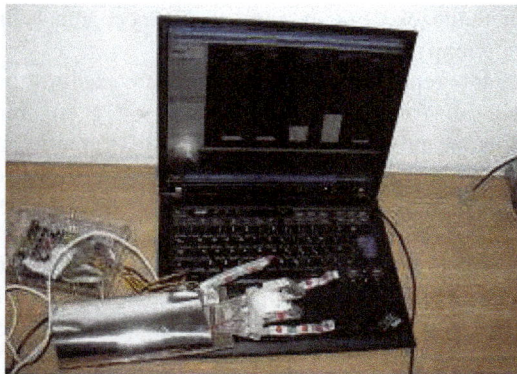

Fig. 2a. First anthropomorphic hand prototype - mechanical structure

2. The Mechanical Structure that Replicate the Human Arm and Hand in the Robotic System

The mechanical robot arm and end effector, as presented in this paper, resembles in principle with the anatomic structure of the complete human arm; it should be able to imitate the main movements of the human arm as shown in fig.1.

The robotic arm consists of a number of joints and links. The mechanical joints are usually restricted to one DOF, which results in simpler control, mechanics and kinematics. A robotic arm attempting to reproduce a human arm consists of 2 moving links connected to a fixed base and three joints, forming a simple kinematic chain. The joint at the base represents the human shoulder, the joint that unites the two links represents the human elbow and the joint situated at the end of this chain connects the end effector, eventually an anthropomorphic hand.

The human hand is considered to have 27 degrees of freedom (DOF): 3 for extension and flexion and one for abduction and adduction, 4 in each finger; the thumb has 5 DOF and there are 6 DOF for the wrist. For the construction of a robotic hand usually some simplifying assumptions are made from the start, like the thumb is considered independent from the other fingers or the adduction/abduction of the finger joints are independent, which correspond to individual control of the fingers.

The mechanical structure of the anthropomorphic hand is

usually a simplification of the human hands structure, being constructed of the same number of joints and links, but having 7 - 20 DOF. Each joint has 1 or 2 DOF.

Upon the completion of the present work, three mechanical structures of the robotic hand were successively adopted and tested. The robotic hand structure resembles the human anatomic structure of the human hand using one DOF joints, which allow efficient grasping of objects maintaining a firm grip. The prototypes described in this section were essentially used in developing the processing algorithms and implementing the control unit able to generate the complex movements of the artificial arm and hand.

The first prototype is presented in fig. 2a without sensors attached, as it first was put in use, and in fig. 2b with pressure sensors attached to the phalanxes and control system.

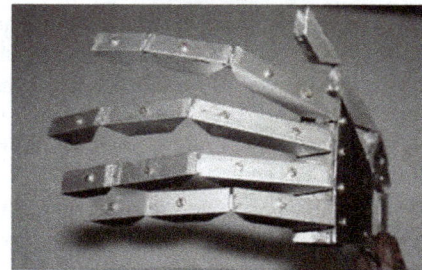

Fig. 2b. First anthropomorphic hand prototype with pressure sensors.

The second developed prototype started from a simple plastic replica of the human hand. Control strings and advanced pressure sensors were attached to the phalanxes, as shown in fig. 3a, resulting in the robotic hand structure presented in fig. 3b.

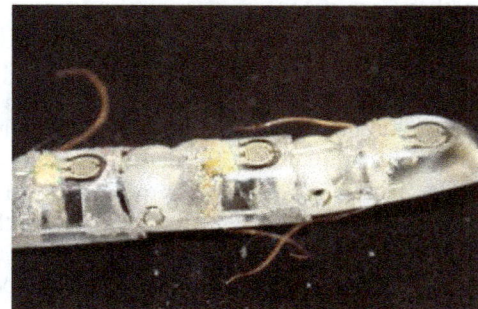

Fig. 3a. Attaching movement strings and pressure sensors to the mechanical structure

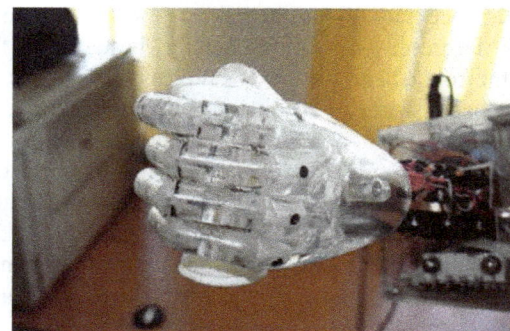

Fig. 3b. Prototype 2 of the robotic hand with haptic feedback

3. Block Diagram of the Robotic System

Our goal is to design a teleoperated robotic arm and anthropomorphic hand with haptic feedback, controlled intuitively by a human operator. The movements of the operator's arm and hand will be exactly reproduced by the robotic system, which provides tactile feedback to the human operator regarding the pressure confronted with. This will enable the possibility of complex and high precision manipulation in unfriendly and hazardous environments, without exposing the human operator to risks. The human operator is equipped with three joint sensors and a control-glove, that will assure performing high-precision tasks with almost no preliminary training required for the operator (the robotic hand having all five articulated and driven fingers and allowing human-like, complex maneuvers).

The block diagram of the implemented robotic system is presented in fig. 4. The robotic arm and hand are remotely controlled by the human operator and imitate his arm and hand movements. The anthropomorphic robotic hand is equipped with a pressure sensor array of 20 sensors distributed on the fingers and palm.

The sensors transmit the force encountered by the hand during manipulation by means of feedback signals to the control unit. The control unit manages to send in real time the sensor signals to a haptic actuator array that comprises also 20 actuators attached to the Complex Interactive Control Glove that is warn by the human operator. The operator perceives the pressure exerted on the robotic hand through vibrations executed by haptic actuators placed on the human hand in the same position as the pressure sensors on the robotic hand. The amplitude of the vibrations is proportional to the pressure level on the robotic hand.

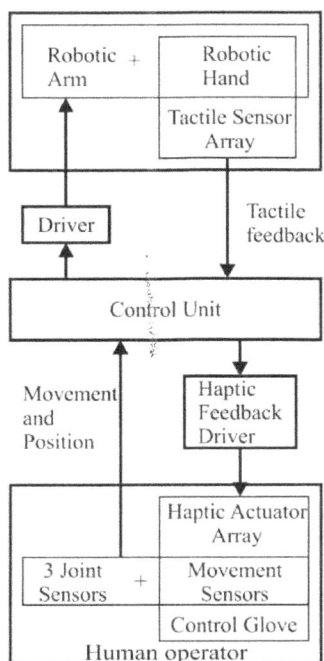

Fig. 4. *Block diagram of the robotic system*

The human operator uses a Complex Interactive Control Glove and three joint sensors to control the movements of the robotic arm and hand. The three joint sensors are situated on the shoulder, the elbow and the wrist and sense their movements. The Complex Interactive Control Glove comprises the tactile actuator array mentioned above and also movement sensors on every finger. The movement and position signals from the joint sensors and the glove are sampled by the control unit, processed and then control signals are sent to the drivers of the robotic arm and hand.

The control function is distributed into sub-systems of lower complexity for each component, simultaneously creating the possibility of implementing the self-management function for achieving the global control. Thus, a hierarchical and distributed architecture for the coordination and control system of the mobile elements in the robotic hand is devised. The main advantage this architecture is the very short reaction time to the control commands the device will receive from the operator, thus ensuring real-time functioning of the mobile elements in the robotic system.

The control unit analyses the movement and position signals from the operators hand and limits the movements of the robotic arm in one of the following situations:

- If the position of the robotic arm and hand that will result consequent to the next movement (performed during a sampling period) will exceed the preprogrammed boundaries. This function is meant to avoid collision between the robot hand and nearby obstacles.
- If the acceleration or speed of the movements executed by the operator exceeds certain values for the three joints and for the fingers. Even if mechanically the arm and hand can perform at the same acceleration or speed as the human operator, some speed levels are considered dangerous for the integrity of the robotic system or for nearby working humans or equipment. Also the possibility of losing grasp on the manipulated object has to be considered due to inertial forces.

The Control unit also analyses the signals from the tactile array and automatically stops the hand grasping movements in one of the following situations:

- If the pressure exerted on an object reaches a preprogrammed value, then this value is maintained to keep the grasping force and is not increased in order to protect the manipulated object from deterioration.
- If the pressure on a specific sensor or sensors exceeds a certain value and also is much higher relative to the pressure on nearby sensors, the grasping force is slightly reduced in order to protect the sensor or the robotic hand from deterioration.

Repetitive movements or operations can be programmed and triggered as desired. During their run, the robotic system functions without the intervention of the human operator performing certain tasks. Thus, the robotic system can function in one of the following three control modes:

- complete human control,
- computer control,

- combined human and computer control.

4. Robotic System Implementation

The first experiments and architectural design were made with the prototypes described in section 2. The final implementation include an artificial hand commercially available equipped with sensor array and various components on the market. The functionality of the control unit is the same with minor parameter adjustment for this model of hand.

A 5 degrees of freedom 10.25" median reach and 13oz lifting capacity robotic arm was used, with a range of motion per axis of 180 degrees. The arm is driven by 4 servo motors located in the base, in the "shoulder", in the "elbow" and in the "wrist".

As end effector for the arm the anthropomorphic MechTE Robot Hand was used, constructed of anodized aircraft aluminum, with 14 points of motion, 5 degrees of freedom, four fingers and thumb open/close. No special force or speed requirements were put on the system, as this is merely an experimental system designed to verify the adopted concepts.

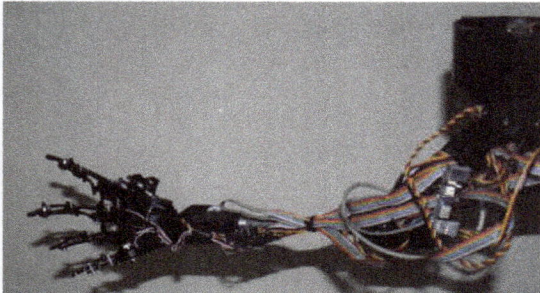

Fig. 5. Robotic arm and hand

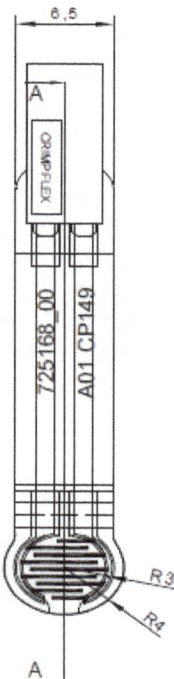

Fig. 6. Pressure sensor CZN-CP6

To the robotic hand there was attached a pressure sensor array with 20 pressure sensors that send signals independently to the control unit, thus assuring the haptic feedback. The pressure sensors are of type CZN-CP6, depicted in fig. 6, with a good response in the force domain similar to the force of a human hand.

The pressure sensors are distributed on the finger's phalanxes and on the palm pressure points, as can be seen in fig. 7.

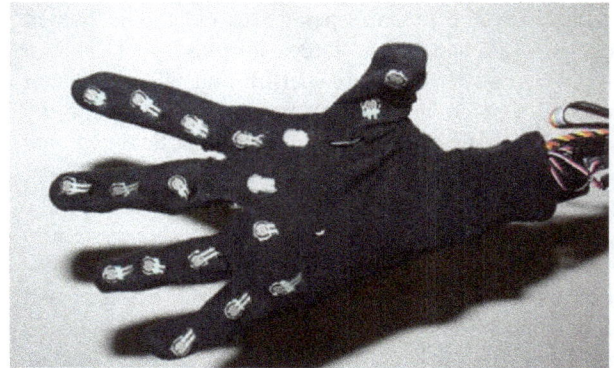

Fig. 7. Pressure sensor array

The Complex Interactive Control Glove shown in fig. 8 is worn by the operator and comprises the haptic feedback actuator array and the movement sensors for each finger. The type of glove used was 5DT Data Glove MRI.

The haptic actuators are of type Pico Vibe 308-100 8mm vibration motor and precision haptic 13mm linear resonant actuators arranged in an array that duplicates the pressure sensor array on the robotic hand.

The signals from the movement sensors are preprocessed by a local control unit attached to the glove, which communicates with the central Control Unit.

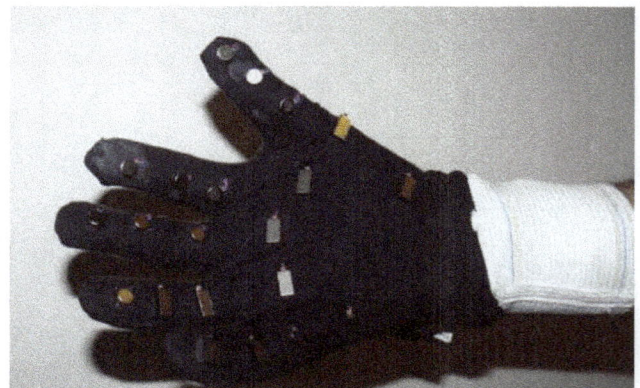

Fig. 8. Complex Interactive Control Glove

The movements of the operator arm is sensed by three joint sensor as presented in fig. 9. The joint sensors are build based on Flex Sensors that offer angle displacement measurement by bending with the robotic arm links. The resistance of the sensors varies proportional with the bending angle and have a very high life cycle.

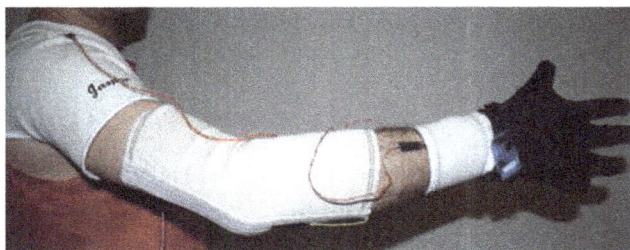

Fig. 9. Joint sensors

As mentioned above, the movements of the arm and hand can be controlled by a human operator, but also by the control unit independently or in collaboration.

5. Experiments and Results

The experiments conducted so far intended to validate the architecture and the processing algorithms. First issue addressed was if the robotic arm and hand really imitate the movements of the operator. Different combinations of movements, sequences of movements varying in distance and rotation were applied in order to verify the efficacy of the control method. The robotic arm and hand are moving as expected.

Relevant images are shown in fig. 10.

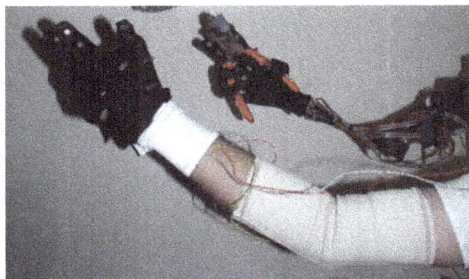

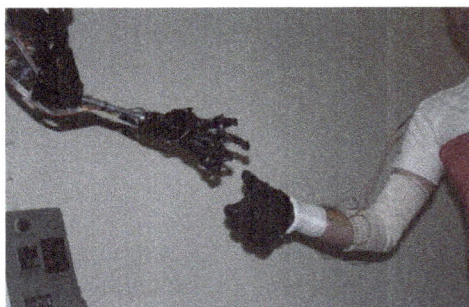

Fig. 10. Experiments conducted on the robotic system

The preliminary experiments conducted on the system showed promising results, as:

- The robotic hand and arm respond in real time to the operator's movements.
- The robotic hand and arm movements are controlled in range, force and speed according to the programmed safety limitations.
- The haptic feedback from the sensor array functions accurately providing the necessary data to the control

unit and clear intuitive feedback to the operator.
- The control of the arm and hand are very intuitive and need no prior training.
- The accuracy of positioning is satisfactory and enables complex maneuvers, like handling pliers.
- The hand does not lose grip on the object during movements.

6. Conclusions

A robotic system comprising a robotic arm and an anthropomorphic hand that are remotely operated by a human operator and providing haptic feedback was designed and implemented. The system is suitable for hazardous environment applications, as the operator remains in a safe location performing complex human-specific handling operations by means of the robotic system.

The teleoperation function uses a Complex Interactive Control Glove and three joint sensors warn by the operator. The telepresence function is performed by an array of pressure sensors mounted on the robotic hand and an array of haptic actuators that convert the pressure in vibrations perceived by the operator's hand on the corresponding locations of the pressure sensors.

The experimental model proves the validity of the telepresence and teleoperation solutions that were adopted and also the control methods implemented in the control unit. The adopted control method is very intuitive and needs little prior training of the operator. The movement precision, the lag in the mechanical and the computer processing of the movement response have been experimentally found satisfactory and the robotic system is proven to be fully operational.

Acknowledgement

This work was supported by a grant of the Romanian National Authority for Scientific Research, Programme for research - Space Technology and Advanced Research - STAR, project number 82/2013.

References

[1] S. Goto, Industrial Robotics: *Theory, Modeling and Control (Force-free control for flexible motion of industrial articulated robot arm)* , Advanced Robotic Systems International, Chapter 30, pp. 813-840, Proliteratur Verlag, 2007.

[2] J. Richer and J. L. Drury, A *Video Game - Based Framework for Analyzing Human - Robot Interaction*, in proceeding s of the ACM conference on Human - Robot Interaction - HRI '06 , 2006, pp . 266.

[3] D. Sakamoto, K. Honda, M. Inami, and T. Igarashi, *Sketch and Run*, in proceedings of the international conference on Human Factors in Computing Systems - CHI '09 , 2009, pp . 197.

[4] C. Glover, B. Russell, A. White, M. Miller, and A. Stoytchev, *An Effective and Intuitive Control Interface for Remote Robot Teleoperation with Complete Haptic Feedback*, in proceedings of the Emerging Technologies Conference - ETC, 2009.

[5]   S. Goto, *Advances in Robot Manipulators (Industrial Robotics: Teleoperation System of Industrial Articulated Robot Arms by Using Forcefree Control)*, Advanced Robotic Systems International, INTECH, 2010.

[6]   S. Goto, T. Naka, Y. Matsuda and N. Egashira, *Teleoperation System of Robot Arms Combined with Remote Control and Visual Servo Control*, Proceedings of the SICE Annual Conference 2010, August 18-21, Taipei, Taiwan, 2010.

[7]   Velagic, J., Coralic, M. and Hebibovic, M. (2004). *The Remote Control of Robot Manipulator for Precise Time-Limited Complex Path Tracking*, Proceedings of the IEEE International Conference on Mechatronics and Robotics (MechRob2004), Volume 2, September 13-15, Aachen, Germany, pp. 841-846

[8]   D. Lee, and M.W. Spong, *Passive Bilateral Teleoperation with Constant Time Delay*, IEEE Transactions on Robotics and Automation , vol. 22, no.2, pp. 269-281, April 2006.

[9]   J. Scholtz, J. Young, J. L. Drury, and H. A. Yanco, *Evaluation of Human - Robot Interaction Awareness in Search and Rescue*, in IEEE international conference on Robotics and Automation - ICRA '04. , 2004, vol. 3, pp. 2327 – 2332

[10] J. L. Drury, J. Scholtz, and H. A. Yanco, *Awareness in Human - Robot Interactions*, in IEEE international conference on Systems, Man and Cybernetics - SMC '03. , 2003, vol. 1, pp. 912 – 918.

7

CFD modeling of the atmospheric boundary layer in short test section wind tunnel

Yassen El-Sayed Yassen, Ahmed Sharaf Abdelhamed

Mechanical Power Engineering, Faculty of Engineering, Port Said University, Port Said, Egypt

Email address:

y_yassen70 @yahoo.com (Y. El-Sayed Yassen), sh_ahmed99@yahoo.com (A. S. Abdelhamed)

Abstract: The aim of this paper is to provide a contribution to algorithms for the numerical simulation of the atmospheric boundary layer (ABL) in short test section wind tunnel, with the lowest pressure loss possible, for large Re, similar to the high values observed in nature. Different turbulent models have been examined for their relative suitability for the atmospheric boundary layer airflow with and without the implementation of buoyancy effects with modified turbulence model constants for the atmosphere. Validation of turbulent models through comparison with wind tunnel experiments is essential for practical applications. It has been observed that the k-ε model is most suitable tool for generation of an ABL in short-chamber wind tunnel. A comparison has been made with the available experimental data, from literature, and the predicted CFD values are very close to the corresponding experimental measurements. The simulation results show the importance of turbulence model constant (C_μ), the non-uniform velocity and turbulence intensity profiles. Also, the significance of y^+ for consistent assessment is confirmed. However, it has been found that the buoyancy force makes significant change in boundary layer thickness without a major impact on computation time.

Keywords: Atmospheric Boundary Layer (ABL), Buoyancy Effect, Turbulence Models, Short Test Section Wind Tunnel, Numerical Simulation, Non-Uniform Velocity

1. Introduction

It would be possible to obtain an appropriate scale of the natural wind structure by covering a considerable length of the wind tunnel's floor with a material of suitable roughness [1,2]. However, the disadvantage of this process is that it requires a length of about 25 m to form a BL with 60 to 120 cm height, which is possible only in tunnels with a long test chamber [3]. Thus, improved techniques for reproduction of the main characteristics of natural winds, as well as the formation of the atmospheric boundary layer (ABL), are needed. These techniques will permit shorter test chambers, so that existing aeronautical tunnels could be used for atmospheric simulations of meteorological interest, with the advantage of flow control and improved data collection [4]. The several studies using the wind tunnel for simulating the characteristics and behavior of the atmosphere can be found in Refs. [5–9]. The accurate computational fluid dynamic (CFD) simulation of the ABL is becoming increasingly important. CFD is a tool which is increasingly being used to study a wide variety of processes in the ABL, where its accurate modeling is an imperative

precondition in computational wind engineering. The application of CFD to study atmospheric dispersion processes in the lower part of the ABL has become an important research subject. Validation is an essential aspect of this research and several comparative studies between CFD and wind tunnel or field measurements have been performed, e.g. [10–13]. In all of these publications, the intermittent nature of the dispersion process in the wind tunnel and field measurements as opposed to the Reynolds Averaged Navier -Stokes (RANS) solution of the CFD simulations is indicated as a reason for the observed discrepancies. These discrepancies could be reported as inaccuracies in the boundary conditions for the flow or the pollutant source and the underestimation of the turbulence kinetic energy [14,15]. The standard k-ε turbulence model, widely employed in the simulation of the ABL due to the availability of appropriate boundary conditions and meteorological data [16–21], will serve as the starting point in investigating of the ABL under the influence of surface heat flux. The re-normalization group (RNG) k-ε model which renormalizes the Navier-Stokes equations to account for small-scale turbulence [22]. While the realizable k-ε model contains a new transport equation for the turbulent kinetic energy dissipation rate (ε) and also, turbulence model constant (C_μ). These are expressed as a function of mean flow and turbulence properties rather than assumed to be constant [23,24]. Other two–equation RANS models are also available, like the k-ω model where the transport equations are the turbulent kinetic energy (k) and the specific dissipation rate (ω). These models have been shown to perform much better than k-ε models in adverse pressure gradients and therefore in predicting separation, but are very sensitive to free-stream/inlet conditions [23,25]. A compromise between the advantages of the k-ε and k-ω models is the shear stress transport (SST) k-ω model which employs the k-ω model near the surface and the k-ε model in the free shear layers through the use of a blending function. Good performance of the SST k-ω model for ABL flow around blunt bodies has been shown [26]. The SST k-ω model has also been adopted for detached eddy simulation (DES) turbulence models, which combine the features of RANS simulation in part of the flow and large eddy simulation (LES) in the separated regions. However, these models solve the unsteady transport equations and are still significantly more computationally expensive than the steady RANS models [23]. Furthermore, the effect of flow in three dimensions over dunes has also been investigated in various studies, which include wind tunnel tests and CFD [27,28]. Most studies to date have focused on simulation of the neutral ABL, where buoyancy effects have mostly been ignored or modeled using a Boussinesq type approach [29]. CFD seems to be the obvious route to quantify these effects, which could make a new contribution to understanding the behavior of the flow fields in short test–section wind tunnel.

The objective of this research is to optimize the flow in the shortest possible extension of the wind tunnel, with the lowest pressure loss possible leading to formation of the ABL and obtain a large Re; similar to the high values observed in nature. Different turbulent models using ANSYS- Fluent14 have been examined for their relative suitability for the atmospheric boundary layer airflow with and without the implementation of buoyancy effects with modified turbulence model constants for the atmosphere. A typical CFD simulation is created in five steps. First, a model of the fluid region is sketched and any solid regions that might be present are defined. Thereafter, a mesh is applied to the sketch. Fluent uses finite–volume methods when calculating the flow field variables, with the mesh elements as the finite volumes. This means that the size of the mesh and the location of its elements determine where the flow field variables are evaluated. Hence, a fine mesh is needed where the flow is changing rapidly, while a coarser mesh can be used at locations in the model where the flow is uniform. The third step is to define the boundary and initial conditions of the problem as well as turbulence models. Then, the numerical calculations can commence, which form the fourth step. Fluent mostly uses second–order accurate numerical methods when evaluating the Navier-Stokes equations, together with various models when calculating for example the turbulence of the flow.

2. Numerical Simulation of the ABL

2.1. Using the Wall Function with a Uniform (Constant) Inlet Velocity

CFD simulations provide a way of predicting the behavior of a fluid without having to perform any experiments, and changes in the problem setup are easily made.

CFD codes employing RANS turbulence generally model the flow under turbulent conditions near walls using a wall function, providing that the flow velocity (constant) at the inlet is chosen to be the mean wind velocity of the wind profile [16]. The roughness of these surfaces is often expressed in terms of the equivalent wall roughness height k_s [18,19]. For the consistent and accurate application of the law of the wall, the dimensionless wall distance y^+ would be in the range of 30 and less 500 [30] , placing a limit on the position of the first grid node from the wall, z_p.

2.2. Using the Roughness Length with Friction Velocity (Non-Uniform Inlet Velocity)

The inlet boundary profile is function of friction velocity $u*$ and the roughness length z_o. The logarithmic velocity profile law is given by [31]:

$$u = \frac{u*}{\kappa} \ln(\frac{z + z_0}{z_0})$$ (1)

Or the more simplified version, the power law:

$$u = u*(\frac{z}{z_0})^\alpha$$ (2)

For $\alpha = 1/7$ for flow of comparatively low Reynolds numbers. The turbulent kinetic energy k can be derived from equations available in the literature [32], for simplicity

$$k = \frac{u*^2}{\sqrt{C_\mu}}$$ (3)

An equation for the turbulent dissipation rate ε is also available in the literature [33]:

$$\varepsilon = \frac{u*^3}{\kappa(z + z_o)}$$ (4)

For the simulated case, the profile at the outlet boundary has exactly the expected profile.

2.3. Using the Effects of Buoyancy on Turbulence in the k- ε Models

When a non-zero gravity field and temperature gradient are present simultaneously, the k-ε models in ANSYS FLUENT 14 account for the generation of k due to buoyancy G_b, and the corresponding contribution to the production of ε [23].

$$\int_A \rho k u_i n_i dA = \int_A \left(\mu + \frac{\mu_t}{\sigma_k}\right)\frac{\partial k}{\partial x_i} n_i dA + \int_V [G_k + G_b - \rho\varepsilon - Y_m]dV$$ (5)

and

$$\int_A \rho \, \varepsilon u_i n_i \, dA = \int_A \left(\mu + \frac{\mu_t}{\sigma_\varepsilon} \right) \frac{\partial \varepsilon}{\partial x_i} n_i \, dA + \int_V \left[C_{\varepsilon 1} \frac{\varepsilon}{k} \left(G_k + C_{\varepsilon 3} G_b \right) - C_{\varepsilon 2} \rho \frac{\varepsilon^2}{k} \right] dV \qquad (6)$$

The generation of turbulence due to buoyancy is given by

$$G_b = \beta g_i \frac{\mu_t}{Pr_t} \frac{\partial T}{\partial x_i} \qquad (7)$$

It can be seen from the transport equations for k that turbulence kinetic energy tends to be augmented ($G_b > 0$) in unstable stratification. For stable stratification, buoyancy tends to suppress the turbulence ($G_b < 0$).

2.4. Using the Change of the k- ε Model Constants

The k-ε model uses five constants in the transport equations, $C_{1\varepsilon}, C_{2\varepsilon}, C_\mu, \sigma_k,$ and σ_ε: C_μ is used to calculate eddy viscosity for the second term of the ε equation.

$$\mu_t = \rho C_\mu \frac{k^2}{\varepsilon} \qquad (8)$$

The standard values of these constants are the default values determined empirically when the k-ε model was first derived by Launder and Spalding [34]. Because the k-ε model

constants are empirically derived, constants modified to fit atmospheric boundary layer data give better results for wind energy research than the standard constants. Alinot and Masson [17] used wind farm data to optimize the k-ε model for atmospheric flow. By extensive algebraic manipulation of the turbulence equations Alinot and Masson derived a set of k-ε constants that produced more accurate results, the values of these constants are in the Table 1.

Table 1. k- ε turbulence model constants.

k- ε Constant	$C_{\varepsilon 1}$	$C_{\varepsilon 2}$	C_μ	σ_k	σ_ε
Standard [34]	1.44	1.92	0.09	1.0	1.3
Alinot-Masson [17]	1.176	1.92	0.03329	1.0	1.3

Other values of the modified model constants for more applications are available in the literature [35].

3. Boundary Conditions

In order to validate the proposed methodology, suitable velocity inlet was used which is similar to the experimental study ($u = 25.5$ m/s). Standard representation of the velocity profile in the ABL is as shown in Table 2.

Table 2. Inlet boundary conditions.

			k-ε (Non-uniform flow)	k-ε (Uniform flow)
Inlet profile		u (m/s)	$u = \dfrac{u^*}{\kappa} \ln(\dfrac{z + z_0}{z_0})$	25.5
		k	$k = \dfrac{u^{*2}}{\sqrt{C_\mu}}$	1
		ε	$\varepsilon = \dfrac{u^{*3}}{\kappa (z + z_o)}$	1
		C_μ	0.09	0.09
Roughness length	Carpet	z_o (m)	0.002	0.002
Roughness height		k_s (m)	$10 z_o / C_s$	$10 z_o / C_s$
Roughness constant		C_s	0.5	0.5

4. Mesh Considerations

4.1 physical Domain

According to [4], the domain represents a rectangular wind tunnel test section with dimensions 0.41×0.41 m^2 with a length of 1.8 m, as shown in Fig. 1a. For initial adjustments, passive devices – screen and spires – were used and for fine tuning a wrinkled carpet were added as shown in Fig. 1b. Three types of screens (2 mm thick) with different meshes were used. (i) thin screen KP = 1, (ii) medium screen KP = 0.75, and (iii) coarse screen KP = 0.05. The carpet used to form the ABL is 900 mm long with roughness of 3 mm. Three spires, each with height of 307.7 mm and base width of 32.6 mm were used.

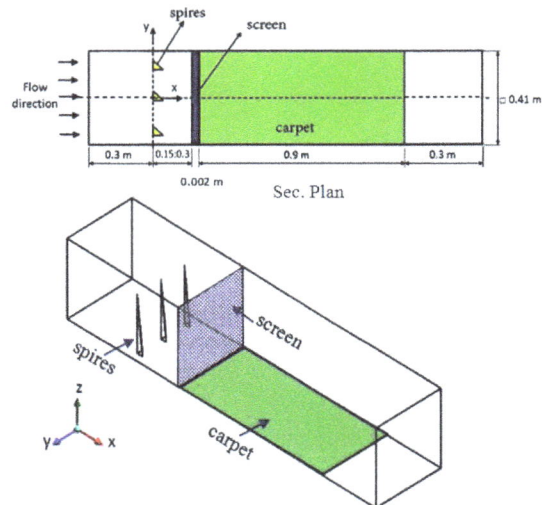

Fig 1a. Physical domain [4].

4.2. Computational Domain

The result of a numerical calculation is dependent on the mesh that is used. A too coarse mesh will give a high error in the result and as the mesh size gets finer this error should decrease. However, if the size of the mesh elements is small enough so that the numerical result is close to the real solution, a further decrease of the element size should not affect the solution significantly since the result is already correct. When this situation occurs the solution is said to be mesh–independent and this should always be achieved when performing a simulation. Mesh design (including the height of the first mesh cell) is critical to obtain a valid and accurate CFD solution, leading to very thin boundary layer. Factors to consider are mesh element shape, surface grid resolution, boundary layer resolution and the overall number of elements. Another requirement of first cell mesh height is that roughness elements can't be higher than the top of the cell. An unstructured Triangle grid was constructed based on refinements with a factor of 2 (397258 elements). The distance between the centre point of the wall-adjacent cells and the wall (carpet) is 0.002 m. The inlet mean wind speed profile and the turbulence intensity profile are taken according to the experimental work [4]. Fig.1(b) shows the grid of the computational domain related to the physical domain [4] as shown in Fig.1(a).

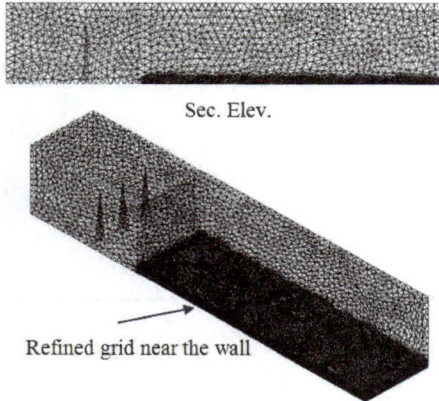

Fig 1b. Computational domain.

4.3. Mesh Independency

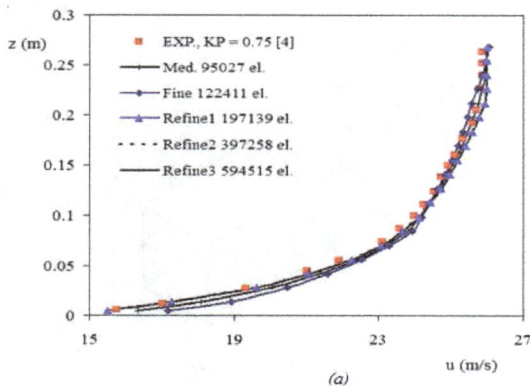

(a)

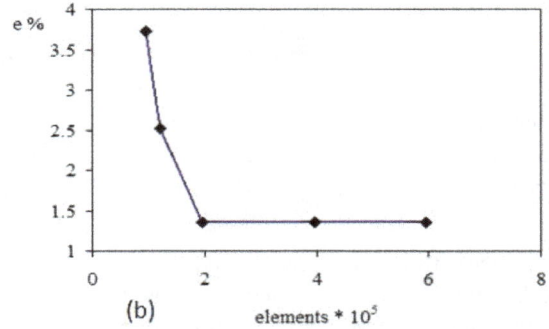

Fig 2. (a) Velocity profile (b) percentage error with different grid elements at x = 1.42 m.

Grid sensitivity study using velocity profile at $x = 1.42$ m, non-uniform flow at inlet, KP = 0.75, and distance 0.3 m from experimental work [4] with different grid sizes as shown in Fig. 2(a).

The grid sensitivity study has led to acceptable grid sizes in the range of refinement 1 (197139 elements) up to refinement 3 (594515 elements), since within this range the velocity profile do not change appreciably as shown in Fig. 2(b).

5. Results and Discussion

In the present study, the predicted results from CFD simulations of the ABL are compared with experimental work [4] for fully developed at $x = 1.42$ m.

The CFD simulations are carried out under different conditions of the different distance between spires and screen with different inlet profiles, the different turbulence models, the different pressure drop coefficient for screen, and different inlet boundary conditions.

Figures 3(a,c) show the wind-speed profile at $x = 1.42$ m for a medium screen KP = 0.75 positioned at 0.15 m and 0.3 m from the spires, at different inlet profiles. It can be seen that the predictive results, based on k-ε turbulence model for non-uniform flow at inlet is very close with experimental. But, some discrepancies are noticeable for the case of uniform (constant) flow at inlet. On comparing the uniform and the non-uniform flow cases, the flow in the case of non-uniform flow enters the duct with imposed shear and hence the boundary layer grows in the downstream direction.

Figures 3(b,d) show the turbulent intensity at $x = 1.42$ m for a medium screen KP = 0.75 positioned at 0.15 m and 0.3 m from the spires, at different inlet profiles. It can be seen that the non-uniform flow at inlet is able to reproduce the experimental measurement in satisfactorily agreement compared in case of uniform flow at inlet.

Table 3 summarizes the outlet values of (τ_w, u^*, y^+, δ, and $e\%$) at different inlet profiles and different distances between spires and screen ($x = 1.42$ m and medium screen KP = 0.75). The experimental results of the height of the ABL increased from 0.18 to 0.2 as the distance between the spires and the medium screen increased (from 0.15 to 0.3 m). The computed boundary layer characteristics from the non-uniform flow at inlet for different distances between spires and screen show better agreement with results from the uniform flow at inlet, compared with the experimental results. Also, the percentage errors are decreased for non-uniform flow at inlet compared with the uniform flow, in particular for the case of non-uniform flow and distance is 0.3m and percentage error (1.36). The boundary layer with spires, screen, and carpet is greater than 7.15 from without spires, screen, and carpet.

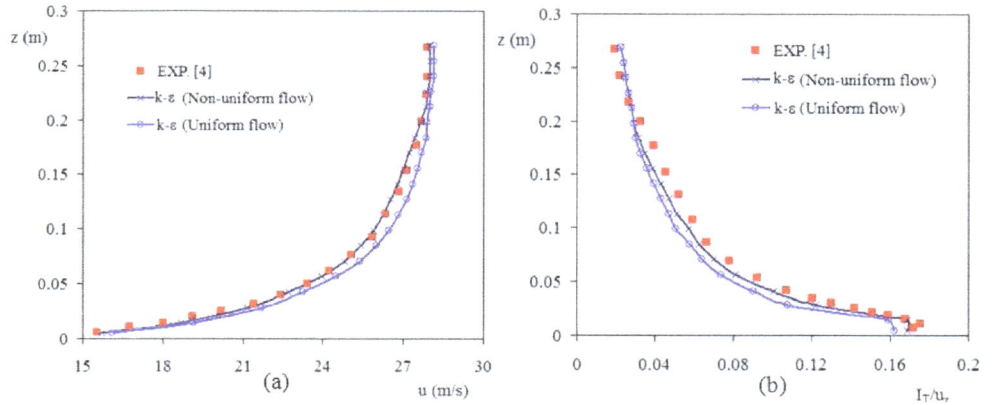

(a and b) Distance between spires and screen = 0.15 m.

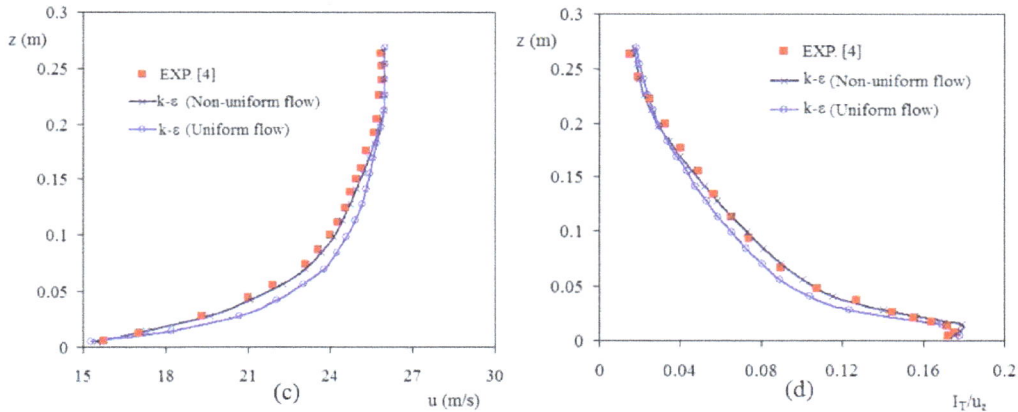

(c and d) Distance between spires and screen = 0.3 m.

Fig 3. *Velocity profile and turbulent intensity at different inlet profile (x = 1.42 m, medium screen (KP = 0.75)).*

Table 3. *Outlet conditions at different inlet profiles and different distance between spires and screen (x = 1.42 m and medium screen KP = 0.75).*

Inlet profile			Non-uniform flow		Uniform flow		Exp. [4]	
Distance between spires and screen (m)			0.15	0.3	0.15	0.3	0.15	0.3
Wall shear stress		τ_w (N/m^2)	0.99	0.99	1.04	0.95		
Friction velocity		u^* (m/s)	0.91	0.91	0.93	0.9		
Wall function		y^+	340	340	348	335		
Boundary Layer thickness without (spires, screen, carpet)	Outlet [x = 1.42 m]	δ (m)	0.028		0.03			
Boundary Layer thickness		δ (m)	0.179	0.199	0.176	0.197	0.18	0.2
Error %		$e\%$	2.91	1.36	5.51	3.74		

Figures 4(a,b) indicate that the numerical solution based on k-ε turbulence model for velocity contour and turbulent intensity at non-uniform flow at inlet, medium screen (KP = 0.75), and distant 0.3 m are capable of representing the experimental measurements in particular, in the ABL (x = 1.42 m).

Figures 5(a,b) show the velocity profile and turbulent intensity at x = 1.42 m for a medium screen KP = 0.75 positioned at 0.3 m from the spires, at several turbulence models for non-uniform flow at inlet. It can be seen that the predictive results, based on k-ε turbulence model are very close with experimental data [4] compared with other turbulence models. Figures 5(c,d) show the boundary–layer thickness and percentage error. The boundary layer thickness for k-ε turbulence model is greater than other turbulence models and the percentage error for k-ε turbulence model is less than other turbulence models. From other hand, the predictive results (boundary layer thickness) using LES in short-chamber wind tunnel would be unexpected, despite its capability of representing the ABL for external flow applications. Therefore, it would be realistic not to be used for internal flow applications. However, the k-ε model is considered to be suitable tool for generation of an ABL in short-chamber wind tunnel.

Figures 6(a,b) show the velocity profile and turbulent intensity at $x = 1.42$ m for different pressure drop for screen positioned at 0.3 m from the spires, for non-uniform flow at inlet. It can be seen that the predictive results, based on k-ε turbulence model for KP = 0.75 are very close with experimental measurements [4]. Figures 6(c,d) show the boundary–layer thickness and percentage error, the boundary layer thickness for k-ε turbulence model for without screen (KP = 0.0) is greater than other KP's. But, the percentage error for k-ε turbulence model for KP = 0.75 is less than other KP's.

Figures 7(a,b) show the velocity profile and turbulent intensity at $x = 1.42$ m for a medium screen KP = 0.75 positioned at 0.3 m from the spires, at different inlet boundary conditions for non-uniform flow at inlet. It can be seen that the predictive results, based on k-ε turbulence model for all cases are in good agreement with experimental results. Figures 7(c,d) show the boundary–layer thickness and percentage error, the boundary layer thickness for k-ε turbulence model with $C_\mu = 0.033$ and k-ε turbulence model with buoyancy are equal to the

boundary–layer thickness for experimental result, and these cases are less than compared other cases.

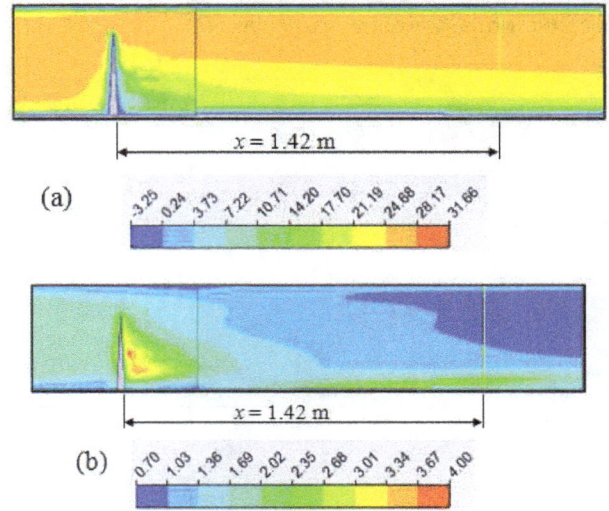

(a)

(b)

Fig 4. (a) Velocity contour (b) turbulent intensity at non-uniform flow, medium screen (KP = 0.75), and distance 0.3 m.

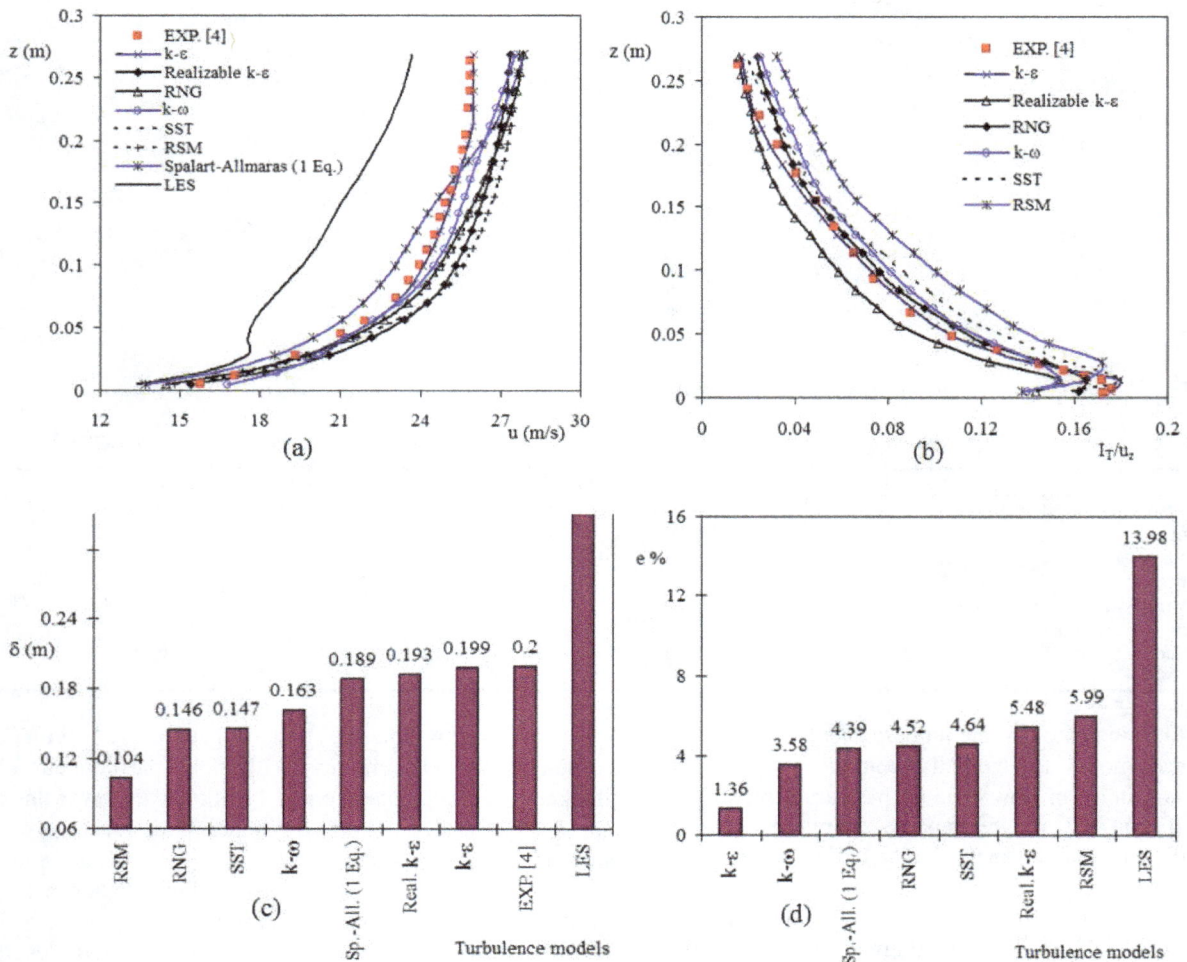

Fig 5. (a) Velocity profile (b) turbulent intensity (c) boundary layer thickness (d) percentage error at different turbulence models (x = 1.42 m, medium screen (KP = 0.75), distance 0.3 m, for non-uniform flow at inlet).

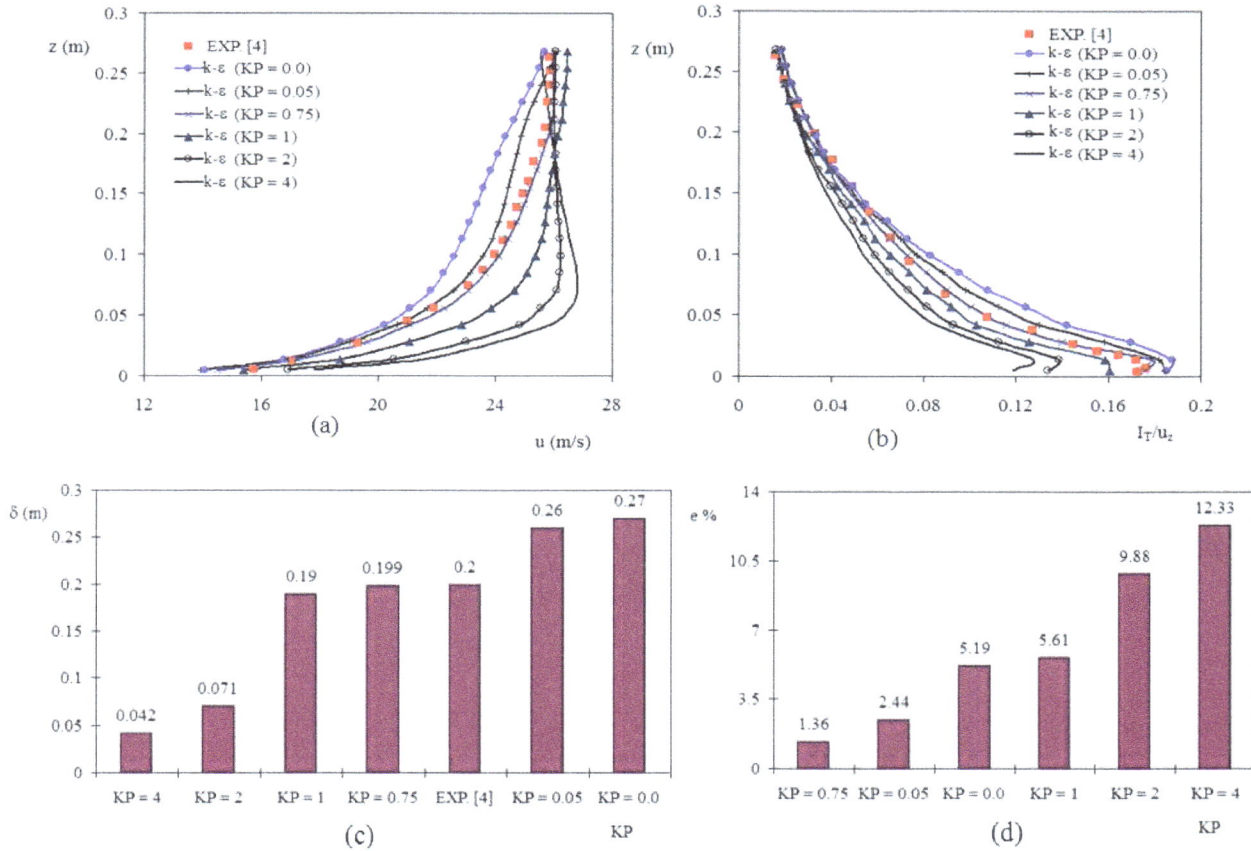

Fig 6. (a) Velocity profile (b) turbulent intensity (c) boundary layer thickness (d) percentage error at different pressure drop coefficient for screen (x = 1.42 m, distance 0.3 m, for non-uniform flow at inlet).

6. Concluding Remarks

In this study the following conclusions are drawn as follows:

- The k-ε model is considered to be a suitable tool for generation of an ABL in short-section wind tunnel leading to reduce the tunnel operation time and cost during the actual experimentation phase.
- Mesh design is critical to obtain a valid and accurate CFD solution.

- Suitable boundary conditions that actually simulate the real flow are required.
- The predicted CFD values are very close to the corresponding experimental measurements.
- The influence of turbulence model constant (C_μ) and the buoyancy force makes significant changes to the flow without a major impact on computation time.
- The dimensionless wall distance $y+$ in the required range is crucial for the consistent and accurate application.

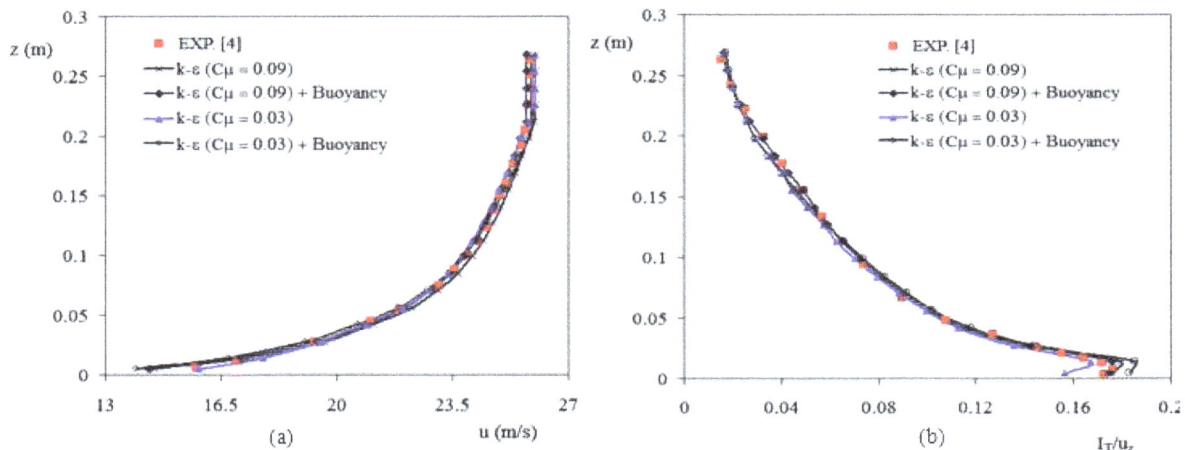

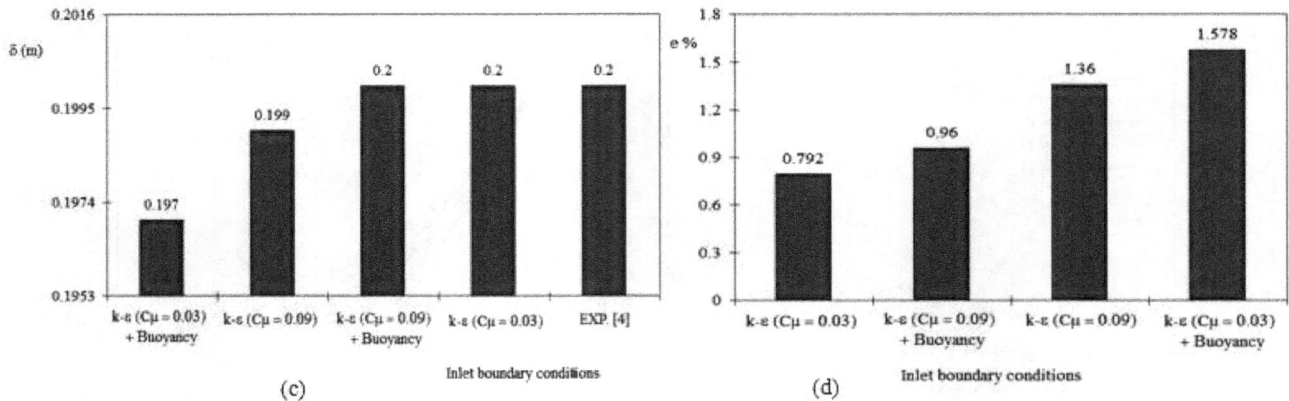

Fig 7. (a) Velocity profile (b) turbulent intensity (c) boundary layer thickness (d) percentage error at different inlet boundary conditions (x = 1.42 m, medium screen (KP = 0.75), distance 0.3 m, and non-uniform flow at inlet).

Nomenclature

C_s	Roughness constant = 0.5
C_μ	k-ε turbulence model constant
$C_{\varepsilon 1}$	k-ε turbulence model constant
$C_{\varepsilon 2}$	k-ε turbulence model constant
$C_{\varepsilon 3}$	k-ε turbulence model constant
e %	Percentage error, $\frac{1}{n}\sum_{1}^{n}\left|\frac{u_{EXP.}-u_{CFD}}{u_{EXP.}}\right|\times100$
G_b	Buoyant production of turbulence
G_k	Shear production of turbulence
g	Gravitational acceleration, m^2/s
I_T	Turbulent intensity
k	Turbulent kinetic energy
KP	Pressure-drop coefficient = $\Delta P/0.5\rho u^2$
k_s	Equivalent wall roughness height, $\approx 10 z_o/C_s$, m
Pr_t	Turbulent Prandtl number = 0.85
Re	Reynolds number = uL/ν
u	Mean velocity, m/s
u_z	Vertical velocity, m/s
u*	Friction velocity ratio = $\sqrt{\tau_w/\rho}$, m/s
y^+	Dimensionless wall distance = z.u*/υ
Y_m	Turbulence production due to compressibility
z	Vertical coordinate
z_o	Roughness length, m
z_p	Height of first cell centroid above wall, m
ΔP	Static Pressure drop across the screen, N/m^2
α	Power law exponents = 1/7
β	Thermal coefficient
δ	Boundary Layer thickness, m
ε	Turbulent kinetic energy dissipation rate
κ	von Karman constant, ≈ 0.41
ρ	Air density, kg/m^3
σ_k	k-ε turbulence model constant
σ_ε	k-ε turbulence model constant
τ_w	Wall shear stress, N/m^2
μ	Laminar fluid viscosity, kg/ms
μ_t	Turbulent fluid viscosity, kg/ms
ν	Kinematic viscosity, m^2/s

References

[1] Jensen, M. and Franck, N., "Model Scale Tests in Turbulent Wind", Part I, Danish Technical Press, Copenhagen, 1963.

[2] Jensen, M. and Franck, N., "Model Scale Tests in Turbulent Wind", Part II, Danish Technical Press, Copenhagen, 1965.

[3] Blessmann, J., "Simulation of the Natural Wind Structure in an Aerodynamic Wind Tunnel" (in Portuguese), Ph.D. thesis, Instituto Tecnológico de Aeronáutica (ITA), S.P., Brazil, 169 p, 1973.

[4] Pires, L.B., Paula, I.B., Fisch, G., Gielow, R., and Girardi, R.,M., "Simulations of the Atmospheric Boundary Layer in a Wind Tunnel with Short Test Section" , J. Aerosp. Technol. Manag., Vol. 5, No. 3, pp. 305-314, 2013.

[5] Novak, M.D., Warland, J.S., Orchansky, A.L., Ketler, R. and Green, S., "Wind Tunnel and Field Measurements of Turbulent Flow in Forests. Part I: Uniformly Thinned Stands", Boundary Layer Meteorology, Vol. 95, pp. 457-495, 2000.

[6] Kwon, K.J., Lee, J.Y. and Sung, B., "PIV Measurements on the Boundary Layer Flow around Naro Space Center", Proceedings of the 5th International Symposium on Particle Image Velocimetry, pp. 22-24, 2003.

[7] Cao, S. and Tamura, T., "Experimental Study on Roughness Effects on Turbulent Boundary Layer Flow over a Two-Dimensional Steep Hill", Journal of Wind Engineering and Industrial Aerodynamics, Vol. 94, pp. 1-19, 2006.

[8] Kozmar, H., "Scale Effects in Wind Tunnel Modeling of an Urban Atmospheric Boundary Layer", Theoretical and Applied Climatology, Vol. 100, pp.153-162, 2009.

[9] Avelar, A.C., Brasileiro, F., Marto, A.G., Marciotto, E. and Fisch, G., "Wind Tunnel Simulation of the Atmospheric Boundary Layer for Study of the Wind Pattern at the Alcantara Space Center", J. Aerosp. Technol. Manag., Vol. 4, No. 4, 2012.

[10] Gorlé C., Van Beeck J., Rambaud P., and Van Tendeloo G., "CFD modelling of small particle dispersion: The influence of the turbulence kinetic energy in the atmospheric boundary layer", Elsevier Ltd., Atmospheric Environment 43, pp. 673–681, 2009.

[11] Leitl, B., Kastner-Klein, P., Rau, M., and Meroney, R.N., "Concentration and flow distributions in the vincinity of u-shaped buildings: wind-tunnel and computational data", J. of Wind Eng. and Ind. Aerody. 67–68, 745–755, 1997.

[12] Chang, C., and Meroney, R., "Concentration and flow distributions in urban street canyons: wind tunnel and computational data", J. of Wind Eng. and Ind. Aerody., 91, 1141–1154, 2003.

[13] Yang, W., Jin, X., Jin, H., Gu, M., and Chen, S., "Application of new inflow boundary conditions for modeling equilibrium atmosphere boundary layer in rans-based turbulence models", In: Proceedings of International Conference on Wind Engineering 2007.

[14] Garcia Sagrado, A., van Beeck, J., Rambaud, P., and Olivari, D., "Numerical and experimental modelling of pollutant dispersion in a street canyon", J. of Wind Eng. and Industrial Aerodynamics 90, 321–339, 2002.

[15] Dixon, N., Boddy, J., Smalley, R., and Tomlin, A.S., "Evaluation of a turbulent flow and dispersion model in a typical street canyon in York", UK. Atmospheric Environment 40, 958–972, 2006.

[16] Pieterse J., E., "CFD Investigation of the Atmospheric Boundary Layer under Different Thermal Stability Conditions", Thesis, MSc., Stellenbosch University, 2013.

[17] Alinot, C. & Masson, C., "Aerodynamic Simulations of Wind Turbines Operating in Atmospheric Boundary Layer With Various Thermal Stratifications", *ASME Conference Proceedings,* vol. 2002, no. 7476X, pp. 206-215, 2002.

[18] Blocken, B., Carmeliet, J. and Stathopoulos, T., "CFD Evaluation of Wind Speed Conditions in Passages Between Parallel Buildings- Effect of Wall-Function Roughness Modifications for the Atmospheric Boundary Layer Flow", Journal of Wind Engineering and Industrial Aerodynamics, vol. 95, no. 9–11, pp. 941-962, 2007a.

[19] Blocken, B., Stathopoulos, T. and Carmeliet, J., "CFD Simulation of the Atmospheric Boundary Layer: Wall Function Problems", Atmospheric Environment, vol. 41, no. 2, pp. 238-252, 2007b.

[20] Hargreaves, D. & Wright, N., "On the Use of the k–ε Model in Commercial CFD Software to Model the Neutral Atmospheric Boundary Layer", Journal of Wind Engineering and Industrial Aerodynamics, vol. 95, no. 5, pp. 355-369, 2007.

[21] Fang, P., Gu M., Tan, J., Zhao, B. and Shao, D., "Modeling the Neutral Atmospheric Boundary Layer Based on the Standard k -ε Turbulent Model: Modified Wall Function", The Seventh Asia-Pacific Conference on Wind Engineering, Taipei, Taiwan, 2009.

[22] Versteeg, H.K. and Malalasekera, W., "An Introduction to Computational Fluid Dynamics: the Finite Volume Method", Prentice Hall, Malaysia, 2007.

[23] ANSYS, Inc., "ANSYS FLUENT, Release 14.0: Installation and User's Guide", SAS IP, Inc., 2013.

[24] Roy, A.K., Babu, N., Bhargava, P.K., "Atmospheric Boundary Layer Airflow theough CFD Simulation on Pyramidal Roof of Square Plan Shape Buildings", VI National Conference on Wind Engineering, 2012.

[25] Rados, K.G., Prospathopoulos, J.M., Stefanatos, N.Ch., Politis, E.S., Chaviaropoulos, P.K. , and Zervos, A., " CFD modeling issues of wind turbine wakes under stable atmospheric conditions", EWEC '09 Proceedings, Marseille, 2009.

[26] Yang, W., Quan, Y., Jin, X., Tamura, Y. and Gu, M., "Influences of Equilibrium Atmosphere Boundary Layer and Turbulence Parameter on Wind Loads of Low-rise Buildings", J. of Wind Eng. and Ind. Aerody., vol. 96, no. 10, pp. 2080-2092, 2008.

[27] Liu, B., Qu, J., Zhang, W. and Qian, G., "Numerical Simulation of Wind Flow over Transverse and Pyramid Dunes", Journal of Wind Engineering and Industrial Aerodynamics, vol. 99, no. 8, pp. 879-888, 2011.

[28] Joubert, E.C., Harms, T.M., Muller, A., Hipondoka, M. and Henschel, J.R., "A CFD Study of Wind Patterns over a Desert Dune and the Effect on Seed Dispersion", Environmental Fluid Mechanics, vol. 12, no. 1, pp. 23-44, 2012.

[29] Alinot, C. & Masson, C., "*k-ε* Model for the Atmospheric Boundary Layer Under Various Thermal Stratifications", *Journal of Solar Energy Engineering,* vol. 127, no. 4, pp. 438-443,2005.

[30] White, F.M., "Viscous Fluid Flow", McGraw-Hill, New York, 1991.

[31] Richards P., Hoxey R., "Appropriate boundary conditions for computational wind engineering models using the *k-ε* turbulence model", J. of Wind Eng. and Industrial Aerodynamics, 46,47, 145-153, 1993.

[32] Wilcox D., "Turbulence modelling for CFD", DCW Industries, California, 1993.

[33] Richards P., Norris S., "Appropriate boundary conditions for computational wind engineering models revisited", J. of Wind Eng. and Indus. Aerody., 99, 257-266, 2011.

[34] Launder, B.E. and Spalding, D. , "The Numerical Computation of Turbulent Flows", Computer Methods in Applied Mechanics and Engineering, vol. 3, no. 2, pp. 269-289, 1974.

[35] Sumner, J., Sibuet, Ch. W., and Masson, Ch., "CFD in Wind Energy: The Virtual, Multiscale Wind Tunnel", Energies, 3, 989-1013, 2010.

8

Determination of Required Torque to Spin Aircraft Wheel at Approach Using ANSYS CFX

Abdurrhman A. Alroqi, Weiji Wang

Department of Engineering and Design, University of Sussex, Brighton, UK

Email address:
aa-alroqi@hotmail.com (A. A. Alroqi), W.J.Wang@sussex.ac.uk (Weiji Wang)

Abstract: Many patents have suggested that spinning the aircraft wheel before touchdown would lessen tyre wear as indicated by landing smoke and rubber deposites on the runway caused by skidding wheel at the point of impact. In this paper, the required torque to spin the aircraft wheel at approach speed has been calculated using ANSYS Workbench CFX, which is used to determine the wheel aerodynamic forces developed by simulation of fluid flows in a virtual environment. The wheel has been tested against different wind speeds, and the aerodynamic forces for the spinning wheel are presented, which include; translational and rotational drags, lift created by vortex, and shaft rolling resistance.

Keywords: Spinning Aircraft Wheel, Aerodynamic Force, Translational Drag, Rotational Drag,
SST Turbulence RANS Model, ANSYS CFX

1. Introduction

Spinning the aircraft wheel before touchdown is a proposed solution to eliminate the smoke generated between the tyres and runway at landing impact [1-9]. The landing smoke phenomenon is described by the following chain events: firstly, the aircraft approaches the runway at a relatively high speed. Therefore, a high velocity difference exists between the wheels and runway, which leads to the tyres being fully locked on the runway at landing until the friction force at the contact surface increases sufficiently for the wheel to 'spin-up' and to reach a constant angular velocity which is equivalent to aircrafts forward speed [10].

During the skidding phase, a high temperature is generated in the tyre tread rubber, sufficient in fact to locally melt the rubber. The melted rubber becomes weak enough for tyre wear to take place [11]. Part of worn rubber sticks to the runway whilst the remaining part is burnt-off forming the distinctive puff of white smoke [12].

In this paper, a case study of Boeing 747-400 main landing wheel has been modelled using ANSYS CFX in order to calculate the required torque to spin the wheel in the aircraft approach phase.

The wheel is tested against three high wind speeds. The lowest wind speed is at least equal to aircraft approach speed in order to simulate the case of a zero heading wind. The model represents the forces created during wheel rotation. Finally, the torque necessary to spin the wheel for required rotation is presented.

1.1. Literature Review

Kothawala, Gatto and Wrobel carried out a computational investigation of the combined effect of Yaw, rotation and Ground proximity on aerodynamics of an isolated wheel using steady and unsteady Reynolds-Averaged Navier-Stokes (RANS & URANS). The diameter and width of the wheel was 0.416m and 0.191m respectively. They tested the rotated wheel against a free stream of air with speeds of 70 and 98 m/s. The wheel rotation speeds were 100, 200, and 327 rad/sec. However, they conclude that the wake on rear of wheel increases with increased rotation speeds [13].

Morelli tested a stationary and rotated wheel against the same wind speed using a wind tunnel. He found that the drag is increases by about 10% when the wheel is rotating. He concluded that this increase of drag was due to negative lift and induced drag [14].

Rahman carried out a computational study on flow around a rotating short cylinder in order to study the effect of rotation on the aerodynamics forces. The cylinder considered is in X-Y plane, the rotation is about the Z axis, and the flow is along the positive x - direction. He found that the clockwise rotation of the cylinder reduced the pressure region above the cylinder with higher pressure at the bottom surface. The difference in pressure producing an upward lift force for clockwise rotation and vice versa for anticlockwise. Also, the lift force is dependent on the cylinder spin ratio, an increase in the rotational velocity producing an increasing lift force [15].

1.2. Case Study Data

The wind speeds used in this simulation are; 80.7, 100, and 120 m/s. The first wind speed (80.7 m/s) is equal to the Boeing 747-400 approach speed whilst the other higher speeds are assumed in case of heading wind speed increase [16]. The wheel geometry data is presented in Table 1 [17, 18].

For required wheel rotation speed, it is important to know the aircraft touchdown speed in order to calculate the equivalent wheel rotation speed with regard to tyre deflection. However, the aircraft will lose about 10 knots (5.14 m/s) during the flare manoeuvre before touchdown, which results in a 75.6 m/s forward speed [19].

The tyre deflects upon landing impact, which means a lesser radius during rotation.

For simplification, the tyre free rolling velocity is assumed to be 121 rad/sec, this is calculated using the tyre radius and aircraft forward speed on the runway. Here we will simulate 50% of the wheel full free rotation, this is because the aircraft wheel is heavy and consequently it may requires large wind

turbines for full spinning. Moreover, designing the turbine for full wheel rotation may lead to excessive free rolling rotation, e.g. in the case of high head wind speed, while using the 50% of free rotation with wind speed equivalent to the aircraft approach speed to design the turbine will guarantee it rotates at the minimum wind speed. This gives the turbine the opportunity to rotate in a safe mode if the head wind were to increase.

Table 1. *Wheel geometry data.*

	Weight (kg)	Radius (mm)	Width (mm)
Tyre	110	622.3	482.6
Rim	74.4	255	-------

2. Theoretical Background

In this model, it is assumed that the wheel is moving through the air with zero angle of attack and the direction of rotation of the wheel is anticlockwise. The wheel will accelerate from zero to the required rotational speed. Fig. 1 shows the forces to be calculated in order to determine the torque is required to spin the wheel. The wind turbine should be physically attached to the wheel rim to consider the tyre deflection effect at touchdown. In this case, the rim mean radius is the force arm to calculate the required torque to be as:

$$T_{required} = F_{required}\, r_{rim} \tag{1}$$

where, $T_{required}$ and $F_{required}$ are the required torque (N.m) and the required force (N) to spin the wheel respectively, and r_{rim} is the mean radius of the rim (m), which depends on the wind turbine position.

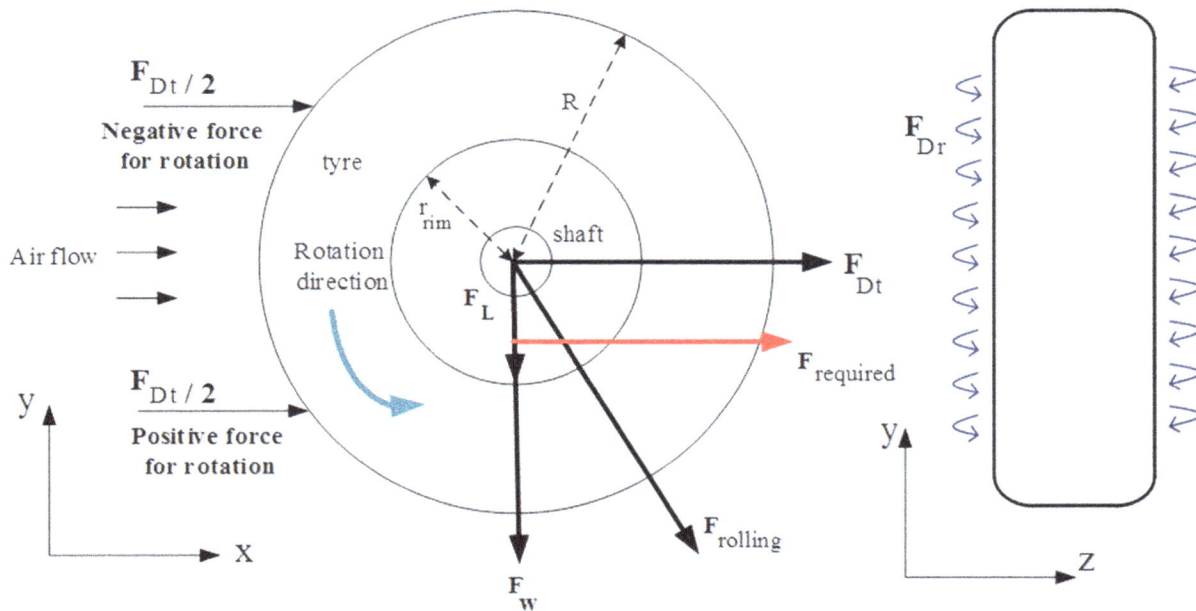

Fig. 1. *External forces acting on the rotating wheel.*

The required force should be at least equal to the sum of all wheel forces including to the aerodynamic forces some of which will be generated during the wheel rotation. Therefore,

the required force will be as:

$$F_{required} = F_{ei} + F_D + F_{rolling} \tag{2}$$

where, F_{ei} is the equivalent inertial force during acceleration (N), F_D is the total drag force, which is inclusive of the translational and rotational drags (N), and $F_{rolling}$ is the shaft rolling resistance (N).

2.1. Drag Force

There are two types of drag force. Firstly, the translational drag in x direction (F_{Dt}), which impacts the wheel frontal area as shown in Fig. 1. Half of this force is positive for rotation and other half is negative "resistance" and because the wheel is symmetrical, therefore, the two parts have an equal and opposite effect in the static condition and the resultant torque on the wheel shaft becomes zero.

On the other hand, the total force, will be applied to the wheel shaft in x direction which will increase the rolling resistance.

The formula of translational drag can be calculated as:

$$F_{Dt} = \frac{1}{2} \rho \, C_{Dt} \, A_f \, U^2 \qquad (3)$$

where, ρ is the air density (kg/m^3), C_{Dt} is the translational drag coefficient, U is the wind speed acts on the wheel (m/sec), and A_f is the wheel frontal area (m^2) [20]. The frontal area is roughly half the wheel circumference area; $A_f = R\pi D$, here R and D are wheel radius and width respectively. The force on the wheel frontal area is different from centre to the top or bottom surfaces.

Once the wheel starts to rotate, half of this force in the positive direction becomes higher than the one in negative direction which is effected by the wheel rotation direction, this force is helpful as it will be added to the required force for spinning the wheel.

The other drag force is that created during wheel rotation and it is called "rotational drag force" which increases in magnitude with increasing wheel angular velocity and occurs around the two side areas of the wheel, acting in the rotation (z) axis. The rotational drag can be calculated by this formula:

$$F_{Dr} = \frac{1}{2} \rho \, C_{Dr} \, A_{side} \, R^2 \omega^2 \qquad (4)$$

where, C_{Dr} is the rotational drag coefficient, A_{side} is the two side areas of the wheel (m^2), R is the wheel radius (m), and ω is the wheel angular velocity (rad/sec) [21].

2.2. Rolling Resistance Force

Three types of force act on the wheel shaft in x and y directions resulting in the total magnitude force, $F_{rolling}$. The three forces are, as follows:

1. Translation drag force (F_{Dt}) presented in (3), is applied to the shaft in (x) direction.
2. The wheel weight force, F_w in $(-y)$ and is simply; $F_w = mg$, where, m is the wheel total mass, and g is the acceleration due to gravity.
3. The lift force, which is created during the wheel rotation.

Kutta-Joukowski lift theorem for a cylinder describes how this force depends on the direction of rotation and acts perpendicular to the air flow direction. As shown in Fig. 2, the

wheel is pulling a thin layer of flow molecules in its rotation direction resulting in a faster flow on the lower surface than on the upper surface, which leads to less pressure than wheel top surface. The wheel upper surface will also pull a thin layer in the opposite direction to the flow which creates a vortex. This vortex has the effect of increasing the upper pressure. The difference in pressure between the wheel top and bottom surfaces is the lift force. In our case, the negative (downward) lift force increases the resistance on the shaft because it acts in same direction as the wheel weight force.

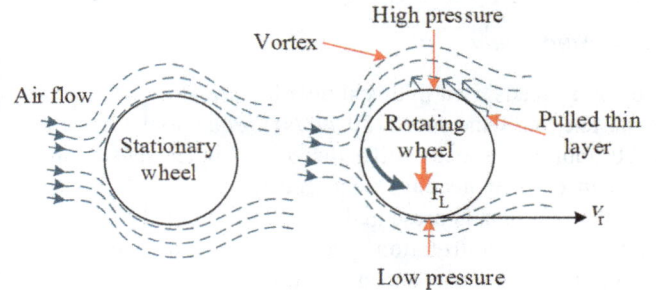

Fig. 2. *Lift force created during wheel rotation.*

The lift force can be calculated as:

$$F_L = \rho \, G \, U \qquad (5)$$

where, G is the vortex strength (m^2/sec) and is given by [22]:

$$G = 2 \pi R v_r \qquad (6)$$

where, v_r is the relative wheel speed (rad/sec), and it is given by: $v_r = R\omega$. The relative speed gives the spin ratio as: $\frac{v_r}{U}$. The increasing of the spin ratio causes an increasing in the lift force [15, 23, 24].

Substituting (6) in (5) with respect to the wheel angular velocity, the lift force will be as:

$$F_L = 2 \rho \pi R^2 \omega \qquad (7)$$

Because all the parameters in (7) are unchanged with the exception of the wheel angular velocity, therefore, the lift force is zero with a stationary wheel, and increases with increasing wheel rotation speed.

Now, simply the total force acting on the wheel shaft will be as:

$$F_{ts} = \sqrt{(F_w + F_L)^2 + F_{Dt}^2} \qquad (8)$$

where, F_{ts} is the total force applied to the shaft. Note: The term $(F_w + F_L)$ will be $(F_w - F_L)$ if the wheel rotate clockwise with the same current flow direction.

The bearing friction coefficient must be considered to estimate the rolling resistance. Therefore, the shaft rolling resistance force will be as [25]:

$$F_{rolling} = C_r \, F_{ts} \qquad (9)$$

where, C_r is the bearing friction coefficient, assumed to be angular contact ball bearing with value of 0.0015 [26].

2.3. Equivalent Inertial Force During Acceleration

This force is internal which is the resistance of the wheel against its acceleration. Using Newton's second law, this force can be expressed as:

$$F_{ei} = m\,\dot{\omega} \qquad (10)$$

where, $\dot{\omega}$ is the wheel angular acceleration (rad/s^2), and can be calculated as:

$$\dot{\omega} = \frac{d\omega}{dt} \qquad (11)$$

where, t is the acceleration time (sec). Rewriting (10) with respect to (11), to be:

$$F_{ei} = m\,\frac{d\omega}{dt} \qquad (12)$$

From (12), this force depends on the time required to complete the acceleration and it decrease to about zero at ($t = \infty$).

Slow acceleration requires less force and Vice versa. However, the wheel aerodynamics forces is fall into the category of experimental science [27]. Therefore, the wheel is modelled using ANSYS CFX to calculate both the forces and the required torque. The wheel is rotated from zero to 60.5 rad/sec during first four seconds with constant acceleration maintaining a constant angular speed for a further two seconds.

The purpose of spinning the wheel is to investigate the wheel aerodynamic forces generated during rotation, these forces are not present on a stationary wheel.

3. Simulation Model

The present work describes a thorough investigation of 3D computations concerning the air flow around the wheel. The calculated results give a clear indication of the air flow distributions for different inlet velocity values. Modelling tasks using CFD programs will allow us to get closer to the real operating conditions.

The 3D wheel geometry is modelled using the ANSYS design modeller together with the data as presented in Table 1. The wheel modelled inside the large domain is as shown in Fig. 3.; the tyre being soft without grooves. The domain dimensions are 20 m x 20 m inlet area by 40 m long to avoid the wall boundary effect and so be representative of the real aircraft wheel conditions during approach in open air.

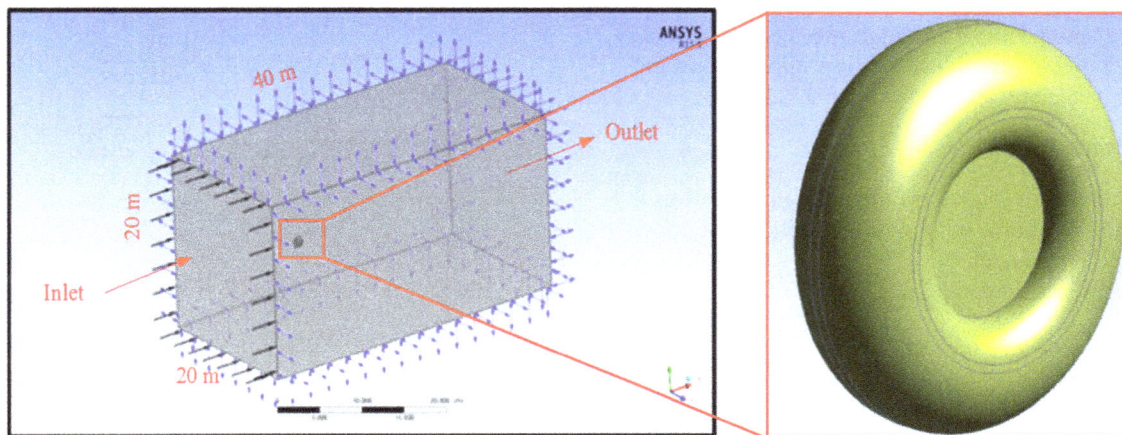

Fig. 3. Wheel domain.

3.1. Mesh Generating

High-quality mesh created for accurate solutions and good convergence. A "Patch Conforming Method" is used to generate the mesh with tetrahedron form elements. In addition, prismatic layers are constructed for the flow near the walls and on the wheel surfaces using "smooth transition" option for more accuracy [28].

The flow regime is subsonic with static temperature at 288 K and the turbulence is zero gradient. Reference and relative Pressures are 1 and $1.013x10^5$ pa. Rough wall surfaces are used as the SST model does not accurately predict the amount and onset of the flow separation from soft surfaces [29]. Fig. 4 shows the mesh model, and Table 2 presents the mesh statistics. The boundary conditions are the same for all simulations except the inlet velocities.

Table 2. Mesh statistics.

	Value
nodes	9197
elements	43703
tetrahedrons	39663
prisms	4040
faces	1952
Orthogonality Angle [30]	27.4^0, acceptable range $> 20^0$
Expansion factor	20^0, acceptable range $< 20^0$
Aspect Ratio	57^0, acceptable range $< 100^0$

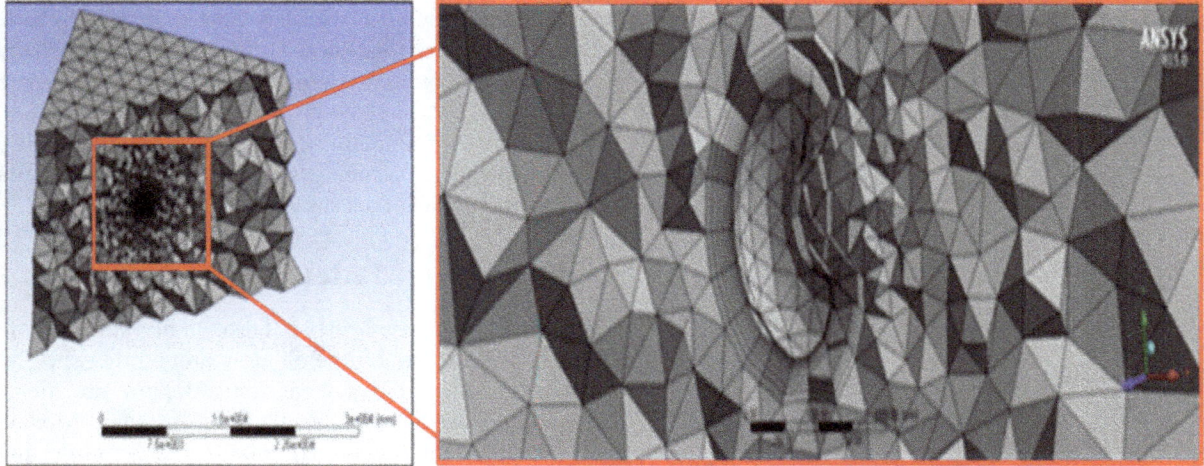

Fig. 4. Mesh model of the Wheel domain.

The flow type and the physical model in the fluid domain is defined as 'steady state' with turbulence conditions using Reynolds-Averaged Navier-Stokes equations. Turbulence is modelled using the SST (Shear Stress Transport) model and heat transfer using the total energy model.

The SST turbulence model is commonly used and is suitable for a wide range of applications.

The total energy model allows for high speed energy effects and is therefore, suitable for high speed flow applications and has better performance at wall boundaries [31, 32].

3.2. SST Turbulence RANS Model

The RANS model allows us to simulate turbulent flow as a steady state. Fig. 5 shows a simple definition for flow velocities in the RANS model.

The flow velocity is calculated as:

$$U(\vec{x}, t) = \bar{U}(\vec{x}) + u'(\vec{x})$$

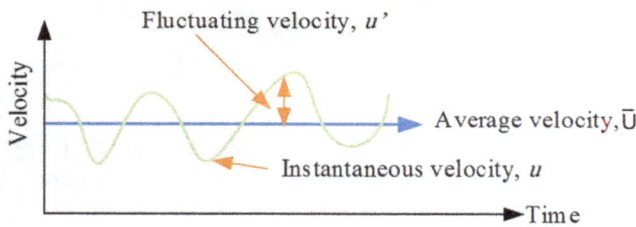

Fig. 5. Velocities definition in RANS model.

Applying the time average procedure to the governing equation which gives the Reynolds-Averaged Navier-Stokes (RANS) equations to be as:

$$\frac{\partial(\rho \bar{u}_i)}{\partial t} + \frac{\partial(\overline{u_i\, u_j})}{\partial x_j}$$

$$= -\frac{\partial \bar{P}}{\partial x_i} + \frac{\partial}{\partial x_j}\left[\mu\left(\frac{\partial \bar{u}_i}{\partial x_j} + \frac{\partial \bar{u}_j}{\partial x_i} - \frac{2}{3}\delta_{ij}\frac{\partial \overline{u_m}}{\partial x_m}\right)\right]$$
$$+ \frac{\partial}{\partial x_j}\left(-\rho\, \overline{u_i' u_j'}\right)$$

here, $-\rho\, \overline{u_i' u_j'} = R_{ij}$ which is Reynolds stress tensor.

SST model:

$$\frac{\partial(\rho k)}{\partial t} + \frac{\partial(\rho U_i k)}{\partial x_i} = \bar{P}_k - \beta^* \rho k \omega + \frac{\partial}{\partial x_i}\left[(\mu + \rho_k \mu_t)\frac{\partial k}{\partial x_i}\right]$$

$$\frac{\partial(\rho \omega)}{\partial t} + \frac{\partial(\rho U_i \omega)}{\partial x_i} = \alpha\frac{1}{v_t}\bar{P}_k - \beta\rho\omega^2 + \frac{\partial}{\partial x_i}\left[(\mu + \sigma_\omega \mu_t)\frac{\partial \omega}{\partial x_i}\right]$$
$$+ 2(1 - F_1)\rho\sigma_{\omega 2}\frac{1}{\omega}\frac{\partial k}{\partial x_i}\frac{\partial \omega}{\partial x_i}$$

$$v_t = \frac{a_1 k}{\max(a_1\omega, SF_2)}; \quad S = \sqrt{2S_{ij}\, S_{ij}}$$

$$P_k = \mu_t \frac{\partial U_i}{\partial x_j}\left(\frac{\partial U_i}{\partial x_j} + \frac{\partial U_j}{\partial x_j}\right),$$

$$\bar{P}_k = \min(P_k, 10.\beta^*\rho k\omega),$$

$$\mu_t = \rho\frac{k}{\omega},$$

$$F_1 = tanh\left\{\left\{\min\left[max\left(\frac{\sqrt{k}}{\beta^*\omega Y}, \frac{500v}{Y^2\omega}\right), \frac{4\rho\sigma_{\omega 2}k}{CD_{k\omega}Y^2}\right\}^4\right\}\right.,$$

$$F_2 = tanh\left[\left[max\left(\frac{2\sqrt{k}}{\beta^* wY}, \frac{500v}{Y^2\omega}\right)\right]^2\right],$$

$$CD_{k\omega} = max\left(2\rho\sigma_{\omega 2}\frac{1}{\omega}\frac{\partial k \partial\omega}{\partial x_i \partial x_i}, 10^{-10}\right)$$

where, k and ω are the turbulence kinetic energy and frequency respectively, Y is the distance to the wall boundary, S is the mean strain tensor rate, $F_{1,2}$ are blending function, which is equal to one. The constants are defined as: $\alpha = \alpha_1 F_1 + \alpha_2(1 - F_1)\dots$ etc., $\beta^* = 0.09$, $\alpha_1 = 5/9$, $\beta_1 = 3/4$, $\sigma_{k1} = 0.85$, $\sigma_{\omega 1} = 0.5$, $\alpha_2 = 0.44$, $\beta_2 = 0.0828$, $\sigma_{k2} = 1$, $\sigma_{\omega 2} = 0.856$ [29, 31].

4. Results and Discussion

A comparison of the air flow around the wheel at three wind speeds; 80.7, 100, and 120 m/s is shown in Fig. 6. As shown in velocity x-y diagrams, a thin layer of flow molecules is pulled by the wheel in the rotation direction. At the top of the wheel, the layer of flow is in the opposite direction to that of the wind which created the vortex, this has the effect of increasing the pressure at the top of the wheel, thus producing a negative lift force. Moreover, the wind speed is higher at the wheel bottom with consequently less pressure than the top wheel area, which leads to the total overall vertical force acting on the wheel to be in a downward direction. The lift force varies directly as the

wind speed, that is the higher the wind speed the higher the lift force developed. The air flow speed behind the wheel is approximately zero for all wind speeds, which is in agreement with [13].

In general, an increase in the wind speed produces a corresponding increase in the aerodynamic forces around the wheel.

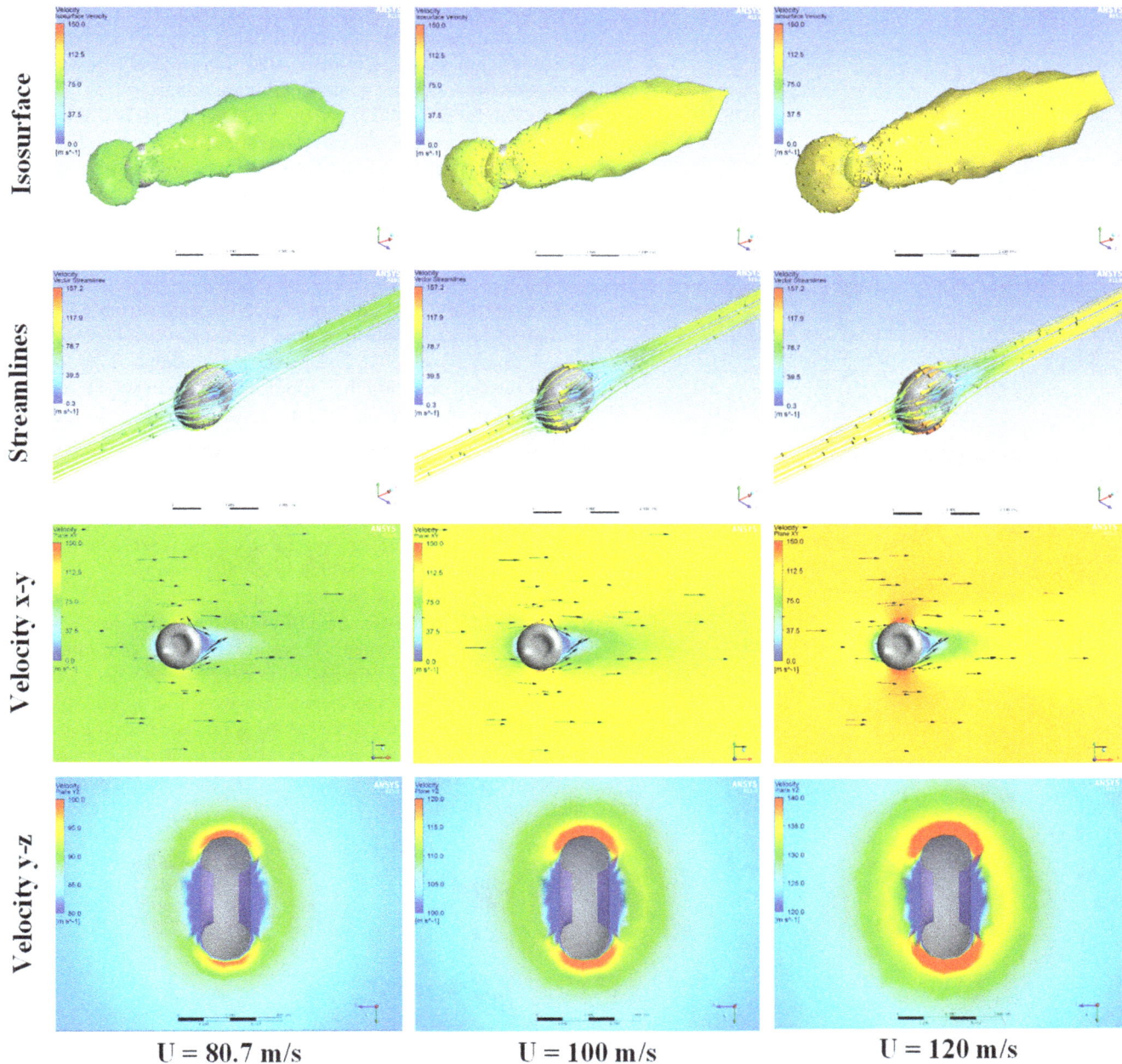

Fig. 6. *A comparison of different flow profiles around the wheel at different wind speeds.*

4.1. Translational Drag Force

The impact of wind flow on the wheel shows different characteristic force behaviour, which is speed dependant. The translational drag force at high wind speeds (100 and 120 m/s) increases over its steady state value immediately to settled

down shortly afterwards. This impact force has the effect of increasing the resistance on the wheel shaft. The translational drag force at the three different wind speeds all show steady values after a period of one second, even while the wheel accelerates. A Comparison of the force curves and distribution profiles are presented in Fig. 7 and Fig. 8 respectively.

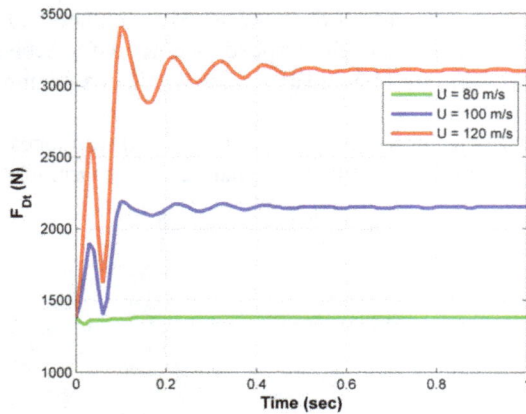

Fig. 7. *A comparison of translation drag forces vs. time with different wind speeds.*

At a wind speed of 80.7 m/s, a drag force of 1383 N is generated by the air flow, which impinges the wheel frontal area at the moment of impact.

Once the wheel starts to rotate, it is seen that the force drops to 1331 N only to increase to 1370 N within 0.09 sec before reaching a steady state value of 1380 N, which is close to the value of a stationary wheel.

At a wind speed of 100 m/s, the force increases immediately from 1383 N to 1900 N within 0.03 seconds. After a further 0.06 seconds it has decreased to 1400 N only to recover and increase again, reaching a steady state value of 2150 N within 0.89 seconds following some slight undulations.

At a wind speed of 120 m/s, the force behaviour is similar to that generated by the 100 m/s wind speed except with higher values. The peak value being 3410 N within 0.1 seconds which settled down to steady state value of 3100 N after 0.97 seconds.

4.2. Side Force - Including Rotational Drag

The rotational and side drag force increases from 64.8 N during wheel acceleration to reach a steady state at ultimate wheel rotation velocity. The resultant rotational drag force acts in the (-z) direction that because the flow acts on the rim side to push the wheel as it larger side area than the other part.

According to rotational drag force formula, it is created during wheel rotation, but as the wheel aerodynamic force is determined by experimental science, this simulation shows that the flow is producing drag force on the wheel sides area because of the tyre shape which is include breadth.

Moreover, the flow is pulled inside the hub and then re-circulated. This re-circulated flow impinges on the free air flow straight past the wheel. Therefore, the force of rotational drag here is inclusive of translation drag on the wheel sides, which is affected by wind speeds. However, Fig. 9 and Fig. 10 shows force curves generated during wheel acceleration and a comparison of wheel profiles at different wind speeds respectively. For all wind speeds, the force starts with the same value of 64.8 N.

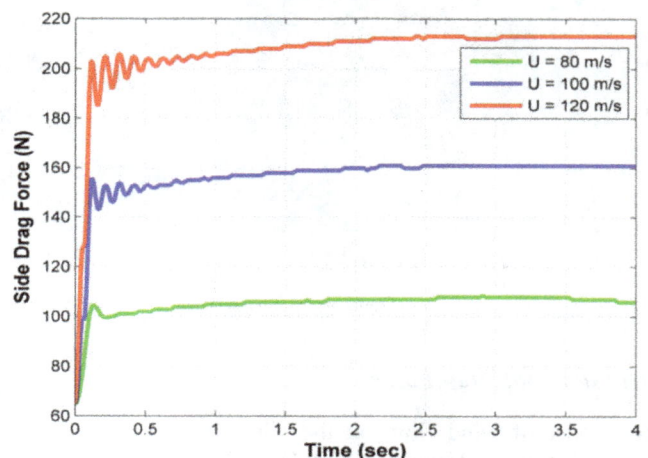

Fig. 8. *Contours of translational drag force with different wind speeds.*

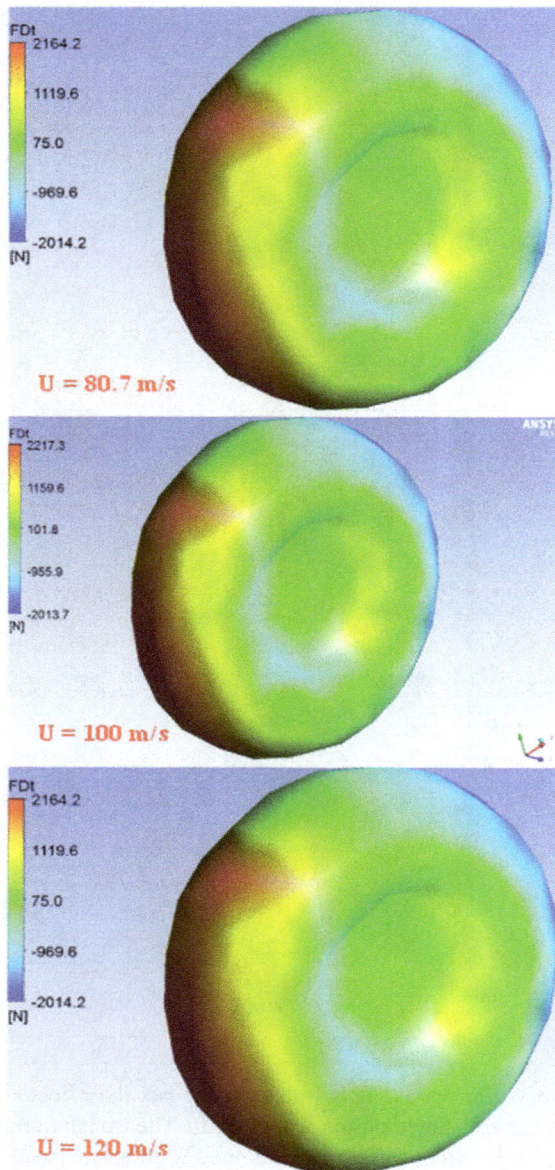

Fig. 9. *A comparison of side drag forces vs. time with different wind speeds.*

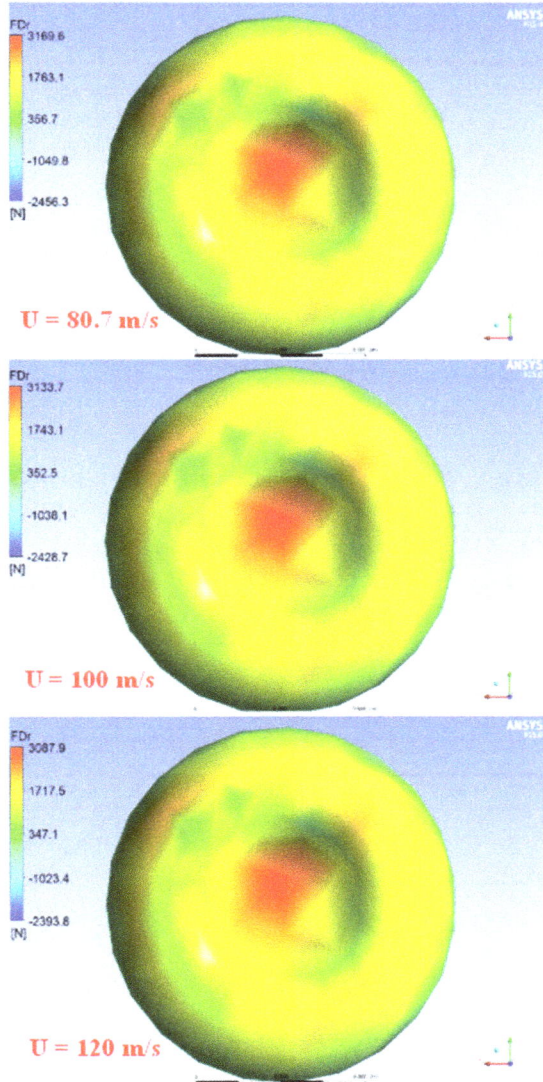

Fig. 10. Contours of side drag force with different wind speeds.

At 80.7 m/s wind speed, the force increased to 105 N within 0.13 seconds before to decreasing to 99.5 N after 0.22 seconds and then rose gradually during the wheel acceleration to attain a steady state at 106 N.

At 100 m/s wind speed, the force increased to its peak value of 156 N within 0.12, which is faster and higher than the corresponding force at 80.7 m/s wind speed. The force is seen to undulate slightly whilst still increasing overall to reach 161 N toward the end of the wheel acceleration.

At 120 m/s wind speed, the associated peak value was 203 N within 0.11 seconds, which is the fastest of the three, and has relatively large waves. The steady state value at the end of acceleration was 213 N.

4.3. Lift Force

The downward (negative) lift force of 90 N for all wind speeds is created immediately at wind impact, and occurs just as the wheel starts to rotate. As rotation progresses different force profiles are produced which depend on the wind speed.

Fig. 11 and Fig. 12 show a comparison of the lift forces for different wind speeds and the associated forces profiles respectively. In each case the force increases during the wheel acceleration stage before reaching a steady state.

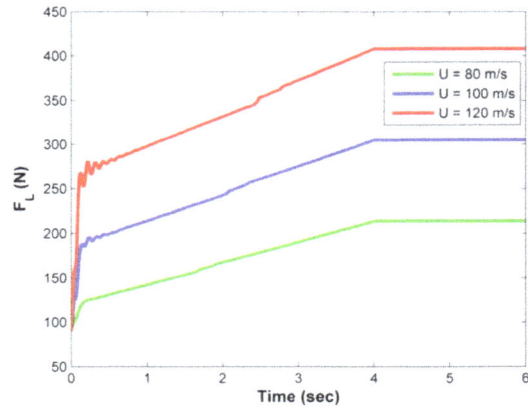

Fig. 11. A comparison of lift forces vs. time with different wind speeds.

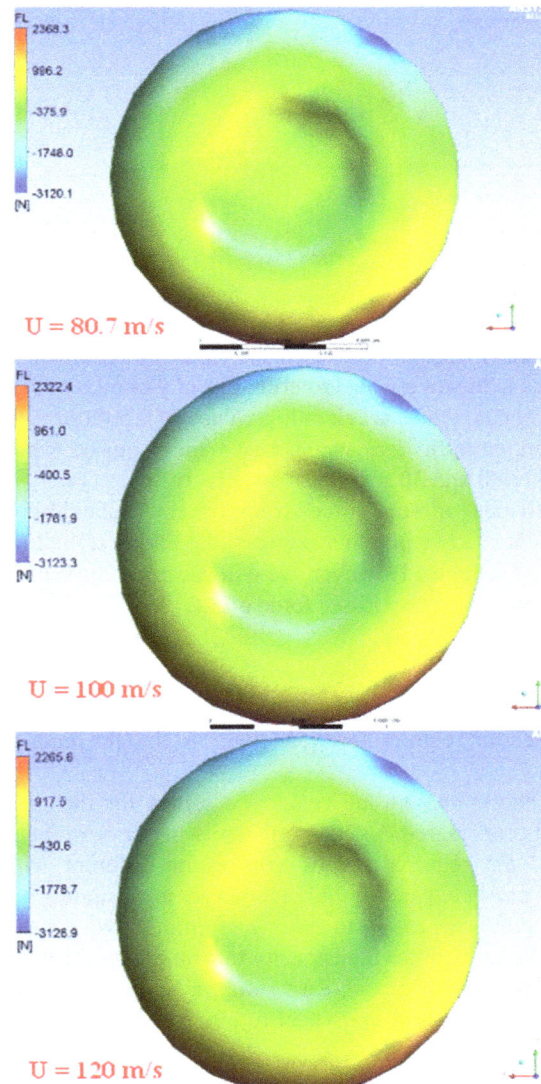

Fig. 12. Contours of lift force with different wind speeds.

At 80.7 m/s wind speed, the force increases to 214 N and the maximum vortex value occurs at 2.21 sec. At 100 m/s, and 120 m/s wind speeds, the steady state value of lift forces are 305 N and 408 N respectively, and the maximum vortexes occur at 2.54 sec and 2.83 sec respectively.

4.4. Equivalent Inertial Force During Acceleration

The aircraft approach period offers sufficient time to accelerate the wheel slowly. However, Fig. 13 shows a comparison of equivalent inertia force with different required wheel angular acceleration.

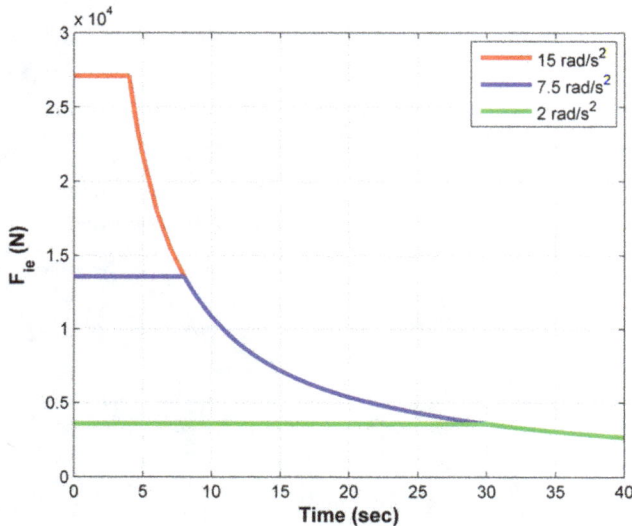

Fig. 13. Equivalent inertial forces vs. time during different accelerations.

All the above results are based on an acceleration of 15 rad/s2 which achieves a wheel rotation velocity of 60.5 rad/sec within 4 seconds. Other acceleration values of 7.5 rad/s^2 and 2 rad/s^2 are investigated, which reach the required wheel rotation within 8 and 30 seconds respectively.

The equivalent inertial force to spin the wheel within 4 seconds is 27.13 kN while this changes to 13.6 kN if the time extended to 8 seconds. However, we will use a spin time of 30 seconds which requires a lesser force of 3.6 kN.

4.5. Required Torque

The shaft rolling resistance force is required in order to calculate the total required force and hence to find the required torque.

Fig. 14 shows the rolling resistance forces for the three wind speeds considered. Using the formula in (1), with 0.8 of rim radius, the required torque curves for different wind speeds are presented in Fig. 15. However, in designing the wind turbine, it is necessary to assume a minimum wind speed to be sure the turbine has the capacity to spin the wheel in the worst-case scenario. For our case study, a torque of 1048.7 N.m is sufficient to spin the wheel to 60.5 rad/sec within 30 seconds.

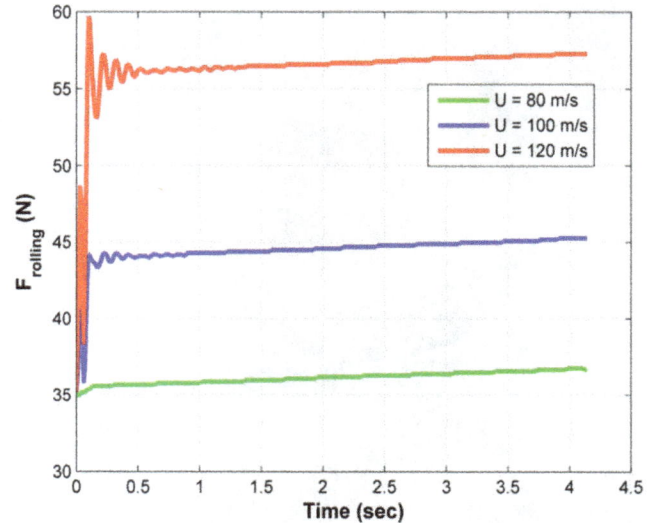

Fig. 14. Shaft rolling resistance vs. time for different wind speeds.

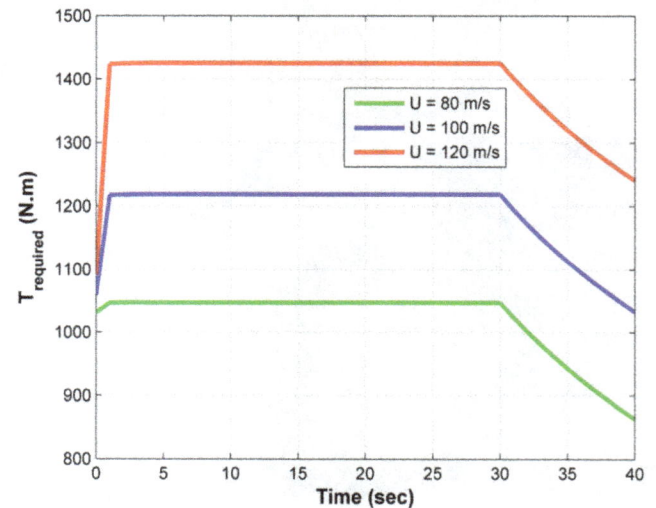

Fig. 15. Required torque vs. time for different wind speeds.

5. Conclusion

An isolated wheel has been tested using ANSYS CFX against different wind speeds. The wheel was accelerated from zero to steady state rolling in order to investigate the aerodynamic forces generated during rotation, from this the torque required to spin the wheel was then determined.

We conclude that the lift force depends on the rotation direction. In the aircraft approach condition, the lift force is negative (downwards) and is additive to the load on the wheel. Also, the shape of the tyre has the effect of increasing the drag force in the rotational axis. The required torque calculation procedure is as presented above.

Finally, as a check, it is recommended that a further study should be carried out to include the use of wind turbines in order to test the ability to spin the wheel against various wind speeds. The acceleration time and torque produced should then be considered and compared against the results obtained above.

References

[1]   Abbasszadeh, M., T., and Abbasszadeh, M., U.S. Patent Application for a "Apparatus for causing an aircraft wheel to rotate", Publication No. US20150021435 A1. Washington, DC: U.S. Patent and Trademark Office. 22 Jan 2015.

[2]   Sweet, R. M., Gilleran, N., Edelson, J. S., Cox, I. W., Cox, R. T., U.S. Patent Application for a "Integrated electric motor and gear in an aircraft wheel", Publication No. US8714481 B2. Washington, DC: U.S. Patent and Trademark Office. 6 May 2014.

[3]   Karl, W., U.S. Patent Application for a "Free Spinning Wheel for Airplanes", Publication No. US20140048648 A1. Washington, DC: U.S. Patent and Trademark Office. 20 Feb 2014.

[4]   Didey, A., U.S. Patent Application for a "Landing gear drive systems", Publication No. WO2014023939 A1. Washington, DC: U.S. Patent and Trademark Office. 13 Feb 2014.

[5]   Khal, S., and Khal, A., U.S. Patent Application for a "Apparatus for Pre-Rotating Aircraft Tires", Publication No. US20130112809 A1. Washington, DC: U.S. Patent and Trademark Office. 9 May 2013.

[6]   Cassetta, F. G., and Perez, L. C., U.S. Patent Application for a "Passive acceleration device for aircraft wheels", Publication No. EP1944233 A1. Washington, DC: U.S. Patent and Trademark Office. 16 Jul 2008.

[7]   Horvath, V., and Szoke, B., U.S. Patent Application for a "Airplane tire saver by protrusion airfoils", Publication No. WO2006130944 A1. Washington, DC: U.S. Patent and Trademark Office. 14 Dec 2006.

[8]   Robert, A., U.S. Patent Application for a "Self rotating airplane tire", Publication No. US3773283 A. Washington, DC: U.S. Patent and Trademark Office. 20 Nov 1973.

[9]   Beazley, R. H., U.S. Patent Application for a "Aircraft wheel spinner and control", Publication No. US2414849 A. Washington, DC: U.S. Patent and Trademark Office. 28 Jan 1947.

[10]   PADOVAN, JOE, AMIR KAZEMPOUR, and YONG H. KIM. "Aircraft Landing-Induced Tire Spinup."Journal of Aircraft, Vol. 28, No. 12 (December 1991): pp. 849–854. doi: 10.2514/3.46108.

[11]   Saibel, Edward A., and Chenglung Tsai. "Tire Wear by Ablation." Wear, Vol. 24, No. 2 (May 1973): pp. 161–176. doi: 10.1016/0043-1648(73)90229-9.

[12]   Bennett, Michael, Simon M. Christie, Angus Graham, Bryony S. Thomas, Vladimir Vishnyakov, Kevin Morris, Daniel M. Peters, Rhys Jones, and Cathy Ansell. "Composition of Smoke Generated by Landing Aircraft." Environ. Sci. Technol. Vol. 45, No. 8 (April 15, 2011): pp.3533–3538. doi: 10.1021/es1027585.

[13]   T. D. Kothalawala, A. Gatto, and L. Wrobel,"Computational Investigation of the Combined Effects of Yaw, Rotation & Ground Proximity on the Aerodynamics of an Isolated Wheel" International Journal of Mechanical, Aerospace, Industrial, Mechatronic and Manufacturing Engineering, Vol: 7, No: 9, 2013, pp. 1789-1795.

[14]   Morelli.A, "Aerodynamic Actions on an Automobile Wheel," Fifth Paper at the First Symposium on Road Vehicle Aerodynamics, City University London, 1969.

[15]   Abu Sadek Saifur Rahman, "Computational study on flow around a rotating short cylinder in order to study the effect of rotation on the aerodynamics of a vehicle", Master Thesis, Texas Tech University, 1996.

[16]   Boeing Commercial Airplane Co., "Approach speeds for Boeing airplanes", 2011. URL: http://www.boeing.com/assets/pdf/commercial/airports/faqs/ar candapproachspeeds.pdf [cited 21 March 2015].

[17]   Lufthansa Technik., "Aircraft tires: more than just rubber on steel", online database, URL: http://www.lufthansa-technik.com/aircraft-tires [cited 11 May 2015].

[18]   Goodyear. (2002). Aircraft data tire book. Akron, OH: The Goodyear Tire & Rubber Co., pp. 32-33.

[19]   Ochi, Y., and K. Kanai. "Automatic Approach and Landing for Propulsion Controlled Aircraft by H/sub ∞/ Control." Proceedings of the 1999 IEEE International Conference on Control Applications, Cat. No.99CH36328, 1999, doi: 10.1109/cca.1999.800951.

[20]   A. Houari, "Determining the drag coefficient of rotational symmetric objects falling through liquids," Eur. J. Phys., vol. 33, No. 4, pp. 947–954, May 2012. doi: 10.1088/0143-0807/33/4/947.

[21]   J. K. Moore, "Aerodynamics of High Performance Bicycle Wheels", Master thesis, University of Canterbury, 2008.

[22]   NASA, "Lift of Rotating Cylinder", Technical note, URL: https://www.grc.nasa.gov/www/k-12/airplane/cyl.html [Cited 02 Feb. 2016]

[23]   Carstensen, S., Mandviwalla, X., Vita, L., and Paulsen, U., "Lift of a Rotating Circular Cylinder in Unsteady Flows", Journal of Ocean and Wind Energy, Vol. 1, No. 1, 2014, pp. 41–49.

[24]   Burns, John A., and Ou, Yuh-Roung., "Effect of Rotation Rate on the Force of a Rotating Cylinder: Simulation and Control" NASA Contractor Report 191442, ICASE Report No. 93-11, 1993, URL: http://ntrs.nasa.gov/search.jsp?R=19930017819 [Cited 17 April 2016]

[25]   Zhang, Darui, Andrej Ivanco, and Zoran Filipi. "Model-Based Estimation of Vehicle Aerodynamic Drag and Rolling Resistance." SAE Int. J. Commer. Veh. 8, no. 2, 2015, pp. 433–439. doi: 10.4271/2015-01-2776.

[26]   Bernard J. Hamrock and William J. Anderson., "Rolling-Element Bearings", NASA Reference Publication 1105, 1983.

[27]   Zhang, Xin, Willem Toet, and Jonathan Zerihan. "Ground Effect Aerodynamics of Race Cars." Appl. Mech. Rev. Vol. 59, No. 1 (2006): pp.33. doi: 10.1115/1.2110263.

[28]   Okumura, K., "CFD Simulation by Automatically Generated Tetrahedral and Prismatic Cells for Engine Intake Duct and Coolant Flow in Three Days," SAE Technical Paper 2000-01-0294, 2000, doi: 10.4271/2000-01-0294.

[29]   Menter, F. R. "Two-Equation Eddy-Viscosity Turbulence Models for Engineering Applications." AIAA Journal, Vol. 32, No. 8, 1994, pp. 1598–1605. doi: 10.2514/3.12149.

[30] ANSYS® Academic Research, Release 15.7, Help System, CFX-Pre Guide, ANSYS, Inc.

[31] Menter, Florian R. "Review of the Shear-Stress Transport Turbulence Model Experience from an Industrial Perspective." International Journal of Computational Fluid Dynamics, Vol. 23, No. 4, 2009. pp. 305–316. doi:10.1080/10618560902773387.

[32] Wilcox, D. C., "Turbulence modeling for CFD", 1st ed, La Canada, CA: DCW Industries Inc., 1998.

9

Erosion of an Axial Transonic Fan due to dust ingestion

Ahmed Fayez EL-Saied, Mohamed Hassan Gobran, Hassan Zohier Hassan

Mechanical Power Eng. Dept., Zagazig University, Zagazig, Egypt

Email address:

amahelal@yahoo.com (A. F. EL-Saied), mhgobran@yahoo.com (M. H. Gobran), hzohier@yahoo.com (H. Z. Hassan)

Abstract: This paper deals with the prediction of the particle dynamic and erosion characteristics due to dust ingestion in an axial flow fan, installed in a high bypass-ratio turbofan engine that operates in a dusty environment. Dynamic behavior comprises the particle trajectory and its impact velocity and location. While the erosion characteristics are resembled by the impact frequency, erosion rate, erosion parameter and the penetration rate. The study was carried out in two flight regimes, namely, takeoff, where the sand particles are prevailing, and cruise, where the fly ashes are dominated. In both cases, the effect of the particle size on its trajectory, impact location, and the erosion characteristics was studied. To simulate the problem in a more realistic manner, a Rosin Rambler particle diameter distribution was assumed at takeoff and cruise conditions. At takeoff, this distribution varies from 50 to 300 μm with a mean diameter of 150 μm sand particles. While at cruise, this distribution varies from 5 to 30 μm with a mean diameter of 15 μm fly ash particles. The computational domain employed was a periodic sector through both the fan and its intake bounding an angle of (360/38) where the number of fan blades is (38). The intake is a stationary domain while the fan is a rotating one and the FLUENT solver is used to solve this problem. Firstly, the flow field was solved in the computational domain using the Navier-Stokes finite- volume supported by the Spalart-Allmaras turbulence model. The governing equations, representing the particle motion through the moving stream of a compressible flow are introduced herein to calculate the particle trajectory. The solution of these equations is carried out based on the Lagrangian approach. Next, empirical equations representing the particle impact characteristics with the walls are introduced to calculate the rebound velocity, the erosion rate, erosion parameter, impact frequency and penetration rate. Moreover, a method to smoothen the irregularity in the calculated scattered data was discussed as well. During takeoff flight regime, the pressure side of fan blade experienced higher particle impact and erosion damage. The highest erosion rate was found at the corner formed by blade tip and trailing edge of pressure side. During cruise conditions, less erosion rates resulted. Maximum erosion rates are found at the leading edge of the pressure side.

Keywords: Turbomachinery Erosion, Transonic Axial Fan, Fan Erosion, Particulate Flow

1. Introduction

Erosion is a main physical problem facing the aircraft engines, especially for those operating in the desert or polluted environment. The tendency of an aircraft engine to ingest ground particles during ground operation is a well- known phenomenon. During high power setting at takeoff, the generated ground vortex has the capability of picking up particles from the ground, which are then ingested into the engine. Thrust reversal can also blow solid particles into the engine intake during landing. During takeoff or landing, helicopters need to use the effect of down-wash to create additional lift. The down-wash associated with hover maneuver and multiple pick-ups with rotor turning may generate a severe dust cloud around the aircraft. This dust cloud circulates up and over the rotor blades to re-enter the circling air from above, inevitably a great deal of dust enters the engine. Moreover, the aircraft engine is subjected to fly ashes during flight operation.

In addition, dust from the desert areas is rich in sodium and potassium, especially particles of diameter up to 100 microns which are the most common. These particles deposit on the surfaces of the hot section of the engine after combining with sulfur and oxygen during the combustion process. Particles of four metals; sodium, potassium, vanadium, and lead will disrupt the protective coating of the hot gas sections leading to faster oxidization of these parts by the hot gases.

Turbomachinery blade erosion has been the subject of many analytical and experimental investigations in the last fifty years. In these studies, the main subject was to correlate the particle impact velocity, impact angle, and impact locations with surface measured erosion parameter for a particular range of conditions. An early study by Montgomery et al. [1] described a test program to relate the losses in a radial turbine operating lifetime to the amount and sizes of ingested particles. This study indicated that, as particle small as $(2-3\mu m)$ it still cause serious losses in the engine operating life time. Tabakoff and Hamed [2], experimentally, derived correlations for the restitution ratio on their study of erosion in radial inflow turbine.

Through an experimental study of an axial flow gas turbine, Tabakoff et al. [3] showed that the maximum stator blade erosion is at the blade trailing edge, and the maximum rotor blade erosion is located near one third the blade height at the leading edge. In an important contribution on the field of gas turbine erosion, Hamed [4] introduced a useful equation to compute the mass erosion parameter. A combined experimental and computational research program was conducted by Hamed et al. [5] to investigate the effect of solid particle impacts on the turbine blade material. The results indicate that many particles impact the vane pressure surface. The vane surface impacts reduce the particles absolute velocity and consequently they impact the rotor blade suction surface. Their prediction indicates a narrow band of high erosion at the vane leading edge and pressure surface erosion increasing towards the trailing edge.

Japikse [6], Wulf et al. [7], and Peterson [8] related the rate of wear to the time of usage. They claimed that, within 6000 -8000 hours of jet engine operation, compressor blades had blunt leading edges with an increased surface roughness.

The damage of the blades is caused mainly by direct material impingement of large particles, and even by recirculation of fine particles due to secondary flow through the blade passage, Neilson and Gilchrist [9].

Tabakoff and Hussein [10], and Elsayed and Brown [11] analyzed the particulate flow in axial turbo-machines. Hamed [12] presented a newer method for particle trajectory calculations to include the influence of the hub and tip shapes, the radial variations and the vane shape on the particle trajectories. She determined the particles rebounding velocity and direction after each impact, and consequently the erosion prediction on the blades by using empirical correlations derived from experiments conducted in a special tunnel. Beacher et al. [13] had improved the model of Hamed [12], involving a more complex representation of the three-dimensional flow fields. Hamed and Fowler [14] used statistical method combined with particle trajectories computations together with the experimental data of metal erosion and particle rebound characteristics in their modified model.

Clevenger and Tabakoff [15] treated the particulate flow problems in radial inflow turbines. Experimentally, they traced the particle trajectories through the different regions of the turbine (scroll, stator and rotor) using high speed camera. These trajectories are used to estimate the erosion rates. Hussein and Tabakoff [16] had developed a code for the three-dimensional particle trajectory calculations through turbines and compressors. The success of their code was attributed to its accurate representation of the blade airfoil shape and the blade-to-blade flow field at the mean radius. Elsayed and Rashed [17] and [18] investigated the air flow and the erosion rate of centrifugal compressors. They estimated the lifetime of centrifugal compressor due to successive operation in polluted environment or sandy areas.

Maxwell [19] conducted a theoretical analysis of particle trajectory in a stationary cascade. Elsayed and Rouleau [20] investigated the erosion of stationary cascades in turbo-expanders. The air flow velocities were measured in many points within the passage using LDV. Next, the particle trajectories were calculated numerically. Erosion not only influences the aerodynamics of turbo-machineries but also affects their structural integrity. Such a combined effect influences the aero-elastic behavior of eroded turbo-machine. The change of the aero-elastic behavior was examined by Elsayed [21].

Suzuki et al. [22] carried out the numerical simulations for the sand erosion phenomena in the rotor/stator interaction of the single stage axial flow compressor. Through this study, authors conclude that: severe sand erosion occurs on the pressure surface at leading edge of the blades and blade tip is deteriorated by sand erosion rather than hub due to high impact velocity.

Corsini et al. [23] presented a numerical study on the evolution of blade leading edge erosion patterns in an axial induced draft fan. The numerical study clarifies the influence of flow structure, blade geometry, particle size, and concentration on erosion pattern. They found that: the blade wear on the leading edge pressure side is larger than that on the suction side, and the leading edge erosion extends over the entire blade. The authors also found that: the erosion sensitive areas were also concentrated over the tip corner due to the combination of the higher peripheral speed and particle migration from the pressure to the suction side.

Brun et al. [24] conducted a semi-empirical model to predict the behavior of solid particles, such as sand, dirt, and dust, in turbo-machines, such as centrifugal compressors, gas turbine engines, or axial compressors. The results allow one to develop a strategy to "harden" a turbo-machine locally against the damaging effects of sand or dirt ingestion.

Carbonetto and Hoch [25] discussed a method of predicting expander flow path erosion and deposition using computational fluid dynamics (CFD) coupled with empirical erosion data. With an understanding of the particle dynamics, new aerodynamically efficient components may be developed to significantly increase the reliability and life of the expander.

Zhang et al. [26] presented a new method to improve the erosion resistance of machine components by biomimetic method. The discrete phase model was used for modeling the solid particles flow, and the Eulerian conservation equations to the continuous phase. The numerical study employs computational fluid dynamics (CFD) software, based on a finite volume method. Gas/solid flow axial fan was simulated to calculate the erosion rate of the particles on the fan blades and comparatively analyzed the erosive wear of the smooth surface. The results show that the groove-shaped biomimetic blade anti-erosion ability is better than that of the other fan blades.

From the above literature review, it is noticed that: most of the researches discussing the erosion problem in turbomachinery were concentrated on the erosion in compressors, turbines, and sometimes small fans. In modern turbofan engines, large fans with special designs are required to achieve high-pressure ratio and mass flux. The deterioration in such fan performance due to the erosion phenomenon has a great impact on its related engine. So, the present work is oriented to study the erosion in a large axial-flow fan installed in a high bypass-ratio turbofan engine at both takeoff and cruise conditions. At takeoff the particle material is considered sand while at cruise it is considered fly ash. For the two flight conditions the effect of particle size is examined. In addition, two particle distributions (at takeoff and cruise) are also examined.

2. Numerical Model

Numerical simulation was conducted to determine the three-dimensional compressible viscous flow field and associated solid trajectories through the fan. The discrete phase model was used for modeling the solid particles flow, and the Eulerian conservation equations to the continuous phase. The flow phase simulations are based on implicit solution of the Reynolds-averaged Navier-Stokes equation in conversation and with Spalart-Allmaras turbulence model with near wall treatment. The numerical study employs computational fluid dynamics (CFD) software, based on a finite volume method. Gas/solid flow axial fan was simulated to calculate the erosion characteristics of the particles on the fan blades and comparatively analyzed the erosive wear of the smooth surface. The simulation incorporates empirical particle-surface and particle-air interaction models. The compressible viscous flow and dispersed particle dynamics simulation were conducted using FLUINT 6.1 software ([27], [28], and [29]).

3. Case Study

The case studied in the present work represents a subsonic intake followed by an axial- flow fan. This fan is the first rotary module in the GENERAL ELECTRIC CF6-50 high bypass-ratio turbofan engines. The fan rotor has a large diameter (2.18 m) and the fan blade is tapered, highly twisted, and has a long span (0.6507 m) with a low hub to tip ratio (0.4033). Figure (1) shows the fan and its intake. The axial length of the intake was found to be 155.55 cm, and the nose of the fan can be simplified as a hemisphere of radius 40.57 cm. The fan and the intake detailed dimensions were measured from the real fan model of the CF6-50 turbofan engine at the maintenance workshop of the Egypt Air Company. The dimensions were measured at four different locations along the blade height. One at the hub, another at the tip, and the

other two sections are in between. These sections help for drawing the overall blade profile. Table (1) gives some important dimensions of the four sections.

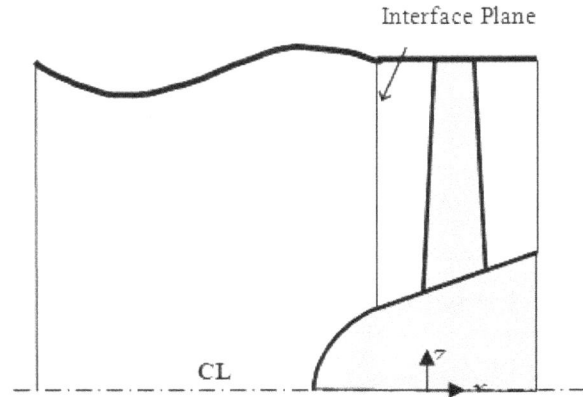

Figure 1. CF6 intake and fan.

Table 1. Blade characteristics at different radius locations

Radial location, (cm)	44 hub	69	89	109 (tip)
Chord length	17.5	19.4	21	24.4
Stagger angle, degree	10	43	49	66
Location of max. thickness, %chord	65.7	58	48	47.8
Max. thickness, mm.	9	7.5	4.2	3.7

4. Computational Domain

The computational domain in the study includes both of the fan and the intake zones. Due to the large dimensions of both the intake and the fan combination, it was very difficult to solve the overall the domain. Since the fan includes 38 blades, a periodic sector of an angle (360/38) was generated separately for both the fan and the intake zones. The two zones were merged together to form the required computational domain. The GAMBIT preprocessor (a software in the Fluent 6.1 package) is used to build the geometry of the solution domain for both the intake and fan zones. It is also used to generate the suitable mesh. Due to complex geometry of both the fan and intake sectors, it was appropriate to use the unstructured tetrahedral grid. Figures (2) and (3) give the mesh at hub and tip sections, respectively and table (2) summarizes the mesh characteristics on the whole domain. Both the fan and the intake zones were meshed separately into two mesh files. The two meshes were merged using the TMERGE utility. Table (3) gives the types of boundary conditions for the computational domain.

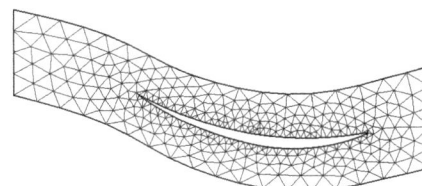

Figure 2. The mesh at the hub section.

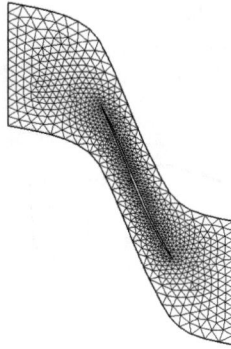

Figure 3. The mesh at the tip section.

Table 2. Grid characteristics

	Intake block	Fan block	Total
Number of cells	49005	281941	330946
Number of faces	99020	585640	692844
Number of nodes	11353	59403	72011

Table 3. Boundary condition

Boundary	Boundary type
Intake inlet	Pressure inlet
Fan exit	Pressure outlet
Intake casing and the fan casing	Stationary wall
Fan hub, fan pressure side, suction side, and the spinner cone.	Rotating walls
Fan inlet and intake outlet	Grid interface plane
Fan and intake periodic sides	Periodic

5. Procedure of Solution

In calculating the particle trajectories and the erosion characteristics on the fan blade, we will use the one- way coupling as discussed before. The procedure of solution can be summarized as follows [28]:

1. The Fluent main program reads the geometry and the air flow field properties were calculated, [29].
2. The particles initial conditions (injection locations and velocity), and properties (density, diameter distribution, and mass flow- rate) are introduced to the solver.
3. The main Fluent 6.1 program starts particle trajectory calculations.
4. If the particle reaches a wall, the main program calls the user defined functions to compute the erosion rate, erosion parameter, penetration rate, and the impact frequency. These values are returned to the element face that the particle impacts.
5. The particle new velocity, after impact, is then calculated and the trajectory calculations continue with this new velocity and other impacts may be considered until the particle leaves the domain.
6. Another particle is traced in the same manner until all particles trajectory are calculated.
7. The erosion data obtained on the boundary wall elements are input to the developed FORTRAN program to begin the localized inverse interpolation calculations.
8. Finally, the obtained results are post processed.

6. Results and Discussion

Many runs of the Fluent program were carried out to predict the particle trajectories and the subsequent erosion phenomena. To understand the erosion problem, equally-space six injection locations are distributed at the intake inlet. These locations represent a part of the total initial injection locations distributed evenly at the intake inlet. These runs are classified according to the flight conditions (cruise and takeoff), different particle diameters (50, 150, and 250 μm for takeoff, and 10, 20, and 30 μm for cruise), and according to the particle size distribution (from 50 to 300 μm for takeoff with the mean diameter at 150 μm, and from 5 to 30 μm for cruise with the mean diameter at 15 μm). At takeoff conditions, the particle material is considered sand while at cruise it is considered fly ash. The particle properties at both takeoff and cruise conditions are shown in table (4). The particle initial injection velocity equals the corresponding flow velocity. The results include the particle trajectories, blade impact frequency, blade erosion rate and erosion parameter and penetration rate.

Table 4. The properties of the particles at both takeoff and cruise conditions. Montgomery [1]

Particle parameter	Takeoff	Cruise
Material	Sand	Fly ashes
Density (kg/m^3)	2650	600
Concentration g/m^3	0.176	0.0176
Particle mass flow rate (gm/s)	97.5	8.83
Particle initial velocity (m/s)	128	133

6.1. Takeoff Condition

Figure (4) shows the trajectory of particle having 50 μm diameters at different locations. One can observe that, the particle injected towards the fan nose (location 1) impacts the nose, then it is acquired a radial outward velocity component till it enters the fan passage where it is influenced by the centrifugal force. The particle then impacts the blade pressure side near the trailing edge at about 40% the blade span with a velocity about 220 m/s and next leaves the domain. Particle injected at location 2 follows the air flow, and then has two successive impacts with the blade pressure side at impact velocity of 180 m/s for each impact. The first impact is at about 20% of the blade span and one third the chord beyond the leading edge, while the second is at about 25% of the blade span near the blade trailing edge. Particle injected from location 3 impacts the blade suction side close to the leading edge with a velocity of 165 m/s. Then, it leaves the domain with a higher circumferential velocity component. Particles injected from locations 4,5 show similar behavior as that of location 3 except with higher impact velocity (230 m/s, 280 m/s, respectively). Particle injected from location 6 near the intake casing is exposed to a slight impact with the intake casing. It continues its motion and follows the air flow until it impacts the blade suction side at about one third of the chord from the leading edge with a velocity of 240 m/s.

P	Blade pressure side
S	Blade suction side.
C	Fan casing.
T	Intake casing.
N	Fan nose
H	Hub.

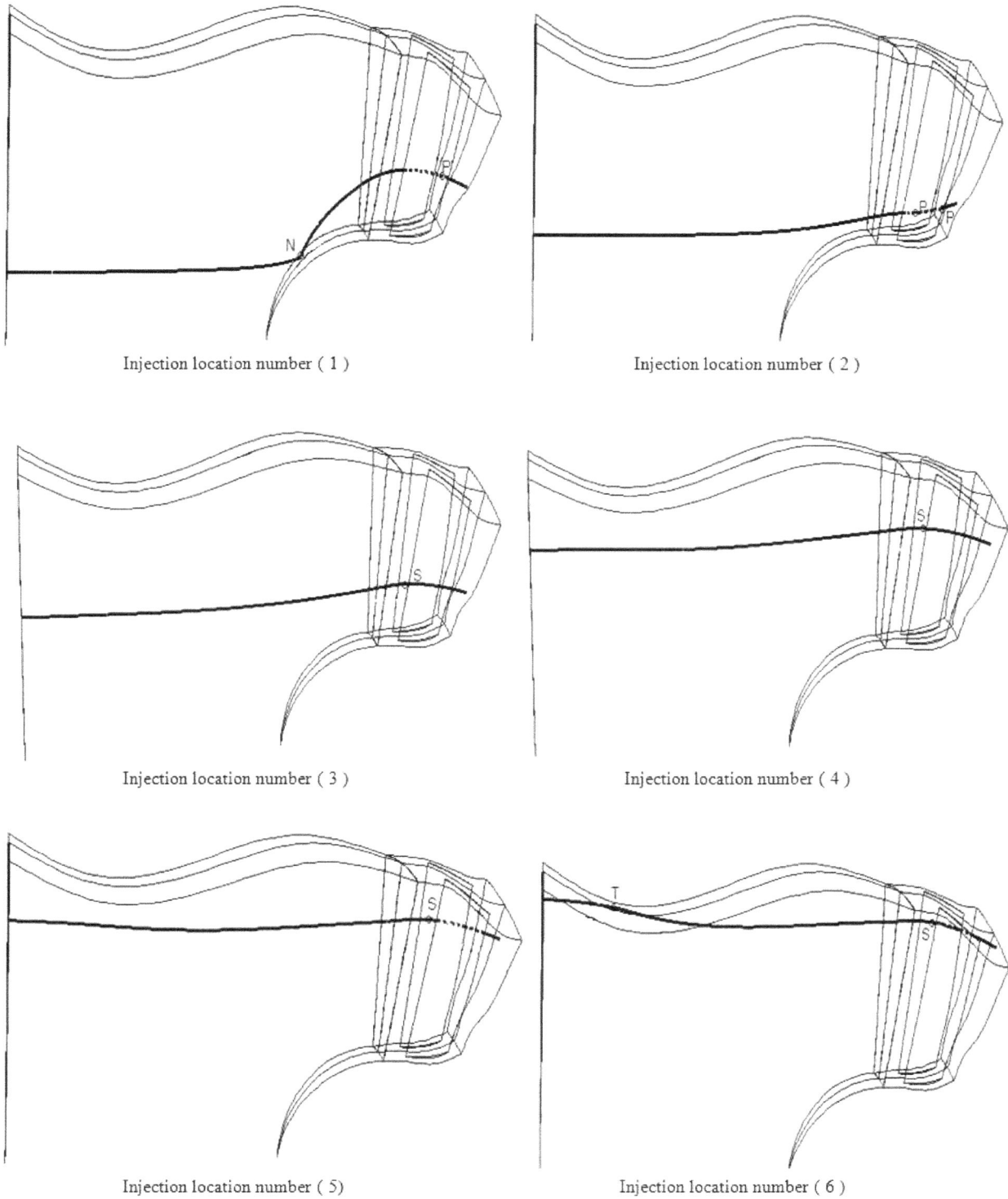

Injection location number (1)

Injection location number (2)

Injection location number (3)

Injection location number (4)

Injection location number (5)

Injection location number (6)

Figure 4. *Trajectory of particles with diameter* $d_p = 50\ \mu m$ *(Takeoff).*

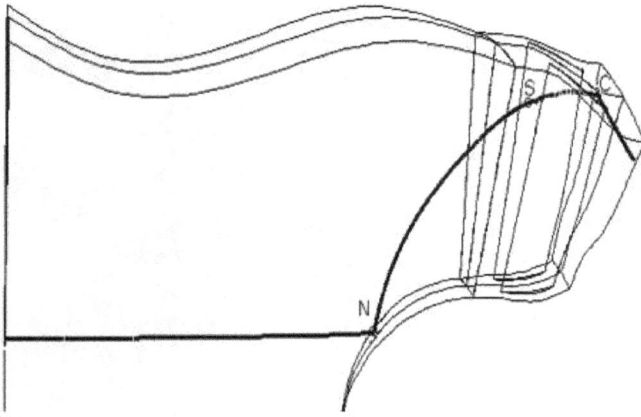

Injection location number (1)

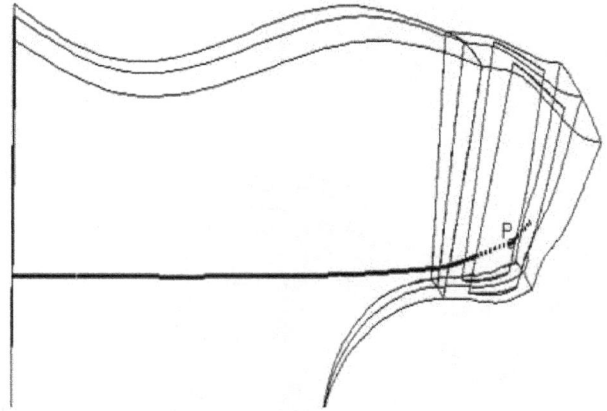

Injection location number (2)

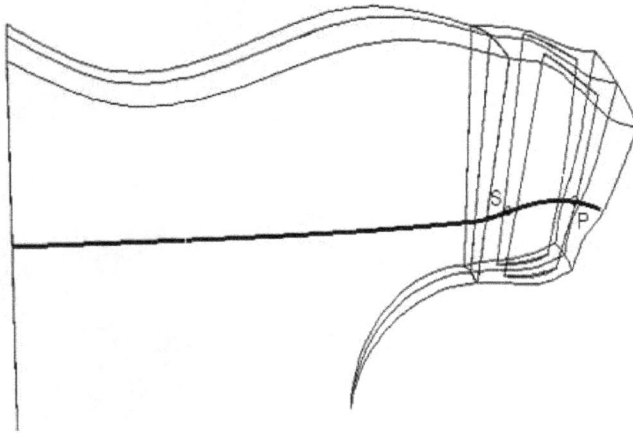

Injection location number (3)

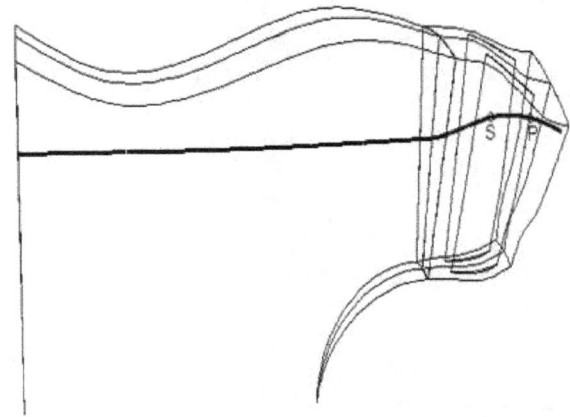

Injection location number (4)

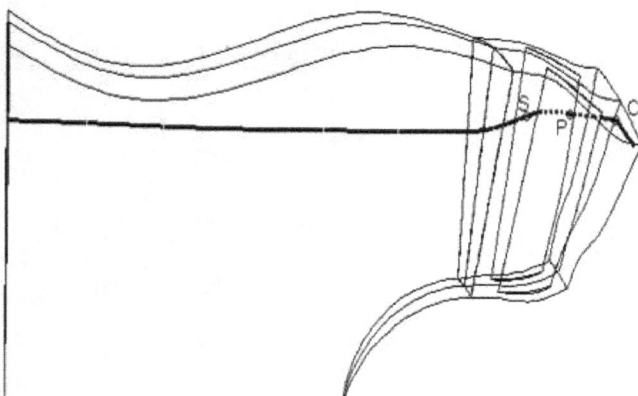

injection location number (5)

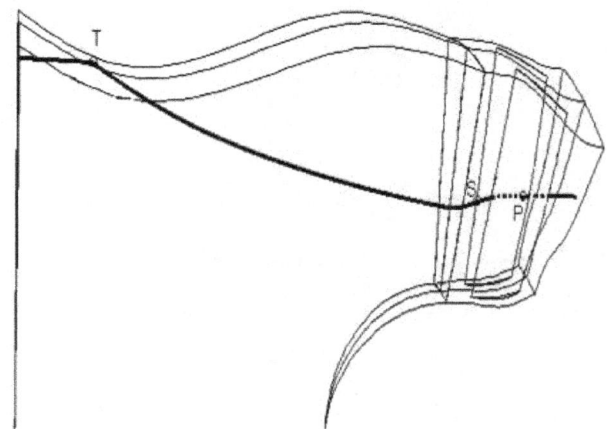

Injection location number (6)

Figure 5. *Trajectory of particles with diameter* $d_p = 150\ \mu m$ *(Takeoff).*

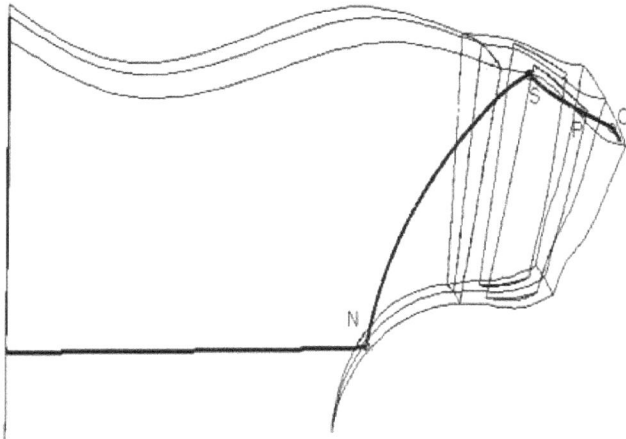

Injection location number (1)

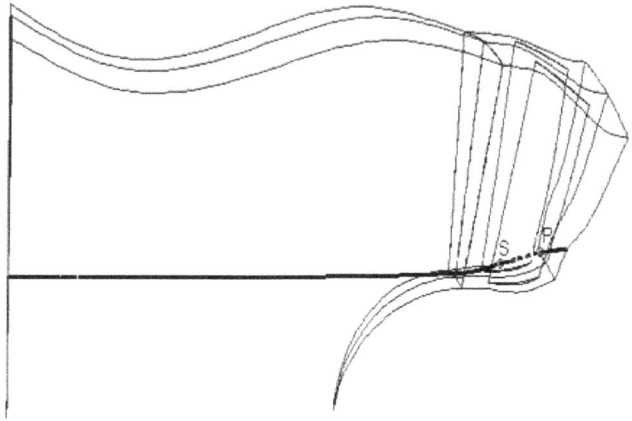

Injection location number (2)

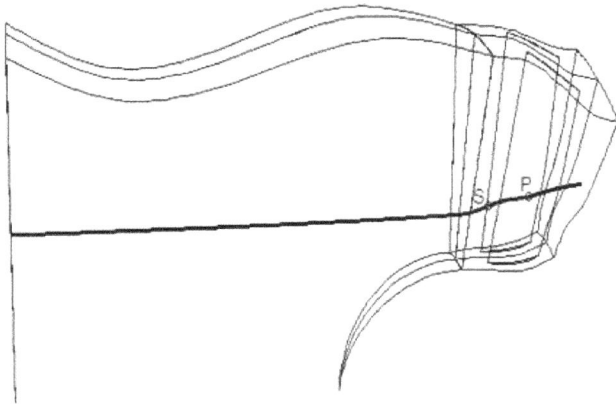

Injection location number (3)

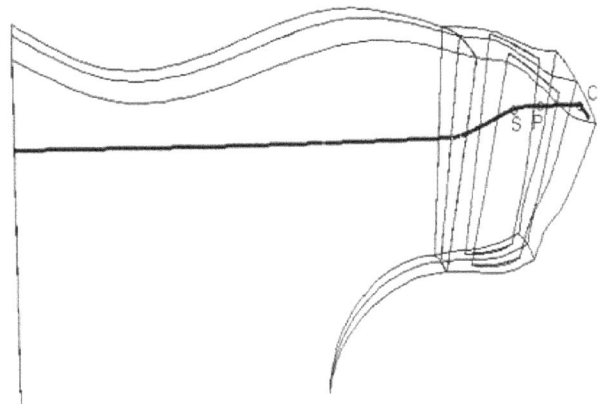

Injection location number (4)

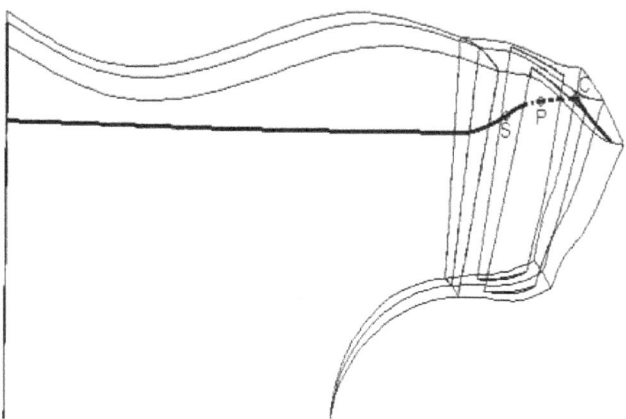

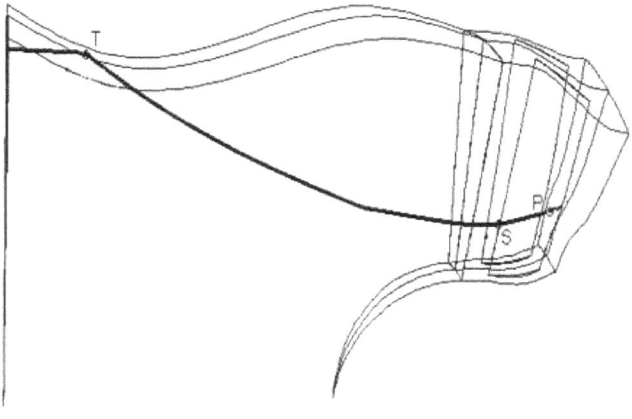

Figure 6. Trajectory of particles with diameter $d_p = 250 \, \mu m$ *(Takeoff).*

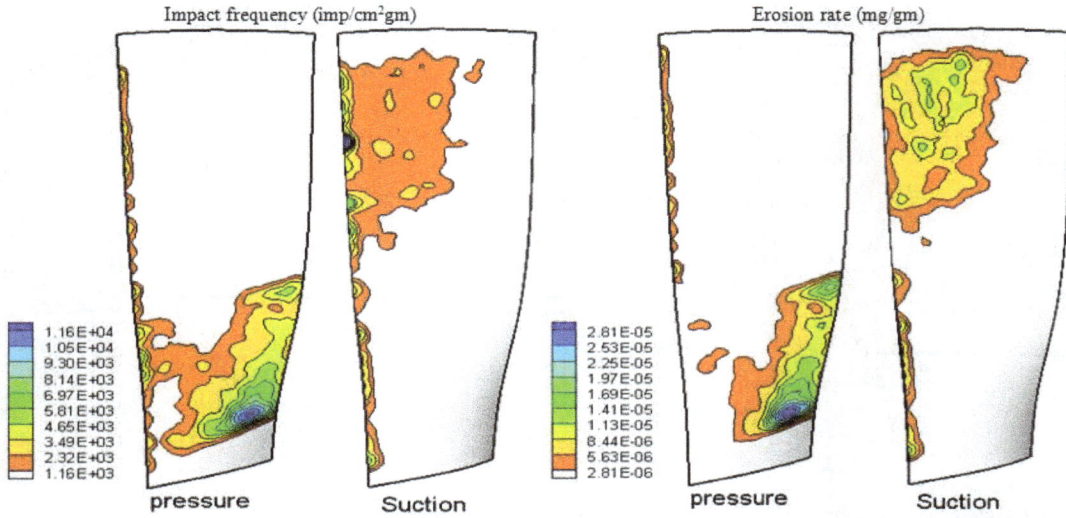

(a) For 50 μm

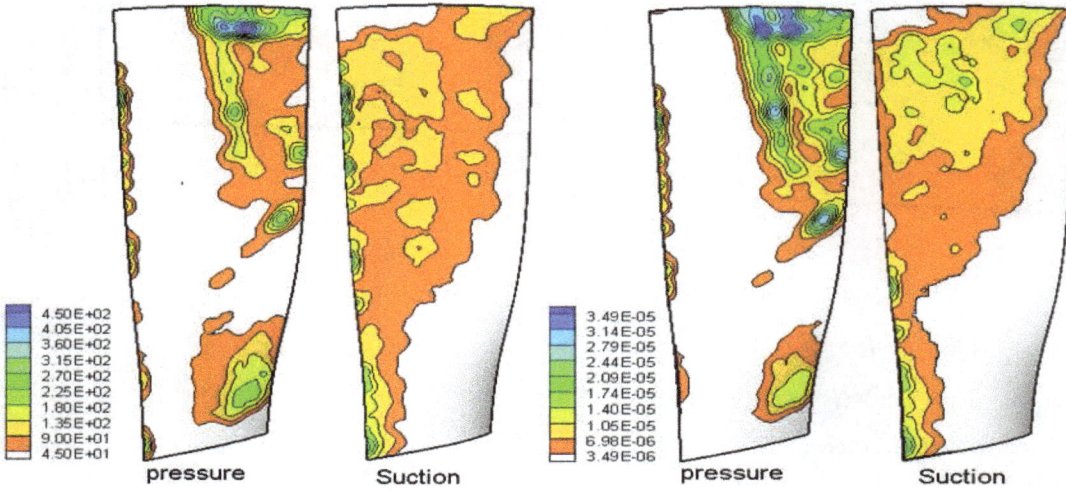

(b) For 150 μm

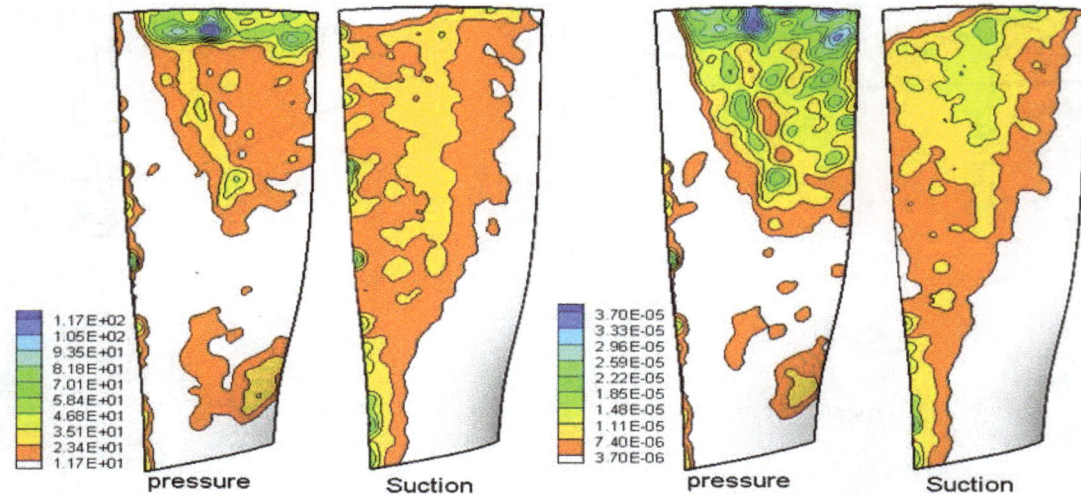

(c) For 250 μm

Figure 7. *The impact frequency and erosion rate, respectively, at take off, for different particle diameters.*

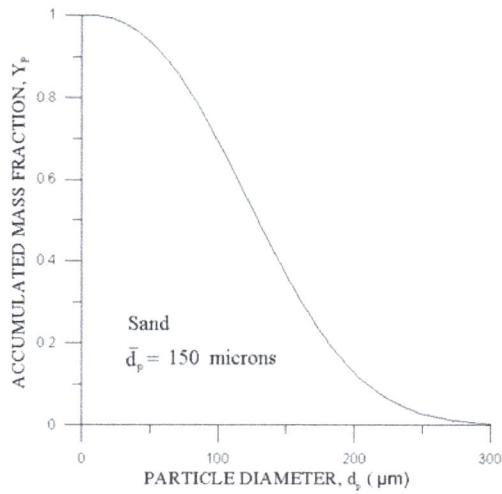

Figure 8. *Rosin-Rammler particle diameter distribution at takeoff.*

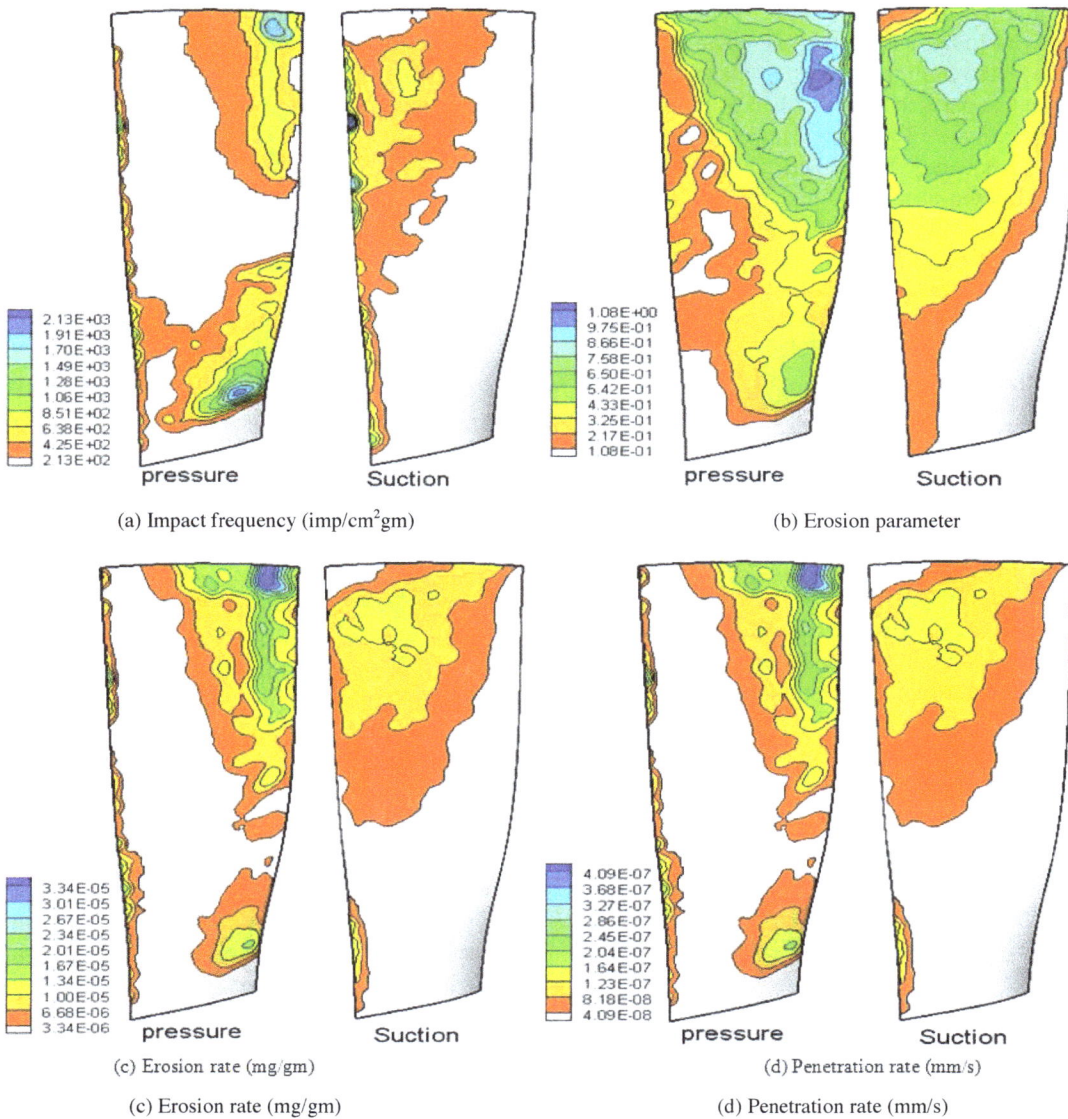

(a) Impact frequency (imp/cm²gm)

(b) Erosion parameter

(c) Erosion rate (mg/gm)

(d) Penetration rate (mm/s)

(c) Erosion rate (mg/gm)

(d) Penetration rate (mm/s)

Figure 9. *Impact frequency, erosion parameter, erosion rate and penetration rate for takeoff Rosin-Rammler distribution.*

The three-dimensional trajectories of particle diameters 150 and, 250 μm are shown in figures (5) and (6), respectively. These particles can be seen to continue their motion unaffected by the flow field due to their high inertia until they impact the blade surface. Most of these particles have two successive impacts. The first impact is observed on the suction side while the second with the blade pressure side. The particle impacts the pressure side with higher velocity and at higher radial location than the suction side.

Figure (7) shows the impact frequency and erosion rate contours on both pressure and suction sides of the fan blade, at takeoff condition and for different particle sizes (50, 150 and 250 μm). For a small particle (50 μm), the impact frequency and erosion rate have maximum values (1.16x104 imp/cm2gm and 2.81x10-5 mg/gm) on the blade pressure side at about 12% span and near the trailing edge. As the particle size increases, the point of maximum impact frequency and erosion rate moves in the blade span-wise direction towards the tip and from trailing to leading edge. The impact frequency decreases with radial position on the blade and it is concentrated near the blade trailing edge up to about 45% of the span. This is due to the particles drift by their impacts with the fan nose. Small numbers of particles impact the blade suction side with low frequency. Moreover, the suction side impacts are generally limited to the leading edge of the outer part of the blade. The leading edge of both pressure and suction sides is exposed to impacts with moderate frequency. It is also noticed that, no particles impact the blade tip either on the suction or the pressure sides. This is due to the effect of the intake geometry which causes the particles to be deflected in radial inward direction.

The Rosin-Rammler particle diameter distribution shown in Figure (8), that representing particle sizes distribution in the Middle East [1], is assumed to predict the erosion rate due to sand particles on the fan blade. The calculated impact frequency for this distribution is shown in figure (9.a). In this figure, one can observe that, the pressure side is exposed to a great number of particle impacts than the suction side. The maximum impact frequency appears at the pressure side on two regions. One of them is at the blade tip between the mid-chord and the trailing edge. This is because the large particles impact the fan nose is migrate outward due to centrifugal force to this region. The other is at about 20% span, near the trailing edge. That is due to small particles that reflected from the fan nose. The particle impacts on the suction side are concentrated at the blade leading edge close to hub and spread towards the trailing edge close to tip. This spreading appears to be deeper with increasing the blade span. Figure (9.b) shows the distribution of the erosion parameter over the blade pressure and suction sides. From this figure, one can conclude that, in general the erosion parameter exhibits greater values on the pressure side than on the suction side. The maximum erosion parameter on the pressure side appears in a region near the blade tip and spreads downward the blade span and towards the blade trailing edge. This region is exposed to higher particle velocities than other regions on the blade surface.

The erosion rate of the fan blade due to the previously mentioned distribution is shown in figure (9.c). Along the blade span, the erosion rate is concentrated near both the trailing and leading edges of the pressure side and near the leading edge of the suction side. On the suction side, the upper half of the blade is exposed to erosion rate ranging from 6.6x10^{-6} to 1.3 x 10^{-5} mg/gm. Other regions on the suction side are less affected by erosion. On the pressure side, the region of maximum erosion rate appears at the corner of the blade between the tip and trailing edge. The value of the erosion rate reaches its maximum in this region, (about 3.34x10^{-5} mg/gm). This is because it is exposed to high frequency and high erosion parameter. Moreover, the region on the blade pressure side at about the upper third of the blade height and about 20% chord from the trailing edge shows an increased erosion rates as well. Also, one can see that, near the leading edge of the pressure side, the erosion rate contours are condensed and indicating a value about 1 x 10^{-5} along the span. The penetration rate contours on the fan blade are shown in figure (8.d). Its distribution is the same as the erosion rate with a maximum value of 4.1x10^{-7} mm/s at the upper and trailing edge corner of the blade pressure side.

6.2. Cruise Condition

The trajectories of ash particles and associated erosion parameter at the cruise conditions are shown in figures (10) through (12). Figure (10) shows the trajectory of the 10 μm diameter ash particles. From this figure, one can observe that the small size and small density of the ash particles makes the associated drag and inertia forces decrease. Thus, the particle trajectories will follow the corresponding streamline. Moreover, these particles acquire different velocities and degrees of turning similar to that of the carrier air flow. Particle injected from location 1 shows no impact with the fan nose because it is drifted by the carrier streamline. There are two successive impacts with the blade pressure side. The first one is at about one third of the chord and the second is near the trailing edge. Particle injected from location 2 shows two impacts with the blade pressure side. One is at the leading edge and the other is at the trailing edge. Particles injected from location 3 and 5 show a single impact with the blade pressure side at the leading edge. Particle injected from location 4 shows a single impact with the pressure side near the trailing edge. Particle injected from location 6 near the intake casing continues its motion and traverses the twisted blades without any impacts. From these trajectories, one can observe that the small ash particles mostly interact with the blade pressure side only even though they may impact more than once. That is because particles entering the fan domain show a tendency to travel from the blade suction side to the pressure side. As the relative velocity of the gas increases from the blade pressure side to the blade suction side, the relative velocity between the gas and particles will attain a minimum value near the pressure surface and maximum value near the suction surface for the same radial coordinates. Consequently, particles are subjected to a positive gradient from the suction surface to the pressure surface. In other words, particles are driven towards the blade pressure side.

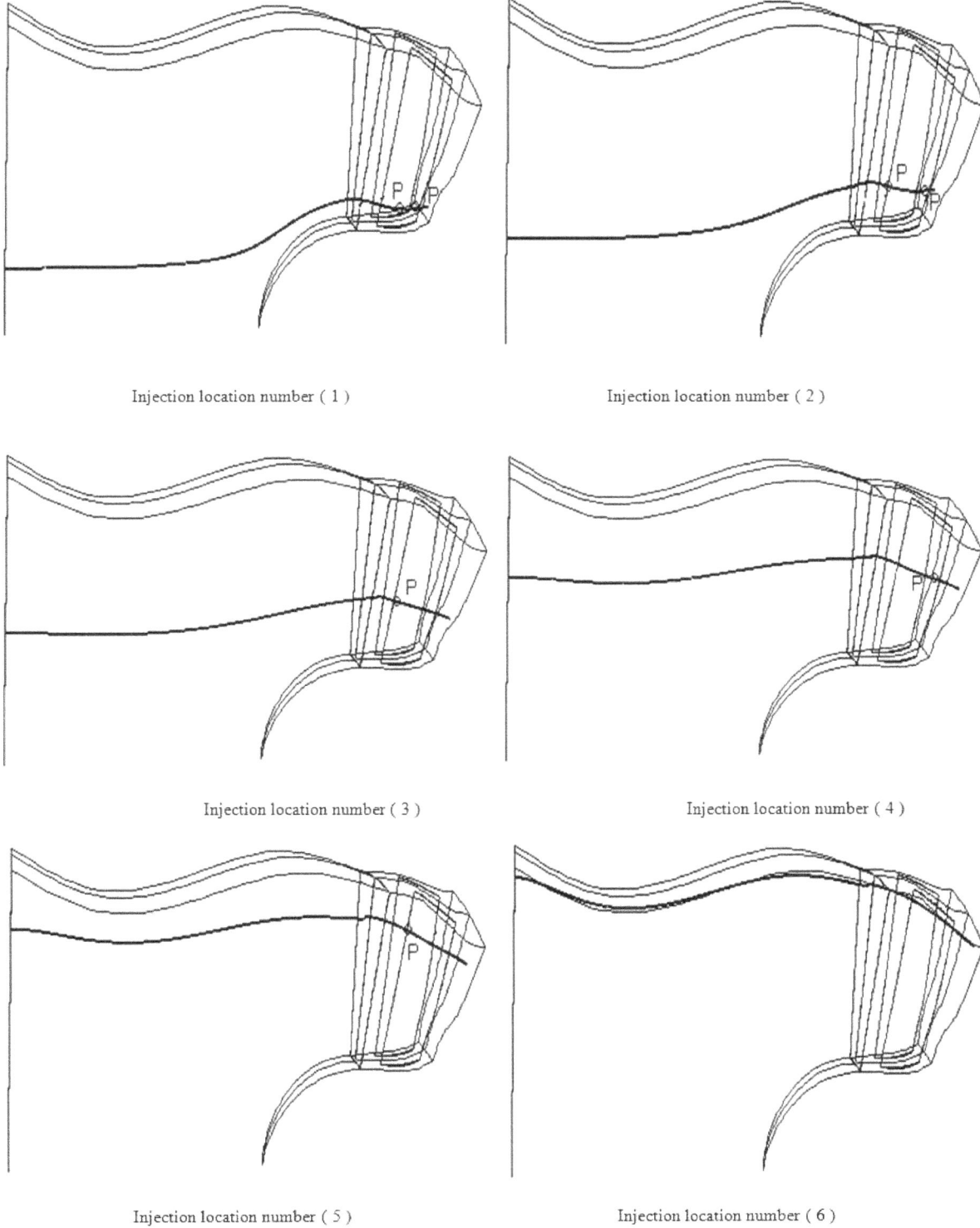

Injection location number (1)

Injection location number (2)

Injection location number (3)

Injection location number (4)

Injection location number (5)

Injection location number (6)

Figure 10. *Trajectory of particles with diameter* $d_p = 10\,\mu m$ *(Cruise).*

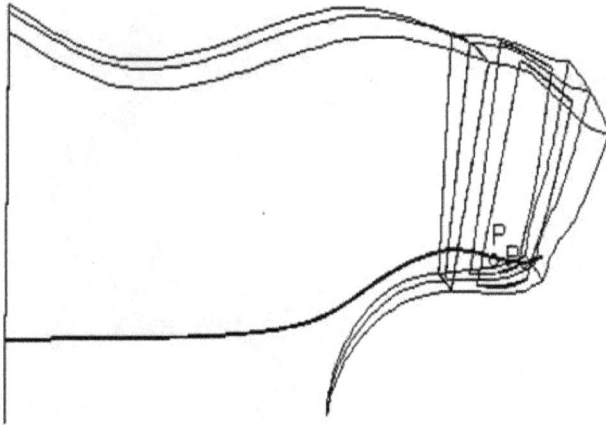

Injection location number (1)

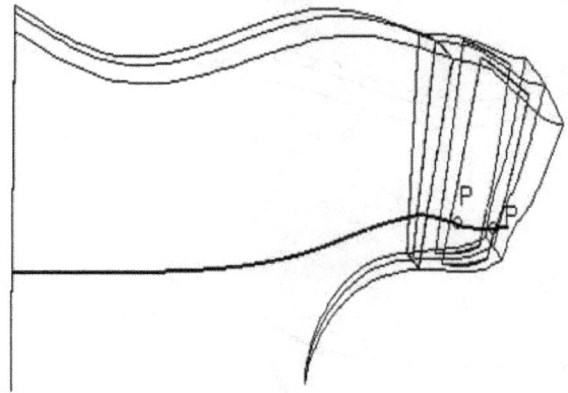

Injection location number (2)

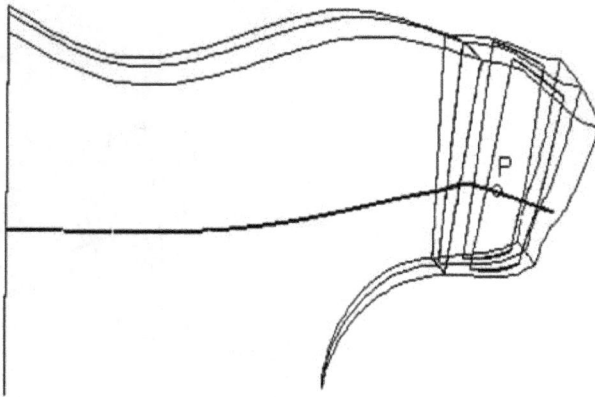

Injection location number (3)

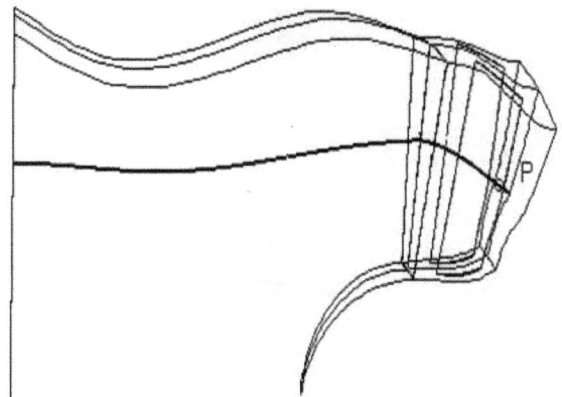

Injection location number (4)

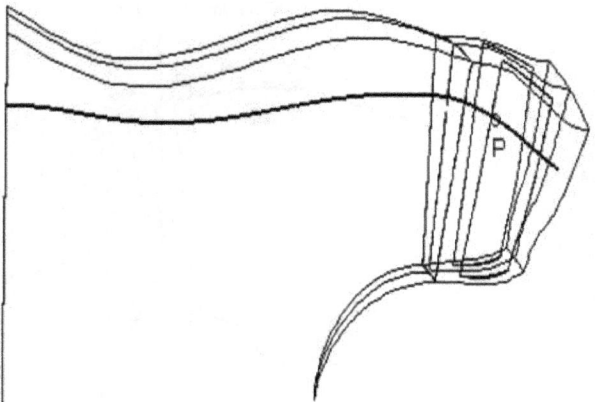

Injection location number (5)

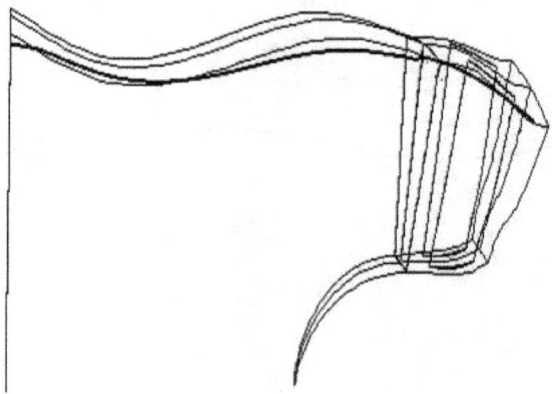

Injection location number (6)

Figure 11. *Trajectory of particles with diameter* $d_p = 20$ μm (*Cruise*).

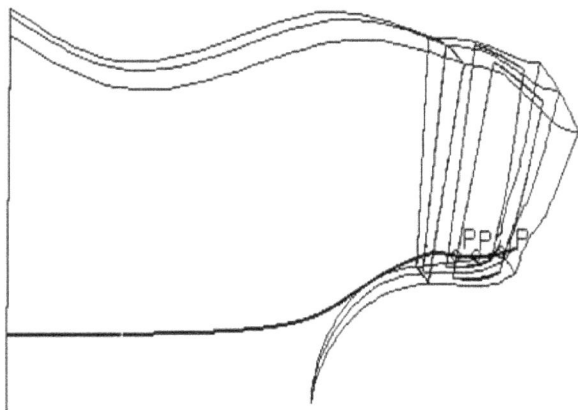

Injection location number (1)

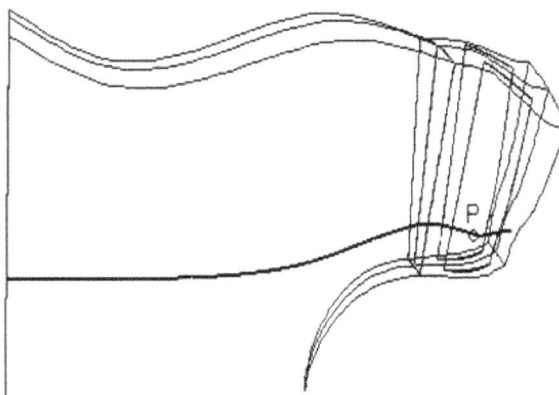

Injection location number (2)

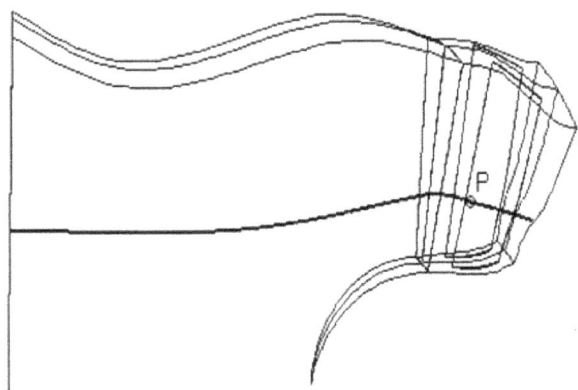

Injection location number (3)

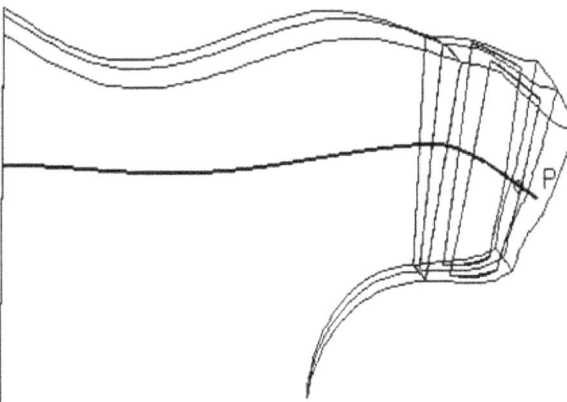

Injection location number (4)

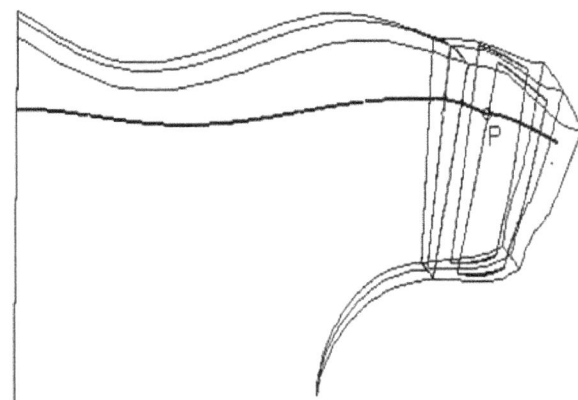

Injection location number (5)

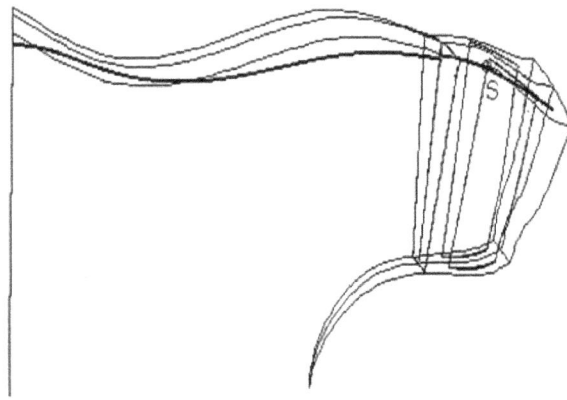

Injection location number (6)

Figure 12. *Trajectory of particles with diameter* $d_p = 30\ \mu m$ *(Cruise).*

Impact frequency (imp/cm²gm)   Erosion rate (mg/gm)

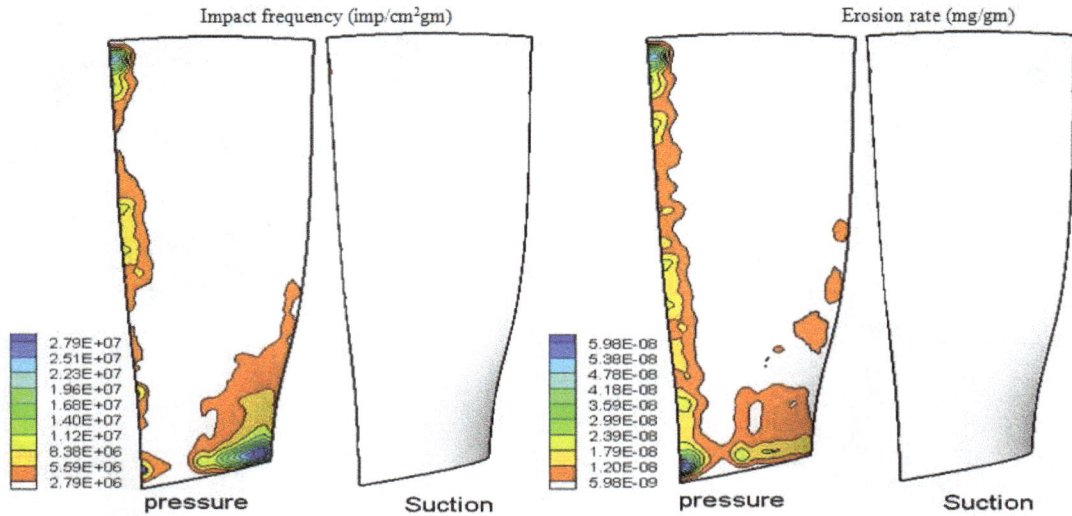

(a) For 10 μm

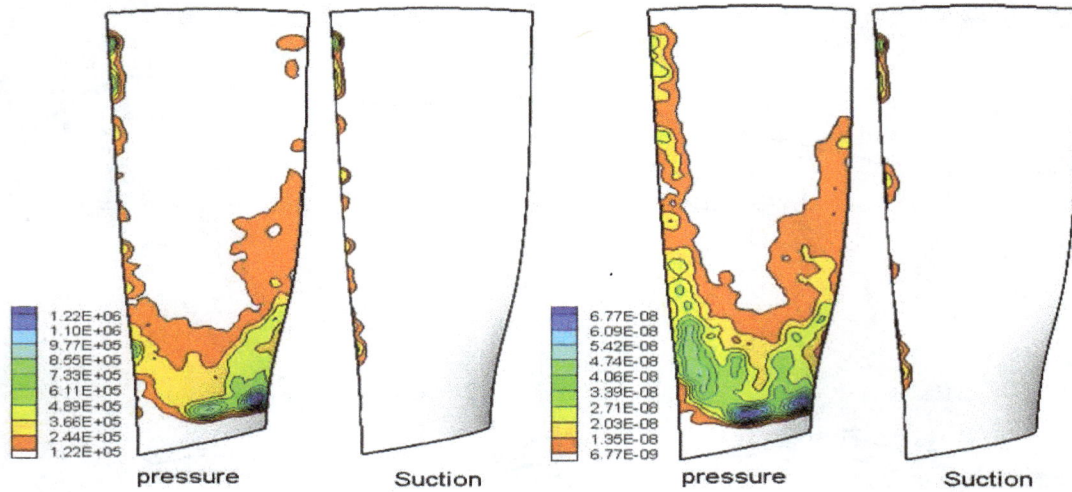

(b) For 20 μm

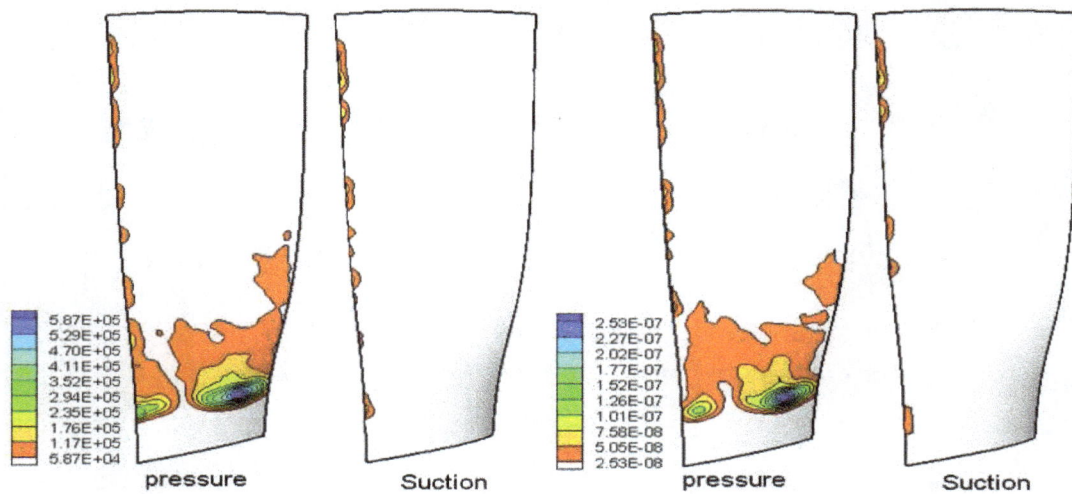

(c) For 30 μm

Figure 13. *The impact frequency and erosion rate, respectively, at cruise, for different particle diameters.*

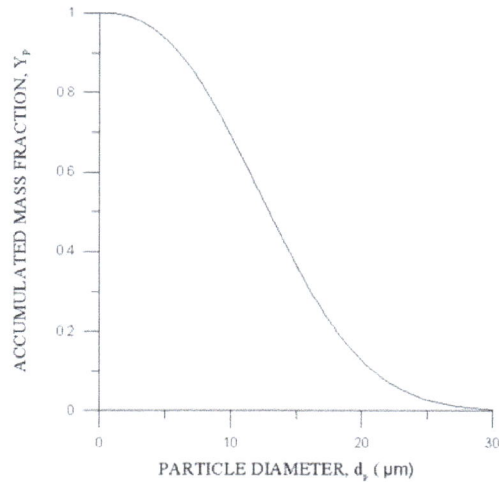

Figure 14. Rosin-Rammler particle diameter distribution at cruise.

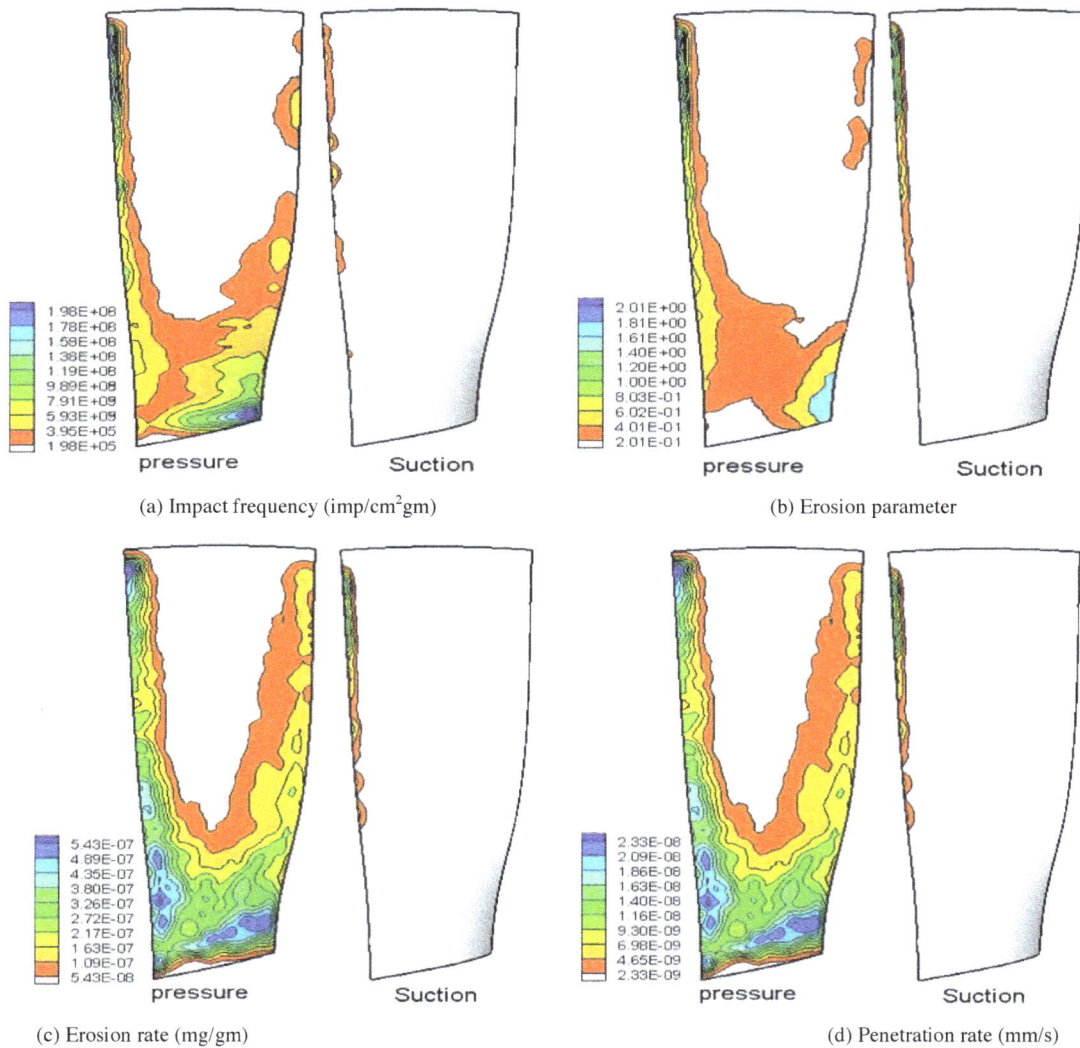

(a) Impact frequency (imp/cm²gm)

(b) Erosion parameter

(c) Erosion rate (mg/gm)

(d) Penetration rate (mm/s)

Figure 15. Impact frequency, erosion parameter, erosion rate and penetration rate at cruise for Rosin-Rammler distribution.

Figures (11) and (12) show the trajectories of the 20 μm and 30 μm diameter ash particles. These trajectories are somewhat similar to that of the 10 μm particles. Most of these particles impact the blade pressure side and a few of them impact the suction side at the leading edge.

Figure (13) shows the impact frequency and erosion rate

contours for the 10, 20 and 30 μm diameters fly ash particles, respectively, at the cruise conditions. From this figure, it is noticed that most of the particles impact the blade pressure side.

For the 10 μm diameter, the maximum impact frequency appears in two regions on the pressure side. One at the corner between the leading edge and the blade tip and the second between the hub and the blade trailing edge (about 2.7×10^7 imp/cm^2/gm). The leading edge of the pressure side is exposed to moderate impact frequency. The leading edge of the pressure side is exposed to maximum erosion rates in two regions, one at the tip and the other at the hub (about 5.9×10^{-8} mg/gm). Other locations on the leading edge are exposed to moderate erosion rates.

For 20 μm diameter particles the maximum frequency appears at about 8% blade span on the pressure side at the inner half of the chord (about1.2×10^6 imp/cm^2/gm). Some impacts with the blade leading edge are also shown at the pressure side. The blade suction side is exposed to few impacts. These impacts are concentrated at the leading edge of the blade suction side. One can observe that the maximum erosion rate is at the same region of maximum impact frequency on the pressure side (about 6.7×10^{-8} mg/gm). The suction side leading edge is exposed to small values of erosion.

For 30 μm diameter particles, the pressure side is exposed to higher impacts and erosion than the suction side. The maximum frequency and erosion appear at about 10% from the blade span (about 5.8×10^5 imp/cm^2/gm and 2.5×10^{-7} mg/gm respectively). There are few impacts with pressure side leading edge causing small erosion rates of about 5×10^{-8} mg/gm. Both the frequency and erosion rate on the suction surface are limited to the leading edge.

The accumulative distribution of the particle size is shown in figure (14). This distribution represents a variation of the particle diameter from 5 μm to 30 μm with the mean diameter at 15 μm. The results of the erosion characteristics of this distribution are shown in figure (14). The impact frequency contours shown in figure (15.a) clarify that the pressure side is exposed to greater number of particles impacts than the suction side. Particles impact the suction side near the blade leading edge with a small frequency. The maximum frequency appears at the blade pressure side at two regions. The first is at the leading edge at about 50% of the upper blade span. The second region exists near the hub at the inner half of the blade chord with a value about 1.9×10^8 imp/cm^2/gm. The erosion mass parameter contours on both pressure and suction sides are shown in figure (15.b). One can observe condensed contours at the leading edge for both pressure and suction sides with moderate values. The maximum value exists at the corner between the hub and the pressure side trailing edge (about 2 mg/gm). The erosion and penetration rates on the fan blade due to the previously mentioned diameter distribution are shown in figures (15.c) and (15.d), respectively. One can observe high erosion and penetration rate values at the pressure side leading edge. The pressure side trailing edge is exposed to moderate values of erosion and penetration rates.

On the blade suction side, the erosion and penetration rates are limited to the blade leading edge at the upper half span.

7. Conclusion

Regarding to the particulate flow, the effect of particle size on the particle trajectory, impact location on the blade, and hence the erosion characteristics were investigated. The particles were sand at takeoff and fly ash at cruise conditions. The diameters of sand particles traced were 50, 150, and 250 μ m, while 10, 20, and 30 μ m were employed for fly ash. The general conclusions are summarized as follows:

7.1. At Takeoff Conditions

a) The deviation of the particle trajectories from the streamlines increased with increasing the particle size. Large size particles trajectories are greatly influenced by the centrifugal force and successively impacted the walls at high speeds and centrifuge faster after impacts.

b) The pressure side was exposed to a great number of particle impacts than the suction side.

c) Most of the large size sand particles had two successive impacts. The first impact was at the suction side while the second was at the blade pressure side.

d) The particle impacted the pressure side with higher velocity and at higher radial location than the suction side. After the first impact, small size particles were more influenced by the air flow field and therefore moved a certain longer distance before next impact.

e) The region of maximum impact frequency and erosion rate appeared on the pressure side at the corner formed by the blade tip and the trailing edge. This is due to the migration of large size particles to this region. The region of maximum erosion rate moved towards the blade leading edge and the blade tip as the particle size is increased.

7.2. At Cruise Conditions

a) The small size and small density of the fly ash particles reduced the associated drag and inertia forces. This caused the particle trajectories to follow the corresponding air flow streamlines. And therefore, these particles acquired velocities and degrees of turning 7 similar to that of the carrier air flow.

b) The ash particles mostly impacted the blade pressure side only. Two successive impacts were noticed at the pressure side, one at the leading edge and the other was at the trailing edge. That was because these small particles are subjected to a positive gradient from the suction surface to the pressure surface and therefore they were driven towards the blade pressure side.

c) The erosion and penetration rates were smaller compared with the takeoff conditions. The region of maximum erosion and penetration rate values appeared at the pressure side leading edge. The pressure side trailing edge was exposed to moderate values of erosion and penetration rates.

References

[1] Montgomery, J. E., and Clark, J. M., Jun., "Dust Erosion Parameters for a Gas Turbine," Soc. of Automotive Engineers Summer Meeting, 1962, Preprint 538A.

[2] Tabakoff, W., and Hamed, A., "Temperature Effect On Particle Dynamics And Erosion In Radial Inflow Turbine," J. Turbomachinery, Vol. 110, APRIL 1988, PP. 258-264.

[3] Tabakoff, W., Hamed, A., and Metwally M., "Effect Of Particle Size Distribution On Particle Dynamics And Blade Erosion In Axial Flow Turbines," J. Engineering For Gas Turbines And Power, October 1991, Vol. 113.

[4] Hamed, A. "An Investigation In The Variance In Particle Surface Interactions And Their Effects In Gas Turbines,". Journal of Engineering For Gas Turbines And Power, APRIL 1992, Vol. 114, PP. 235-241.

[5] Hamed, A., Tabakoff, W., Richard B. Rivir, Kaushik Das, and Puneet Arora, "Turbine Blade Surface Deterioration By Erosion", Journal of Turbomachinery, July 2005, Vol. 127, Issue 3, pp. 445-452.

[6] Japikse, D., "Review- Progress in Numerical Turbo-machinery Analysis" ASME J. of Fluids Engineering, pp 592-606, 1976.

[7] Wulf, R. H., Kramer, W. H., and Paas, J. E., "CF6-6D Jet Engine Performance Deterioration" NASA/CR-159786, NASA, 1980.

[8] Peterson, R. C., "Design Features for Performance Retention in the CFM56 Engine" Turbomachinery Performance Deterioration FED, Vol. 37, the AIAA/ASME 4th Joint Fluid Mechanics Plasma Dynamics and Lasers Conference, Atlanta, Georgia, 1986.

[9] Neilson J. H. and Gilchrist A., "Erosion By A Stream Of Solid Particles ", Wear, ∏ , 1968 .

[10] Tabakoff, W., and Hussein, M. F., "Computation and Plotting of Solid Particle Flow In Rotating Cascades" Computers and Fluids, Vol. 2, 1974.

[11] Elsayed, A. F., and Brown, A., "Computer Prediction of Erosion Damage in Gas Turbine" ASME Paper 87-GT-127, 32nd ASME Int. Gas Turbine Conf., California, USA, 1987.

[12] Hamed, A., "Particle Dynamics of Inlet Flow Fields with Swirling Vanes" Journal of Aircraft, Vol. 19, No. 9, pp 707-712, 1982.

[13] Beacher B., Tabakoff W. and Hamed A. "Improved Particle Trajectory Calculations through Turbo-machinery Affected by Coal Ash Particles.",ASME Journal of Engineering for Power, vol. 104, pp 64-68, 1982.

[14] Hamed, A., and Fowler, S., "Erosion Pattern of Twisted Blades by Particle Laden Flows" ASME J. of Engineering for Power, Vol.105, pp839-843, 1983.

[15] Clevenger, W. B., Jr., and Tabakoff, W., "Similarity Parameters for Comparing Erosive Particle Trajectories in Hot Air and Cold Air Radial Inflow Turbines", ASME Paper 74-GT-65, 1974.

[16] Hussein, M. F., and Tabakoff, W., "Computer Program for Calculations of Particle Trajectories through a Rotating Cascade" Univ. of Cincinnati, Cincinnati, Ohio, USA, Tech. Report 76-47, 1976.

[17] Elsayed, A. F., and Rashed, M. I. I., "Erosion in Centrifugal Compressors" Paper No.55, Proc. 5th Int. Conf. on Erosion by Liquid and Solid Impact (ELSI V), Newnham Collage, Cambridge Univ., Uk, 1979.

[18] Elsayed, A. F., and Rashed, M. I. I., "Computation of Gas Flow in Centrifugal Compressors Used in Helicopters." Proc. of 5th Int. symposium on Air Breathing Engines, Bangalore, India, 1980.

[19] Maxwell, B. R., "Particle Flow in Blade Passage of Turbo-machinery with Application of Laser-Doppler Velocimetry", NASA CR-134543, 1974.

[20] Elsayed, A. F., and Rouleau, W. T., "Three Dimensional Viscous Particulate Flow in a Typical Turbo-expander", Int. J. of Energy Systems, Vol.5, No.2, 1985.

[21] Elsayed A. F., "Aerodynamics/Aero-elastic Behavior of Eroded Axial Turbines" 2nd Int. Symposium on Transport Phenomena, Dynamics and Design of Rotating Machinery, Honolulu, Hawaii, USA, 1988.

[22] Suzuki, M., Inaba, K., and Akoto Yamamoto, M." Numerical Simulation of Sand Erosion Phenomena in Rotor/Stator Interaction of Compressor", Proceedings of the 8th International Symposium on Experimental and Computational Aerothermodynamics of Internal Flows, Lyon, July 2007.

[23] Corsini, A., Marchegian, A., Rispoli, F., Venturini, P., and Sheard, A. "Predicting Blade Leading Edge Erosion in an Axial Induced Draft Fan", Journal of Engineering for Gas Turbine and Power, published online January 30, 2012.

[24] Brun, K., Nored , M., and Kurz, R. " Analysis of Solid Particle Surface Impact Behavior in Turbo-machines to Assess Blade Erosion and Fouling", Proceedings of the Forty-First Turbo-machinery Symposium September 24-27, 2012, Houston, Texas.

[25] Carbonetto, B. and Hoch, V." Advances in Erosion Prediction of an Axial Flow Expander", Proceeding of The 28th Turbo-machinery Symposium.

[26] Zhang, J., Han, Z., Cao, H., Yin, W., Niu, S., and Wang, H. "Numerical Analysis of Erosion Caused by Biomimetic Axial Fan Blade", Hindawi Publishing Corporation, Advances in Materials Science and Engineering, Volume 2013, Article ID 254305, 9 pages.

[27] Fluent 6.1 Documentation (User Guide Manual).

[28] Zohier, H. Z., "Solid Particulate Flow in an Axial Transonic Fan In a High Bypass Turbofan Engine", M.Sc. Thesis, Zagazig University, Egypt, 2006.

[29] Elsayed, A. F, Gobran M. H., and Zohier H., "Three-Dimensional Flow in A Transonic Axial Flow Fan of A High Bypass Ratio Turbofan Engine", 11th International Conference On Aerospace Science and Aviation Technology, Military College, Cairo, Egypt, 17-19, May 2005.

10

Organization of Uniform Dispersal for Group of Small Satellites After Their Separation and Acceptable Spread at Stages of Their Further Approaches

Alexander Degtyarev, Irina Vorobiova, Anatoliy Sheptun

Yuzhnoye SDO, Dnepropetrovsk, Ukraine

Email address:

info@yuzhnoye.com (I. Vorobiova)

Abstract: Mutual-relative motion of satellites constellation after their placing into initial orbit is characterized the satellites reiterated approaches to minimal distances. Determined separation conditions for each satellite pair in constellation, which realization leads to minimal satellites approaches. It was sustained, that distances between satellites can reach extreme minimal values due to minor angular deviations of orbit planes and always occurring flight altitudes long-period oscillations. Received analytical relations for estimation of angular deviations and distances between satellites orbit planes at the phases of periodical approaches. It was approved that in some vicinity of orbit numbers can occur following almost simultaneously: small differences of angular distances from the equator for any satellite pair in constellation; minimal angular deviations of satellites orbit planes (in general they have different precession after separation of different directions and with various velocities). Analytical relationships were numerically confirmed with using software for satellites movement prediction at the phases of their approaches to the minimum distances. The area of satellites separation parameters was built, with its implementation in a small vicinity of flight orbits the satellites approach to distances less than 100 m. The minimum distance in the specified area according to the single calculations is ~10 m.

Keywords: Group Launches, Minimal Distances, Satellites Spatial Separation, Angular Spatial Separation of Orbit Planes, Estimation of Maximal Satellites Approaches, Maximal Satellites Approaches

1. Introduction

1.1. Basic Assumptions

The tendency of reducing weights and dimension of near space satellites, considerable rocket capabilities and new technologies allow ensuring injection of fair quantities of satellites by one launch. Issues concerning organization of uniform dispersal for constellation of small satellites after their separation and acceptable spread at stages of their further approaches are topical in the rocket building industry. Physical movement foundations in the terrestrial gravitational field determine satellites periodical approaches. Such issues occur two times per orbit in case satellites separation along the normal to orbital velocity and more rarely (up to several thousands of orbits) in case separation along flight direction (fig. 1 shows satellites approaches during their autonomous flight with differences

between their separation velocities 0.75 m/s).

In case of small number of satellites in constellation their separation velocities are differ from each other by 1-2 m/s. The mutual relative motion analysis of satellites constellation and their dispersal during first flight orbits after separation is sufficient. Such analysis method is acceptable for practice. However, relative separation velocities are reducing with growth the number of satellites in constellation (10-20 and more). As a result difficulties of ensuring organization of uniform dispersal for small satellites constellation occur during their first flight orbits after separation. The necessity of satellite approaches estimation during long time intervals occurs at the same time.

The conditions of uniform dispersal for satellite constellation (the number of objects is 12, fig. 2) during their first flight orbits were determined and analytical relations for analysis satellites approaches during long time intervals were preliminary obtained. They were presented in articles [2, 5] and in the reports [3, 4] at 64th (Beijing) and 65th (Toronto) International Astronautical Congresses. Numerical evaluations of satellites approaches for their single separation along the flight direction were also presented. There were noticed satellite approaches on the distances near 100 m. The results of such calculations forces to carrying out of more detailed researches about mentioned issues.

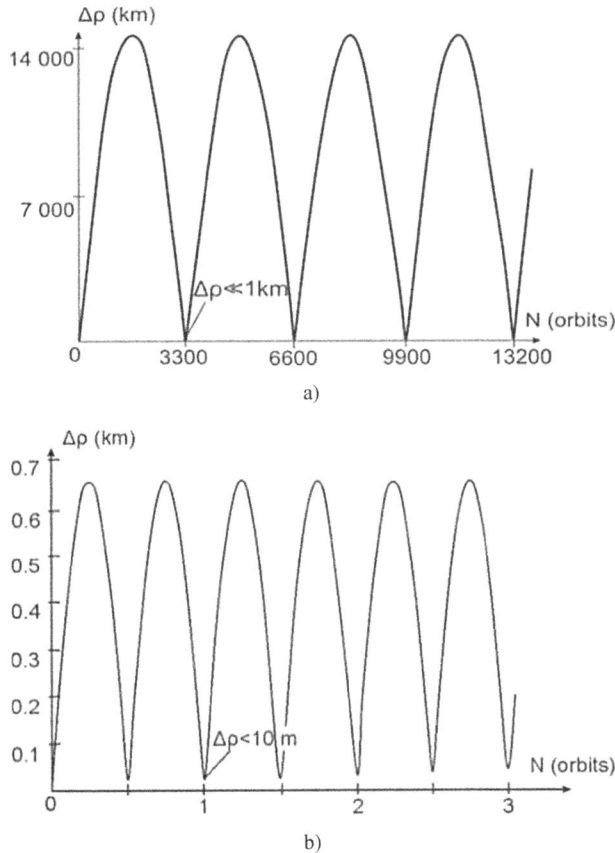

Fig. 1. *The typical picture of changing of distances between satellites: a) separation along normal to orbital velocity vector b) separation along flight direction.*

The conditions which correspond to satellites approaching on minimal distances for a long time intervals have not researched before.

1.2. Problem Statement

The research task is definition of conditions which implementation leads to reducing distances between satellites at the phases of their maximal approaches and obtaining data of numerical evaluation of these distances. Current researches continue ballistic analysis of mutual-related motion for satellites constellation (12 objects) after their separation from side surface of dispenser normal to longitudinal axis of launch vehicle (fig. 2), which was started in [2-5]. It was presumed that longitudinal axis of dispenser specified along radius-vector. It should be noted that dispencer longitudinal axis OX_1 was set along nadir direction.

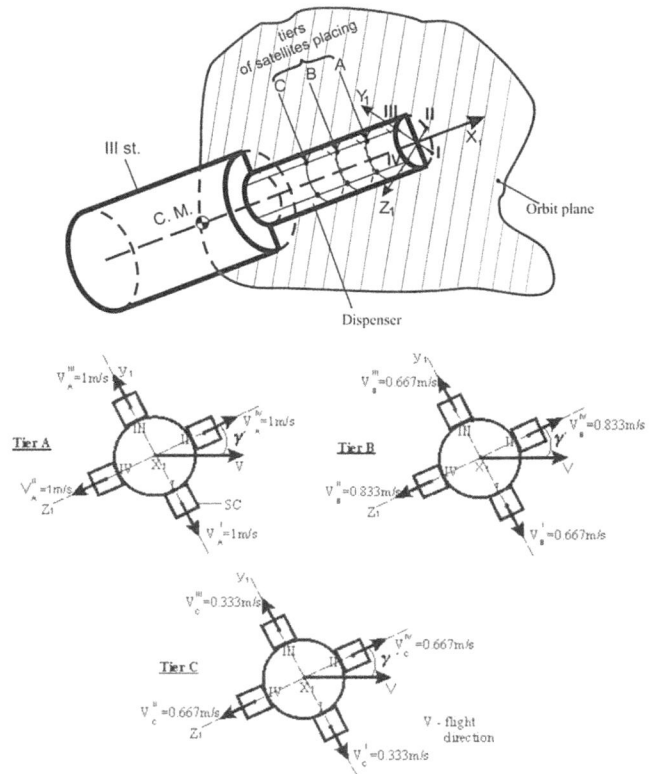

Fig. 2. *The scheme of satellites location on dispenser's surface.*

The conditions of satellites constellation separation after putting them into initial orbit which ensure uniform dispersal with step of ~2.6 km along orbit for each flight orbit (fig. 3, table 1) were determinedin [2].

Table 1. *Directions and satellites separation velocities (m/s).*

Platformindex	Stabilizationplanes	Satellites separation velocities			
		I	II	III	IV
A		V	V	V	V
B		0.677·V	0.833·V	0.677·V	0.833·V
C		0.333·V	0.677·V	0.333·V	0.677·V

The estimation of satellite pairs approaching in constellation were carried out on the base of calculation results of satellites movement after their separation, in case of putting them into sun-synchronous orbits with altitude ~650 km and inclination 98° by one launch.

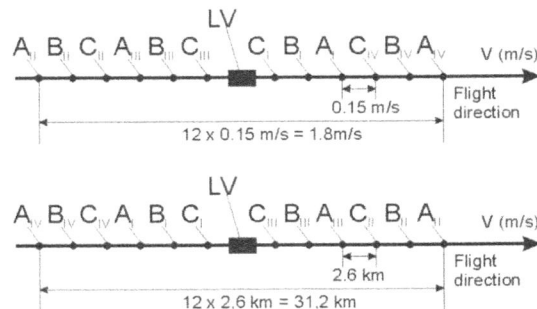

Fig. 3. *Satellites separation scheme and their location in flight ($\overline{V} = 1$ m/s)a) after separationb) by the end of first flight orbit.*

Application Satellite Tool Kit was used for required calculations.

2. Analytical Relations

It is examined mutual-relative motion of small satellites constellation with equal ballistic coefficients after putting them into frequently used orbits with low aerodynamic drag ($h>600$ km) by one launch. General characteristic of each satellite flight orbit after separation is small differences of semi axis dimensions and inclination (in most cases to 1 km and 10^{-4} rad correspondingly). As a result parameters of satellites orbit planes precession motion differ insignificantly. Satellites approaches on minimal distances during long time intervals are also possible.

Location of two of satellites orbits after their separation and at the expiration some number of flight orbits $N \sim N_{i,j}$ showed at fig. 4, 5.

The distances between satellites decrease and reach their minimal values in some vicinity of flight orbits $N \sim N_{i,j}$ after reaching of the maximum distances between any pair (i, j) of satellites in constellation ($\rho_{i,j} = r_i + r_j \sim 14$ thousands of km, where $r_{i,j}$ – are orbital radiuses):

$$N \sim N_{i,j} = \frac{T_i}{|T_i - T_j|} \sim \frac{1}{3} \frac{V_0}{\Delta V_{V_{i,j}}},$$

(T_i, T_j – orbital periods of two (i, j) satellites, V_0 – flight velocity at the initial orbit, $\Delta V_{V_{i,j}}$ – the difference of separation velocities projections of satellites pair along flight direction).

General number l of satellite pair combinations determines as $l = \frac{n(n-1)}{2} = 78$ in case of number of flight objects equal to $n=13$ (constellation of 12 satellites and the dispenser).

The distances between satellites in constellation at phases of minimal distances in the vicinity of orbit number $N \sim N_{i,j}$ are determined by angular spacing $\Delta(\delta\psi_1)_{i,j}^{N_{i,j}}$ of orbital planes ($\delta\psi_1$ – orbital plane rotation angle per one flight orbit), by changing of average altitudes owing to atmospheric resistance, flight altitudes periodic changes, etc.

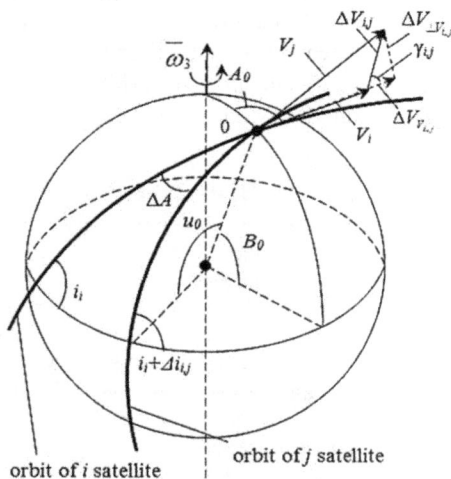

Fig. 4. *Location of two satellites orbits after their separation.*

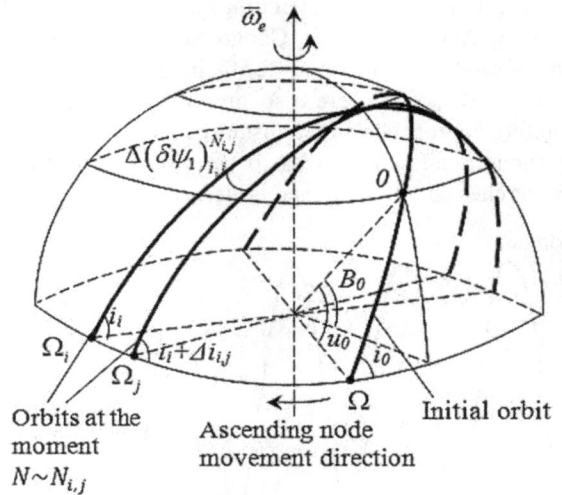

Fig. 5. *Location of two satellites orbits at the expiration some number of flight orbits $N \sim N_{i,j}$.*

Relations for estimation of orbital planes angular spacing $\Delta(\delta\psi_1)_{i,j}^{N_{i,j}}$, linear shifts along lateral direction $\Delta Z_{i,j}^{N_{i,j}}$ and changings of orbital average altitudes $\Delta H_{i,j}^{N_{i,j}}$ for two of satellites in constellation in the vicinity of orbit number $N \sim N_{i,j}$ are presented below:

$$\Delta(\delta\psi_1)_{i,j}^{N_{i,j}} = \pi \left(\frac{a_e}{r_0}\right)^2 \alpha \left[\frac{7}{6}\sin 2i_0 - \delta i_{i,j} N_{i,j} \cos 2i_0\right]$$

$$\Delta Z_{i,j}^{N_{i,j}} = r_0 \Delta(\delta\psi_1)_{N_{i,j}} \cos u_0 \qquad (1)$$

$$\Delta H_{i,j}^{N_{i,j}} = \frac{8}{3}\pi B_k r_0^3 \frac{\partial\rho(h_0)}{\partial h}$$

(index «0» refers to initial orbit parameters)

Using expressions for $\Delta i_{i,j}$, $\Delta r_{i,j}$, $N_{i,j}$ from [1-5] and spherical geometry formulas the first relation in (1) was transformed into:

$$\Delta(\delta\psi_1)_{i,j}^{N_{i,j}} = \frac{\pi}{3}\left(\frac{a_e}{r_0}\right)^2 \alpha \sin 2i_0 \left[\frac{7}{2} - \frac{\Delta V_{\perp V_{i,j}}}{\Delta V_{V_{i,j}}} ctg 2i_0 \cos u_0\right] \qquad (2)$$

($\Delta V_{\perp V_{i,j}}$ – the difference of separation velocities projections of any satellites pair (i, j) in constellation along the normal to flight direction V_0; u_0 –point on the orbit, which corresponds to angular distance from the equator at the moment of satellites constellation separation (fig. 4)).

First summand in (2) characterizes influence of different average flight altitudes of two satellites onto orbital precession parameters and as a result onto angular dispersal of orbital planes in the vicinity of flight orbit number $N \sim N_{i,j}$. Second summand in (2) characterizes influence of different initial orbit inclinations after satellites separation. Also next statements follow from (2):

- Angular dispersal of any satellite pair (i, j) in constellation are caused by the difference of separation velocities projections $\Delta V_{V_{i,j}}$ in the vicinity of flight orbit number $N \sim N_{i,j}$. Such dispersal are equal to each other and in accepted calculation conditions are:

$$\Delta(\delta\psi_1)_{i,j}^{N_{i,j}} = \frac{7}{6}\pi\left(\frac{a_e}{r_0}\right)^2 \alpha \sin 2i_0 \sim 9.5 \text{ ang. min}$$

- The usage of satellites separation angularly to initial orbit plane ($\Delta i_{i,j} \neq 0$) leads to changing (increase or decrease) angular dispersal of two satellites orbits. Assimilation to zero the expression in brackets in (2) leads to satellites orbit planes correspondence accurate within $\Delta i_{i,j}$ in cafe of realization next condition:

$$\frac{\Delta V_{\perp V_{i,j}}}{\Delta V_{V_{i,j}}} = \frac{7}{2}\frac{tg2i_0}{\cos u_0} \tag{3}$$

Relation obtained below follows from (3) for sun-synchronous orbits ($i_0 = 98°$) and satellites separation at the equator $u_0 = 0$:

$$\frac{\Delta V_{\perp V_{i,j}}}{\Delta V_{V_{i,j}}} = 1.0023 \tag{4}$$

Relation (3) affects more onto increasing of two satellites (i, j) approach frequency to minimal distances in the vicinity of flight orbit number $N \sim N_{i,j}$. In case of condition (3) fulfillment the angle between satellites orbit planes does not exceed of angular minutes splits in both directions during valuable orbits number related to 'approaching event' time. This implies that frequency of events of satellites approaching to minimal distances increases.

For example, for sun-synchronous orbits ($i \sim 98°$) with 600-700 km altitude in case of condition (3) fulfillment the angular distance between satellites orbit planes does not exceed of 0.2 - 0.3 angular minutes during 200 flight orbits.

There were determined separation conditions (separation velocities and inclinations) for any satellite pair (i, j) in constellation, which realization leads to minimal satellites approaches. In particular it was approved that in some vicinity of orbit numbers can occur almost simultaneously following:

- Small differences of angular distances from the Equator for any satellite pair (i, j) in constellation;
- Minimal angular deviations of satellites orbit planes (in general they have different precession after separation of divers directions with various velocities).

Satellites approaches occur during further flight after their first approaching event. The number of such events $k_{i,j}$ for each satellite pair determines as integer part of relation $k_{i,j} = [\frac{N_\Sigma}{N_{i,j}}]$, where N_Σ – the number of flight orbits during active satellite exploitation time. This relation for accepted conditions of satellite separation approximately is of the form $k_{i,j} = [10,5 \cdot \Delta V_{V_{i,j}}]$. The range of $k_{i,j}$ value changing (during 5 year exploitation period, which corresponds to $N_\Sigma \sim 27000$ orbits) for every possible satellite pairs in constellation is from 1 to 19 approaching events. General approaches number for all 78 satellite pairs during mentioned time interval is 551.

Angular orbit planes dispersal of two (i, j) satellites at the moment their k_{ij} approach to minimal distances can be written as $\Delta(\delta\psi_1)_{i,j}^{kN_{i,j}} = k \cdot \Delta(\delta\psi_1)_{i,j}^{N_{i,j}}$ besides condition of angular orbit dispersal minimization [3] will be constant.

It should be noted that orbit inclinations would not change with dispenser longitudinal axis angular placing normal to orbit plane and maintaining of conditions showed at fig. 2 and table 1. As a result conditions (3) for minimal angular deviation are not implemented in some vicinity of flight orbit $N \sim N_{i,j}$. With increase of satellites approaching events number the expression of inaccuracy angular orbit planes dispersal $\Delta(\delta\psi_1)_{i,j}^{kN_{i,j}}$, which can be represented as a linear subjection from number k_{ij} of approaching events, decreases. As a result the accuracy of formula (4) for k_{ij} of approaching event decreases.

3. Numerical Estimation of Minimal Distances

The calculations were carried out for sun-synchronous orbits with ~650 km altitude for constellation composed of 12 satellites[2-5], which were uniform set on the circle of dispenser in points I, II, III, IV at tiers A, B, C. Satellites separation was accepted at the equator ($u_0 = 0$) normal to dispenser longitudinal axis (the angular orientation of it is presented at fig. 2). Velocities and separation directions implement satellite uniform dispersal during first flight orbits and their numerical values were accepted according to data from fig. 2.

The calculations of flight trajectory for approach evaluation of satellite pair (i, j) in constellation in some vicinity of number $N \sim N_{i,j} = \frac{1}{3}\frac{V_0}{\Delta V_{V_{i,j}}}$ were carried out considering change of initial movement conditions due to satellites separation from dispenser. The differences of relative separation velocities $\Delta V_{V_{i,j}}$ and of inclinations $\Delta i_{i,j}$ of satellite pair (i, j) in constellation, which were accepted for calculations according to initial data (table 1, V=1 m/s), are showed in table 2. Also there were showed data of full orbit numbers ($N_{i,j}$) to the moment of first satellite pair (i, j) approach and count $k_{i,j}$ of approach events.

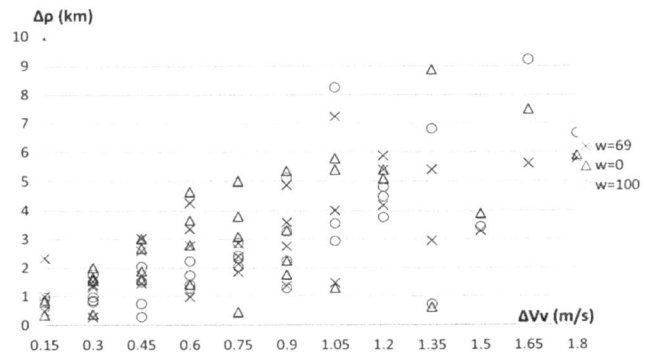

Fig. 6. *Distances between satellites at the moment of their first approach in some vicinity of orbit number* $N \sim N_{i,j}$ *(nominal separation conditions).*

The calculations for three values pericentre argument $\omega_0 = 0°$; $69°$; $100°$ were carried out for pericentre argument initial value influence of initial orbit with excentricity equal $e_0=0.00115$ ($\omega_0=69°$ and $e_0=0.00115$ with $h_0\sim650$ km altitude satisfy stable near-Earth orbits, which parameters ω_0 and e_0 remain constant in ascending nodes for long time intervals).

The numbers of distances $\rho_{i,j}$ between satellite pair (i,j) in some vicinity of number $N\sim N_{i,j}$ on the assumption of

nominal separation conditions are showed at fig. 6.

Next conclusion can be drawn from fig. 6, that low border of minimal distance between satellites with nominal separation conditions is at 0,5 km level in difference range of relative separation velocities $\Delta V_{V_{i,j}}$ =0,15-1,35 m/s. Nominal values of satellites minimal distances increase with time.

Table 2. Combination of satellite pair (i, j) separation parameters differences.

$\Delta V_{V_{i,j}}$ (m/s)	0.15	0.3	0.45	0.6	0.75	0.9	1.05	1.2	1.35	1.5	1.65	1.8
$\Delta i_{i,j}$ (rad)	10^{-5} $4\cdot10^{-5}$ $1.6\cdot10^{-5}$	$2\cdot10^{-5}$ $8\cdot10^{-5}$ $1.2\cdot10^{-4}$ $1.7\cdot10^{-4}$	$8\cdot10^{-5}$ $1.2\cdot10^{-4}$ $1.8\cdot10^{-4}$	$4\cdot10^{-5}$ $9\cdot10^{-5}$ $1.4\cdot10^{-4}$ $1.6\cdot10^{-4}$	0 $5\cdot10^{-5}$ 10^{-4} $1.6\cdot10^{-4}$	10^{-5} $4\cdot10^{-5}$ $6\cdot10^{-5}$ $2.4\cdot10^{-4}$	$2\cdot10^{-5}$ $3\cdot10^{-5}$ $8\cdot10^{-5}$	$2\cdot10^{-5}$ $7\cdot10^{-5}$ $8\cdot10^{-5}$	$6\cdot10^{-5}$ $9\cdot10^{-5}$	10^{-4}	$1.1\cdot10^{-4}$	$1.2\cdot10^{-4}$
$N_{i,j}$	~16500	~8250	~5500	~4100	~3300	~2750	~2350	~2050	~1820	~1650	~1500	~1380
k_{ij}	1	3	4	6	8	9	11	13	14	16	18	19

Relation (3) equal "1" (minor difference from condition (4)) for two satellite pair A_2-C_3 and A_1-C_4 (fig. 7) among 78 satellite pair combinations, which were created after satellites separation from dispenser.

Orbit planes of satellite A_2-C_3, A_1-C_4 in some vicinity of orbit number $N\sim N_{i,j}$ have minor angular separation (to 20-30 ang. sec). The values of differences of relative separation velocities projections along flight direction $\Delta V_{V_{i,j}}$ and along the normal to OX_1 axis of dispenser $\Delta V_{\perp V_{i,j}}$ for these satellite pairs are $\Delta V_{V_{i,j}} = \Delta V_{\perp V_{i,j}} = 0.75$ m/s. Distance between mentioned satellites pairs is about ~0.5 km by the moment of their maximal approach in some vicinity of orbit number $N\sim N_{i,j}$ (fig. 6).

Values of inaccuracies for satellite separation parameters $\delta(\Delta V_{V_{i,j}})$ and $\delta(\Delta i_{i,j})$ were determined due to carried out calculation results. These inaccuracies correspond to maximal satellites approaches to distances 9 and 12 m (fig. 8 points "●"). Thus 1/3 of objects from all satellite number in constellation (fig. 2, 7) can approach to distances about ~10 m.

Area $D((\Delta V_{V_{i,j}}), \delta(\Delta i_{V_{i,j}}))$ of inaccuracies is also showed at fig. 8. Its bounds correspond to satellites approach to 100 m (it was conditionally accepted that distance 100 m between satellites is acceptable boundary point of their autonomous flight including existing separation inaccuracies, physical nature of movement, calculation model imperfective and other factors).

Behavior of altitude inclinations ΔH, geodetic latitudes ΔB and longitudes $\Delta\lambda$ at the phase of satellite approach to minimal distances are showed at fig. 9 a-c (the moment of satellite pair maximal approach is marked with «●»)

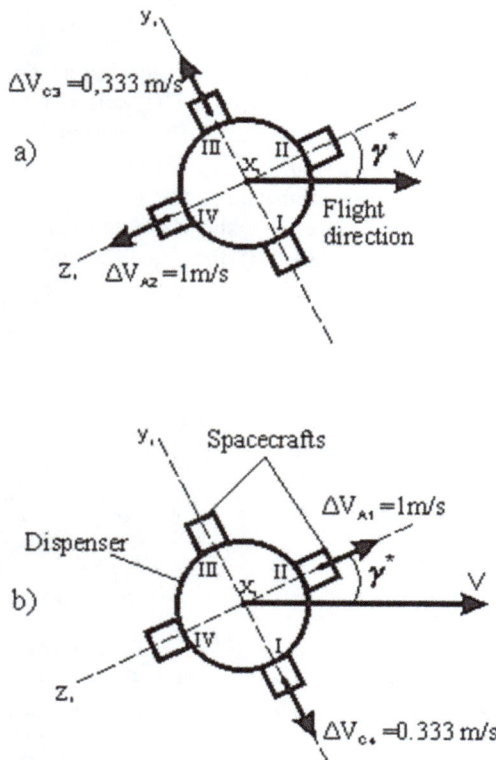

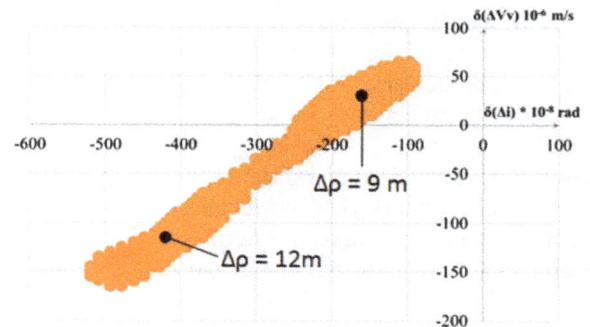

Fig. 8. *Area of satellite approach critical values.*

* Origin of coordinates corresponds to calculating parameters of satellite separation $\Delta V_{V_{i,j}} = 0{,}75$ m/s and $\Delta i_{i,j} = 10^{-4}$ rad.

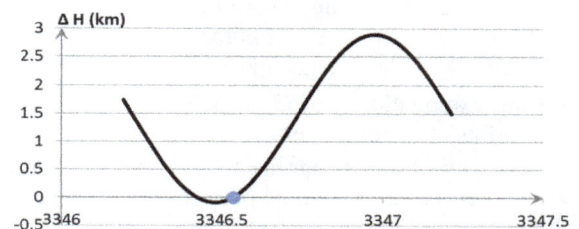

Fig. 7. *Conditions of satellite pairs separation* ($\Delta V_{V_{i,j}} = 0.75 \frac{m}{s}$, $\Delta V_{\perp V_{i,j}} = 0.75\ m/s$) *a) pair* A_2-C_3. *b) pair* A_1-C_4.

a)

b)

c)

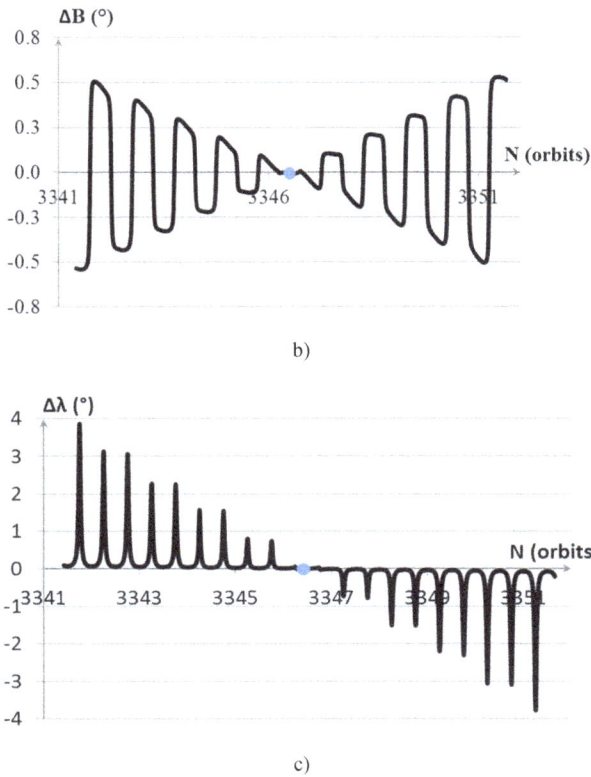

Fig. 9. Range of orbital parameters change of satellites pair A_2-C_3a) by altitude ΔHb) by latitude ΔBc) by longitude Δλ.

4. The Alternative Separation Scheme for Big Constellation of Satellites

The implementation of satellites separation scheme by fig. 2 is based on the usage of mechanical, electrical, pneumatic and other separation appliances. It is obvious, that such appliances have significant scattering of satellites separation operation (about 10% from nominal) because of absence of power adjusting hardware. As a result there will be marked change of satellites separation conditions and deviation of further distances between them at the beginning of their autonomous flight and in further. In this case providing of satellites non-closing in constellation (or part of them) above mentioned bound of minimal distances between satellites in some vicinity of number $N \sim N_{i,j}$ sometimes can lead some difficulties.

Scheme at fig. 10 can be used as alternative one (satellites are separated serially normal to dispenser longitudinal axis in case of rocket moving with minor longitudinal acceleration along flight direction $\dot{w}_{x_1} \sim 0{,}02 - 0{,}1 \ m/s^2$).

The advantage of this scheme is possibility of every further satellite separation by the control system instructions after achieving determined increase of flight velocity. Required characteristics of satellites constellation mutual-relative motion at the first phase after separation (including realization of tracing their flight trajectories on the safety distance (y_{acc}) from gas flows, which are formed by racing of thruster) and at further phases of their approaches are provided by following:

- Longitudinal rocket acceleration$\dot{w}_{x_1}$;
- Durations of satellites separation time intervals$\tau_{i,j}$;
- Satellites separation velocities normal to longitudinal rocket axis $\dot{y}_{0_i}$.

Imposition of velocity increment $\Delta V_{V_{i,j}}$ restriction for two (i, j) satellites, which are separated serially, provides their non-closing in some vicinity of orbit number $N \sim N_{i,j}$ to distances that are lower than determined distance bounds. For satellites safety flight in the gas flows zone this restriction can be determined by relation (5) (fig. 10):

$$\dot{y}_{0_i} \geq y_{acc}\sqrt{\frac{\dot{w}_{x_1}}{2x_{0_i}}} \quad (5)$$

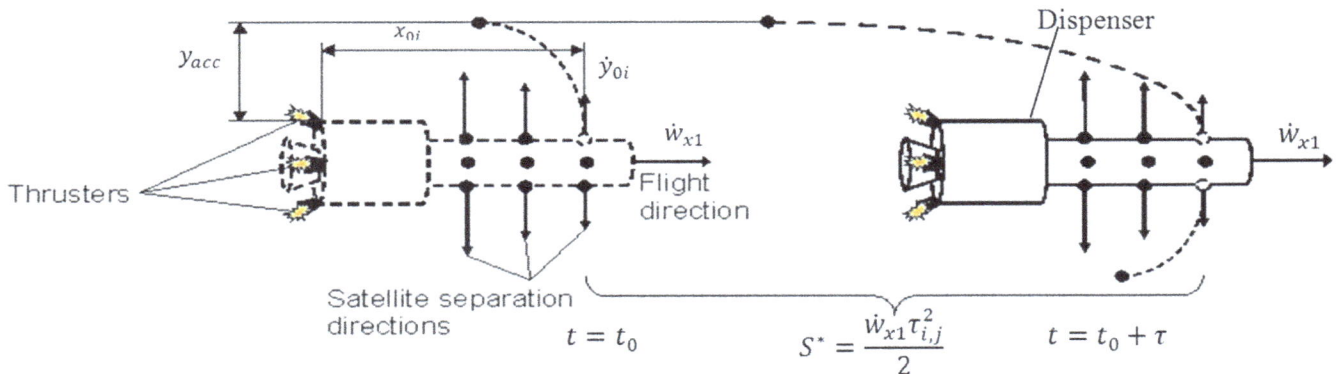

Fig. 10. Alternative separation scheme for big constellation of satellites.

References

[1]   Эльясберг, П. Е. Введение в теорию полета искусственных спутников Земли // Павел Ефимович Эльясберг // – М.: Наука, 1965. – 540с.

[2]   Degtyarev A. Research of trajectories of spacecraft cluster injection into different orbits in one «Cyclone-4» vehicle launch//Degtyarev A., Degtyarenko P., Sirenko V., Sheptun A.//Space research in Ukraine 2010-2012.–Kyiv.

[3]   Degtyarev A. Feasibility study of spacecraft cluster lunches with lunch vehicle into various basic orbits // Degtyarev A., Sheptun A., Mashtak I. // IAC-13-D2.P.16, 64th International Astronautical Congress, China, 2013.

[4]   Vorobiova I. Launch vehicle energy performance during single injection of the satellites into individual orbits // Vorobiova I., Sheptun A. // IAC-14-D2.2.9, 65th International Astronautical Congress, Canada, 2014.

[5]   ВоробьеваИ. А. Проектно-баллистические исследования по групповым запускам спутников в одном пуске ракеты носителя на разнесенные базовые // Воробьева И. А., Шептун А.Д. // Научный журнал ХУПС, № 3, 2014.

11

High-efficiency mission planning and distribution system of the land observation satellites

Qiu Hu

China Center for Resource Satellite Data & Application, Beijing 100830, China

Email address:

tigerqiuchina@hotmail.com, tigerqiuchina@gmail.com

Abstract: The mission planning of earth satellite observation generally refers to the optimized use of satellites' payloads. According to relevant experiences in project construction, this paper puts forth the concept of the integrated hub-style mission planning on the basis of consumers' satisfaction and data's timeliness. This concept takes the ground processing center as the integrated service point, and data demands of difference levels of users as the driving force. It fully considers the practical problems like station resources construction lag and data transmission link frequency interference conflicts, reasonably plan the satellite payloads data reception stations resources, monitor and control resources and data processing resources, so as to implement the remote sensing data from a data request, overall planning, telecontrol, data acquisition, data processing and the unified distribution in the whole process of mission planning model. This model effectively solves the problems like the difficulty to match single satellite programming model with user demands, and the low efficiency of resource use. Besides, the paper seeks to explore relevant elements, technology and techniques which are necessary in such mission planning. The author thinks that from the vintage point of systems engineering, the commercialization of satellite observation products requires the full-cycle integrated mission planning that covers the period from the beginning of a user's order to the distribution of final data products, among which dispatching satellites' payloads is only one of important links. During the cycle, activities and constraints related to land support systems and other elements should also be taken into consideration. The comprehensive consideration including points of data acquisition, processing, archiving and distribution will make mission planning more complicated, which needs to be studied theoretically and practically. The full-cycle and integrated hub-style mission planning coordinates with national policy of civil remote sensing satellites. It is of great significance to deliver remote sensing products with high quality and high timeliness to every professional user through remote sensing data planning and major processing points, which demonstrates in two aspects. The first aspect is that it helps to achieve the win-win situation where different users get what they want and the state reduces investment in this field. The other one is that it is facile to monitor various satellite data and in return promote technology advancement of satellite processing and construction, which eventually stimulates the development and progress of aerospace remote sensing technology.

Keywords: Mission Planning, Land Observation Satellite, Hub-Style, Full-Cycle, Frequency Interference, Overall Planning

1. Introduction

Because of its superiority in data acquisition, the observation satellite is playing a crucial role in all aspects of social development. With the maturity of satellite manufacture technology and constant development of aerospace equipment, China is applying Earth observation satellites in a more commercialized way. Thus, with the help of mission planning, how to maximize resources utility and users' satisfaction through overall management of producing of land observation products? At present, both at home and abroad satellite mission planning technology has received mass attention and

has achieved admirable success in this field though, the concept of mission planning has been confined to the optimized operation of satellite payloads[1-3]. This paper from the aspect of systematic engineering, holds that the commercialization of satellite observation products requires the full-cycle integrated mission planning from users' requests to the final data distributed to users, of which the scheduling of satellite payloads is just one of important links. Besides, we should consider activities and constraints associated with the ground support systems and other factors including the deadline of distributing products to customers, the meteorology of observing regions, the constraints on the capability of satellite payloads, the limit of ground data transmission capability, the task allocation and the effective use of the distributed ground receiving systems, the flexibility of handling changes and unexpected events, and the frequency interference between different ground links. The whole management of data collection, processing, archiving and distribution make mission planning even more complicated, which needs to be explored further theoretically and practically, so that a number of elements should be considered, and a set of advanced techniques should be referred to.

2. The Full-Cycle Mission Planning

2.1. Fulfillment of Full-Cycle Satellite Application Mission

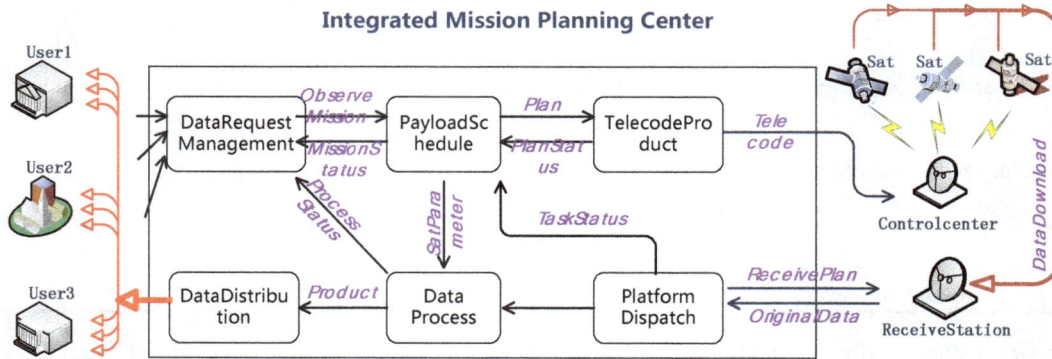

Figure 1. *Full-cycle satellite application mission system model*

A fulfilled satellite application mission is a circulating from the submission of users' requests to the distribution of data products. As shown in the chart 1, the process is as follows.

(1) Provide users with satellite observation planning through the Internet;
(2) Receive users' observation request through the Internet and the like;
(3) Preliminarily examine the feasibility of users' requests;
(4) Put users' requests into the payloads planning optimization module to avoid the conflict in resource use;
(5) Transform the planning results into remote commands towards satellite payloads;
(6) Check availability of other resources related to data acquisition, such as TDRSS satellites;
(7) The planning of ground support activities, including moderation made by the availability of resources and priority of different missions, to the scheduling of ground receiving activities, the data processing, and the scheduling of data distribution;
(8) Acquire, receive and process data;
(9) Send situations feedback to users;
(10) Archive and distribute data products;
(11) Users receive ordered products.

Among the steps above, (4) and (7), where resources use conflicts are most likely to happen thus mission planning is needed, are of the greatest importance. Traditional satellite mission planning always focuses on step (4). The result of planning corresponds to the sequence of satellites' observation activities, which is described in the nest section.

2.2. The Traditional Concept of Satellite Mission Planning

Traditional satellite planning, usually interpreted as optimal use of satellite payloads, aims to schedule the space-borne sensor resources and specific time for each observation activity on the basis of the constraints on satellite payloads and requirements for observing missions, in order to maximize users' satisfaction. Relevant constraints should be considered by traditional planning include:

(1) The constraint of satellite system's capability Elements, including energy and power, the number and the type of sensors, capacity of data storage, and the time to change the angle of side-sway, would hinder the performance of observations.
(2) The constraint of the time window When satellites observe objects or transmit initial data down to the land reception station, all tasks should be finished within limited time window. Some objects could not be observed.
(3) The constraint of weather conditions For example, visible light shooting has strict restraints on cloud thickness.
(4) Special requirements for certain missions Special constraints like periodical or paired observations are needed in some missions.

Traditional mission planning could only decide when to acquire initial data, while it could not determine whether the acquired data could be transferred to land stations without delay, or when the data could be processed and distributed to users successfully. As land observation satellites undergo the transition from experimental use to more practical and commercial use, users have more strict requirements on the quality of data products and the efficiency of collecting those products, for example, at some users' request, the time span, from high resolution imaging of the targets to the submission of data products should be less than a few hours. Typically the stricter the requirements are, the more profits are achieved by observing missions. Finally, application systems of land observation satellites shall provide consumers with continuous information service instead of occasional data products. However, traditional satellite mission planning cannot fulfill this kind of task. To solve the problem, this paper introduces the concept of full-cycle mission planning.

2.3. The Intension of Full-Cycle Mission Planning

From the perspective of systematic engineering, full-cycle mission planning considers the whole process of satellite application task, including data acquisition, downlink, processing and distribution, as integral. By scheduling all activities in a integrated way, the planning endeavors to maximize users' satisfaction and satellites' use efficiency. Highlighting the quality of service, the full-cycle mission planning fully considers users' requirements on products, which contributes to commercialization of satellite data products and users' satisfaction.

Compared to traditional mission planning, some other elements should be added in the full-cycle mission planning:

1. The change of distribution system While the ground support system is previously designed, the distribution system can be changed all the time. In that way, the service for consumers and the planning should be more flexible.

2. The mechanism of ground support system The ground support system usually comprises a number of land stations with similar functions and processing centers. Its planning also requires coordination and distributed handling, and also involves specific constraints.

3. The connection to the external data communication network Data distribution requires a communication network covering large areas, which is also a part of the full-cycle. Therefore, special requirements and constraints towards the connection between ground support system and communication network are involved.

4. Automatic operation and operating cost saving The trend of modern technology development is to save operating cost by promoting automation of operations, while weighing automation against technical risks. To achieve the goal of full-cycle mission planning, we can make full use of the progressive technology satellite payloads, for example, the solid-state storage devices with large capacity can increase the flexibility of data recording and downlinking, efficient power usage can prolong the service cycle of certain equipments, TDSSR satellites can improve amount of sent-back data, and technology of mass data downlinking at high speed improves the ability of missions accomplishment. At the same time, the scope of traditional mission planning shall be extended. Planning technology needed to be innovated, and only by combining hard technology and soft technology can maximize the profits. The following section will mainly describe the full-cycle mission planning from the perspective of soft technology.

3. Elements to Realize Full-Cycle Mission Planning

3.1. User Interface

User interface is the direct channel of communication between users and satellite application systems. Fine user interface can enhance users' satisfaction of the service. Firstly, user interface should provide users with convenient consulting product database. When current database can meet users' requests, the interface then should permit them to download the data. When existing products fail to meet users' needs, the interface could supply standard formats and effective guidance for user's inputting new observation requests, so that the full-cycle mission planning can analyze the request and extracting mission-related information. Secondly, in view of different locations of the data users, the interface should apply distributed features, that is, the service meant for users' terminal should have a unified view that does not change with locations. What is more, to satisfy the need of near-real-time products, more customer service centers can be established in accordance with the layout of data storage, processing centers and ground stations. Data products are distributed by external communication network. Users can constantly submit orders or receive data products in a recycled way by the same network [4].

In addition, when the system is handling their requests, the interface should allow users to keep track of the progress. In particular, the cancellation of order and false information must be fed back to users in time. Thus, the interface should encompass monitoring mechanism of contingency in data distribution as well as the identification of the infeasible requests.

3.2. The Identification of Missions and Constraints

Identifying all detailed requirement and related constraints is the premise of full-cycle mission planning. Users' observation orders may vary, so different satellite resources and different constraints are involve. For example, some users demands visible light imaging, indicating that the image should be taken by visible light imaging satellites under favorable sunlight conditions; some users may require all-weather observation, where SAR satellites are used. Some global scale missions require regular target observation and data downlink in every circle of satellites' flight to obtain continuous data products, where satellites with low resolution can manage. But in regional missions, it is satellites with high resolution that implement observation and downlink the data to ground station or TDRSS satellites. Requirements for the service are always changing, among which the near-real -time trait becomes the main constraint in the planning of data acquisition, processing and distribution.

3.3. The Planning of Satellite Payloads

It is difficult to figure out a unified model and an efficient scheduling algorithm of multi-type satellites and their payloads based on constraints in satellite payloads planning. To give prominence of the advantages of full-cycle mission planning, it is necessary to schedule the various satellites in a unified and coordinated way. As for the different operating constraints of various satellites, such as visible light imaging satellites and SAR imaging satellites, they have to be described in a unified form in the integrated planning schedule model.

There are other aspects in scheduling algorithm need considering. If we simply follow the principle of First Come First Served, we may obtain a feasible solution, but it may not maximize the profits or users' satisfaction. The planning of satellite payloads is a combinatorial optimization problem with NP trait, which suggests that it is difficult to attend both the time constraint and the optimal solution of the algorithm. Therefore, the scheduling algorithm ought to strike a balance between solution performance and solution time to satisfy users' requirements in terms of the priority and time span of different observing missions.

3.4. The Planning of Ground Support Systems

According to its structure, mission planning of ground support system can be divided into two stages--data processing and control center stage and distributed ground stations stage. Stage two after preliminarily processing the data downlinked from satellites, sends the data to the former stage, and then the former processes data again and delivers final products to users. The planning cycle of stage one is rather longer (1~2 days), while the cycle of stage two depends on the circle of the satellite. The planning of two stages have to solve all conflicts in resource utilization related to data processing and distribution, as well as allow addition, revision and deletion of specific data products. A good number of departments and links are involved in the problem of ground support system planning, of which the scheduling of data processing is quite crucial. Because the same processing chains need to tackle offline missions, such as the additional processing of products and the disposition of overstocked missions, at the interval of finishing near-real-time data processing missions.

Furthermore, other elements should be considered in the scheduling of data processing, for instance, the test of capacity and feasibility of data processing chains, the allocation of the operating loads to different processors, and the automation of mission planning. The large number of products also contributes to the complexity of ground support planning, let alone the combination of processing and distribution of the products, which resulted from the different types of data processing activities (including near-real-time processing, offline processing, and overstocked mission handling), the site of data process(generally there are several data processing centers or stations.), the type of products(including regular products of global missions or occasional products of regional mission), the media for data distribution(such as solid-state recorders, wireless broadcast communication links, and multicast links), and the targets of data distribution(users or processing and control centers). The aspects of ground support

systems mentioned above add to the difficulty of full-cycle satellite land observation mission planning.

3.5. Data Distribution

Data distribution is the last link of full-cycle mission plan. Users' satisfaction largely depends on whether they could receive ordered products within given time. Since the data processing center and the user who submits requests are not always at the same place, both near-real-time data products and regular mission products are delivered through particular distributed data communication networks. What is more, the mass amount of satellite observing data could easily lead to network congestion or interrupt, thus may delay users obtaining the products and decrease their satisfaction. Hence compression technique of massive information is significant to lessen the total amount without much information loss. After compression, data products are saved in remote servers and system will provide operating port to remote users.

In light of the limited bandwidth and transmission rate of current network system, proper transmission strategy and agreement should be reached to accelerate mass information transmission. Generally, data of smaller amount can be transmitted through broadcast communication links with relative low speeds. As for the massive data, multicast data links have more advantages. Also, if massive data is required to be transmitted within a rather short period, other high-speedy transmission channels can be utilized.

3.6. Relevant Technology and Techniques

A practical full-cycle mission planning software has to be flexible enough to facilitate the addition, moderation, and deletion of rules and constraints. Meanwhile, the software needs to maintain interaction with planners, while the final decision to solve the conflicts comes to planners. The expert system requires identification and authentication mechanism, so it is not applicable to planning and scheduling. Traditional operational researches methods like mathematic programming could not manage such a complicated problem either, whether in modeling or seeking the proper solution.

Given the development of planning and scheduling software outside of China, the algorithm of constraint-based scheduling is the most feasible. Constraint-based scheduling algorithm transforms scheduling problem into Constraint Satisfaction Problem(CSP) in artificial intelligence and applies relative Constraint Programming(CP) to modeling and scheduling. CSP problem can be defined by the triple combination of a variable set, a variable value set and a constraint set of variable values, and its solution should be a value combination of variables that meets all constraints. If required, the given target should be optimized. CP inaugurates a highly effective way to solve CSP--its programming language can describe a host of CSP models, and its constraints processing algorithms and predefined search algorithm library guarantee the efficiency of problem solving. For example, American GREAS satellite mission planning system employs CP tools, ILOG Solver and ILOG Scheduler, developed by French company OLOG, providing moulds of all missions, resources and constraints. Here, missions refer to the activities and operations to be conducted, resources are human, satellites, sensors and communication bands used in implementation, and constraints are those defining the time of mission implement, relationship between missions, and the capacity and availability of resources. Although GREAS system does not include the planning for ground support system, its solving approach is a splendid reference to full-cycle mission planning. Another point to be stressed--CP technique is different from mathematic programming in that the models do not necessarily correspond to the solving algorithms. In other words, the same model may have a variety of methods to figure out the solution, including accurate search like branch and bound algorithm, and also inaccurate heuristic search such as taboo search, and genetic algorithm. Accordingly we can make full use of CP approach to build flexible models in solving complex planning and scheduling problem. Meanwhile, we can design by ourselves or borrow other inaccurate search algorithms with outstanding time behavior to get satisfactory mission planning based on the exact description of problem traits.

3.7. Prevention of Frequency Interference

At present, most data transmission systems of remote sensing satellites use the X band (8025-8400MHz) for data downlink, they march in sun-synchronous polar orbit, their adopt the local time of descending node is 10:30, and the data is received by the same ground receiving station network. These similar parameters make frequency interference more likely to occur between different data link channel.

In the past years, one operator corresponds to one specific satellite, however, this mode could not solve the problem of frequency interference effectively. Nowadays, the "hub-style" satellite mission planning can formulate plans in advance, and then settle the problem with the optimal solution.

In accordance with the interference protection threshold of International Telecommunication Union(ITU), the calculation and analysis of interference margin of the link can make suggestions to the judgment and prevention of data link frequency interference.

4. Conclusion

That the application of observing satellites converts from experimental use to practical use requires the full-cycle mission planning of satellite systems and ground support systems as integral. Satellite is just one of the links in service to users and the feature of current missions embodies in the stricter requirement on the efficiency and quality of data products. New problem about management of the relevant activities around ground support system emerges as well. Since it is a long cycling chain from requests submission to products delivery, the monitoring and optimization of ground data processing and resource allocation become necessary.

Only by viewing the cycle chain of production as a whole can we satisfy users' different requirements of desirable products. When practicing full-cycle mission planning, constraint-based scheduling is a favorable method, while more detailed techniques shall be further studied.

In the last a few decades, remote sensing satellites of China have witnessed significant changes, from single number to more than a dozen, from universal usage to specialized division. The full-cycle hub-style mission planning corresponds to national policy of civil remote sensing satellites, that is, unified planning and centralized handling. It is of great significance to timely deliver products of high quality to every professional user through remote sensing data planning and major processing points--first, this mode technically resolves the planning and optimization problem of numerous satellites; second, it helps to achieve the win-win situation where different users get what they want and the cost is reduced; third, it keeps supervising and monitoring various satellite data, and in turn propels the development of satellite processing and manufacturing techniques, and systematically stimulates the development and progress of aerospace remote sensing technology.

References

[1] Payload Design Description, RapidEye internal document, RE-DD-DJO-2000,Nov.2007

[2] E.Stoll,R.Schulze,B.D'Souza, M.Oxfort,"The impact of collision avoidance maneuvers on satellite constellation management", in proc. Of European Space Surveillance Conference, Madrid, Spain,2011

[3] Nicola Bianchessi,Jean-Francois Cordeau, Jacques Desrosiers,Gilbert Laporte, Vincent Raymond. A Heuristic for the Multi-Satellite, Multi-Orbit and Multi-User Management of Earth Observation Satellites. European Journal of Operational Research.vol.177,pp.750-762,2005

[4] Xu Xue-ren,Gong Peng,Study on Optimization of Satellite.Journal of Remoter Sensing,vol.56,pp:962-968,2007

[5] H. Muraoka, R. .Cohen, T. Ohno, and N. Doi. Aster observation scheduling algorithm. In Proceedings of the International Symposium Space Mission Operations and Ground Data Systems.1998

[6] J. Pemberton. Towards scheduling over-constrained remote sensing satellites. In Proceedings of the 2nd International Workshop on Planning and Scheduling for Space,2000.

[7] D. Smith, J.Frank, and A, Jonsson. Bridging the gap between planning and scheduling. Knowledge Engineering Review,15(1),2000.

[8] B-S Lee, J-S.Lee,B-S.Lee,and J-W.Eun, "Mission Analysis and Planning System for Korea Multipurpose Satellite-I",ETRI Journal, Vol.21,No,3,1999,pp.29-40.

[9] Zitzler E.Evolutionary algorithms for multiobjective optimization: methods and applications [D]. Zurich: Swiss Federal Institute of Technology,2002

12

Computational investigation of aerodynamic characteristics and drag reduction of a bus model

Eyad Amen Mohamed[*], **Muhammad Naeem Radhwi, Ahmed Farouk Abdel Gawad**

Mech. Eng. Dept., College of Eng. & Islamic Archit., Umm Al-Qura Univ., Makkah, Saudi Arabia

Email address:

eytworld@gmail.com (E. A. Mohamed), mnradhwi@uqu.edu.sa (M. N. Radhwi), afaroukg@yahoo.com (A. F. A. Gawad)

Abstract: It is well-known that buses comprise an important part of mass transportation and that there are many types of buses. At present, the bus transportation is cheaper and easier to use than other means of transportation. However, buses have some disadvantages such as air pollution due to engine exhaust. This study is an attempt to reduce the gas emissions from buses by reducing the aerodynamic drag. Several ideas were applied to achieve this goal including slight modification of the outer shape of the bus. Thus, six different cases were investigated. A computational model was developed to conduct this study. It was found that reduction in aerodynamic drag up to 14% can be reached, which corresponds to 8.4 % reduction in fuel consumption. Also, Neuro-Fuzzy technique was used to predict the aerodynamic drag of the bus in different cases.

Keywords: Computational Investigation, Aerodynamic Characteristics, Drag Reduction, Bus Model

1. Introduction

1.1. Background

Nowadays, the waste of energy and the environmental pollution are some of the major global concerns for all science disciplines especially engineering. There are a lot of researchers who studied the aerodynamic behavior around heavy vehicles and tried to control their harmful emissions. Thus, they considered how to find out a better way to improve the vehicle performance by modifying the shape and weight of the vehicle.

Buses are one type of the heavy vehicles that consume much fuel. They are road vehicles designed to carry passengers in different applications. Buses can have a capacity as high as 300 passengers. The most common type of buses is the single-decker rigid bus. The larger loads are carried by double-decker buses and articulated buses. The smaller loads are carried by midi-buses and minibuses. Coaches are used for longer distance services.

Bus manufacturing is increasingly globalised with the same design appearing around the world. Buses may be used for scheduled bus transport, scheduled coach transport, school transport, private hire, tourism, *etc*. Promotional buses may be used for political campaigns and others are privately operated for a wide range of purposes.

Historically, Horse-drawn buses were used from the 1820s, followed by steam buses in the 1830s, and electric trolleybuses in 1882. The first internal combustion engine buses were used in 1895 [1]. Recently, there has been growing interest in hybrid electric buses, fuel cell buses, electric buses as well as ones powered by compressed natural gas or bio-diesel.

1.2. Previous Investigations

Generally, there is somehow shortage in the investigations that consider aerodynamics of buses in comparison to other heavy vehicles, *e.g.*, trucks.

Newland [1] aimed to develop a transit bus fuel consumption function based upon relationships found in the literature between bus fuel consumption and various bus operating characteristics especially their variable passenger loads.

Roy and Srinivasan [2] studied the aerodynamics of trucks and other high-sided vehicles that are of significant interest in reducing road accidents due to wind loading and in improving fuel economy. They concentrated on the associated drag due to the exterior rear-view mirrors. They stated that modifying truck geometry can reduce drag and improve fuel economy.

Diebler and Smoth [3] developed experimentally a ground research vehicle (*GRV*) to study the base drag on large-scale vehicles at subsonic speeds. They concentrated on base drag of trucks, buses, motor homes, reentry vehicles, and other large-scale vehicles. They presented preliminary results of both the effort to formulate a new base drag model and the investigation into a method of reducing total drag by manipulating forebody drag.

Yamin [4] used computational fluid dynamics (CFD) technique to simulate external flow analysis of a coach. His results suggested that the steady state CFD simulation can be used to boost the aerodynamic development of a coach.

Abdel Gawad and Abdel Aziz [5] investigated experimentally and numerically the effect of front shape of buses on the characteristics of the flow field and heat transfer from the rear of the bus in driving tunnels. Their study covered three bus models with flat-, inclined-, and curved-front shapes. They found that the front shape of the bus affects its aerodynamic stability in driving tunnels. Also, they stated that the cooling of the inclined- and curved-front vehicles is better than the cooling of the flat-front bus by about 20%.

François et al. [6] studied experimentally the aerodynamics characteristics and response of a double deck bus, which is a bus type very used in the Argentinean routes, submitted mainly to cross-wind. They measured pressure distributions over the frontal and lateral part of the bus and also drag and lateral forces related to the position of centre of gravity.

Yelmule and Kale [7] considered experimentally and numerically the aerodynamics of open-window buses where airflow due to motion provides comfort. They stated that an overall drag reduction of about 30% at 100 km/h can be reached by modifying the bus exterior body.

Mohamed-Kassim and Filippone [8] analyzed the fuel-saving potentials of drag-reducing devices retrofitted on heavy vehicles. They considered realistic on-road operations by simulating typical driving routes on long-haul and urban distributions; variations in vehicle weight. Their results show that the performance of these aerodynamic devices depend both on their functions and how the vehicles are operated such that vehicles on long-haul routes generally save twice as much fuel as those driven in urban areas.

Patil [9] performed aerodynamic flow simulation on one of conventional bus to demonstrate the possibility of improving the performance with benefits of aerodynamic features around the bus by reducing drag, which improves the fuel consumption. They optimized one of the conventional bus models and tried to reduce drag by adding spoilers and panels at rear portion along with front face modification. Their results showed that drag can be decreased without altering the internal passenger space and by least investment.

Also, the issue of fuel consumption was covered by many authors [10] and [11].

1.3. Present Investigation

The present study focuses on the aerodynamic characteristics of buses especially drag, either form or friction, which influences directly the fuel consumption.

A computational model was developed using the commercial code ANSYS-Fluent 13 to predict the aerodynamic performance of buses.

Modifications of the external body and/or surface of the bus to reduce the aerodynamic drag are proposed. The authors carefully considered that the proposed modifications do not affect the safety and operation of the bus. Also, the modifications do not change the main body/structure of the bus. Actually, modifications can be applied with considerably low cost and fairly technical skills.

The computations were carried out for different values of Reynolds number.

2. Governing Equations and Turbulence Modeling

2.1. Governing Equations

The equations that govern the fluid flow around a model are time-averaged continuity and momentum equations which, for the steady, incompressible flow, are given by, respectively:

$$\frac{\partial U_i}{\partial x_i} = 0 \quad i = 1, 2, 3 \tag{1}$$

$$U_j \frac{\partial U_i}{\partial x_j} = \frac{\partial}{\partial x_j} \left(\nu \frac{\partial U_i}{\partial x_j} - \overline{u_i u_j} \right) - \frac{1}{\rho} \frac{\partial P}{\partial x_i} \quad i,j = 1, 2, 3. \tag{2}$$

In the above, U_i is the mean-velocity vector with components U, V and W in x, y and z directions, respectively, P is the static pressure, ρ is the fluid density and ν is its kinematic viscosity. Repeated indices imply summation. The turbulence model involves calculation of the individual Reynolds stresses ($\overline{u_i u_j}$) using transport equations. The individual Reynolds stresses are then used to obtain closure of the Reynolds-averaged momentum equation (Eq. 2).

2.2. Turbulence Modeling (Realizable k-ε Turbulence Model)

The realizable k-ε turbulence model was used in the present study. The realizable k-ε model differs from the standard k-ε model in two important ways:

- The realizable **k-ε** model contains an alternative formulation for the turbulent viscosity.
- A modified transport equation for the dissipation rate, ε, has been derived from an exact equation for the transport of the mean-square vorticity fluctuation.

For further details about the realizable k-ε turbulence model, one may refer to [12].

2.3. Drag Calculations

The results focus on the drag coefficient, i.e., pressure (form), friction, and total drag coefficients.

2.3.1. Pressure (form) Drag

The coefficient of pressure drag, C_{D_p}, is calculated by Eq. 3 as follows,

$$C_{D_p} = \frac{\Delta P}{(0.5 \times \rho \times U_\infty^2)} \tag{3}$$

Then, the force of pressure drag is calculated as

$$D_p = C_{D_p} \times 0.5 \times \rho \times U_\infty^2 \times A_F \qquad (4)$$

Where, D_p is the drag force due to pressure, A_F is the frontal (projected) area of the bus = $H \times W$, ρ is the flow density, U_∞ is the bus speed, and ΔP is the pressure difference between the front and rear surfaces of the bus.

2.3.2. Friction Drag

The force of friction drag is calculated for the two side surfaces and roof of the bus using the following equation:

$$D_f = C_{D_f} \times \frac{1}{2}\rho \times U_\infty^2 * A_{RS} \qquad (5)$$

Where, C_{D_f} is the coefficient of friction drag, A_{RS} is the area summation of roof and side surfaces = $A_R + A_S$, A_R is the roof area = $L \times W$, and A_S is the side surfaces = $2 \times L \times H$.

Generally, the actual operating Reynolds number ($Re = U\infty$ L/ν) is greater than the critical Reynolds number ($Re_{cr} = 5 \times 10^5$ for a flat surface) for all test cases, which means the flow is turbulent. Thus, the coefficient of friction drag is calculated as [13]:

$$C_{D_f} = \frac{0.031}{Re_L^{\frac{1}{7}}} \qquad (6)$$

Where, L is the bus length and ν is the flow kinematic viscosity.

2.3.3. Total Drag

The value of the total drag force, D_T, is calculated as:

$$D_T = D_p + D_f \qquad (7)$$

Then, the coefficient of total drag, C_{D_T}, is calculated as:

$$C_{D_T} = \frac{D_T}{(0.5 \times \rho \times U_\infty^2) A_F} \qquad (8)$$

3. Original and Modified Models

The original model represents an actual bus that was produced by Mercedes Benz, Type: Coach Travego M [14], Fig. 1.

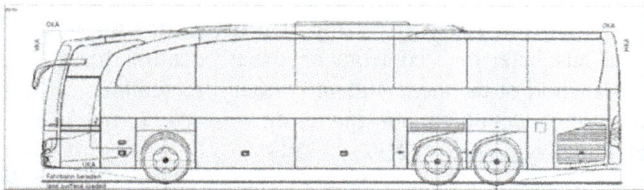

Figure 1. Side view of the bus [14].

This original model is considered as the comparing reference of the present study. The total length of the bus is 13 m, the width is 2.55 m, and the height is 3.1567 m as shown in Fig. 1. The fuel tank capacity is about 475 litres.

Some modifications were proposed to the original model. Each modification produced a new model. A specific name was given to each new model for classification. Table 1 shows the names and shapes of the different models.

Table 1. Names and shapes of the different models

No.	Name	View
1	Original	
2	MCOBS1	
3	MCOBS2	
4	MCOBS3	
5	MCOBS4	
6	MCOBS5	
7	MCOBS6	

The names and shapes that appear in table 1 can be explained as follows:

Original: It is the actual shape of the bus without modifications.

MCOBS1: A curved device is added at the rear of the bus to direct the air flow downward directly behind the bus. It has two supports.

MCOBS2: Similar to MCOBS1 with closing the two left

and right ends of the curved device to grantee that all air is directed downward without side escape.

MCOBS3: The bus is equipped with the same device of MCOBS2. Also, two small ducts ($4500{\times}50{\times}300mm$) are added on both sides of the bus to drive air, with relatively high pressure, to the low-pressure zone behind the bus.

MCOBS4: Only two small ducts are added on both sides of the bus. There is no curved device.

MCOBS5: The bus is equipped with a curved device similar to MCOBS1. The front surface of the bus is modified to have a suitable curvature.

MCOBS6: Similar to MCOBS5 but the rear surface has also a curvature similar to the one of the front surface.

4. Computational Aspects

4.1. Tested Velocities

The computations were mainly carried out at 100 km/h (27.22 m/s) for all cases. However, to evaluate the effect of bus velocity on the aerodynamic characteristics and drag, other three values of velocity were examined for the case of MCOBS5. Table 2 shows the four values of velocity and the corresponding values of Reynolds number.

Table 2. Tested bus velocities.

NO.	Speed (Km/h)	Speed (m/s)	Reynolds number
1	70	19.44	16.73×10^6
2	100	27.22	23.9×10^6
3	120	33.33	28.68×10^6
4	150	41.66	35.85×10^6

4.2. Computational Domain and Boundary Conditions

As seen in Figure 2, the computational domain is a rectangle that contains the bus. The dimensions of the domain were selected to ensure free development of the air flow around the bus.

The boundary conditions of the domain can be listed as: (*i*) Uniform velocity at the inlet surface. (*ii*) Zero pressure-gradient at the outlet surface. (*iii*) Solid condition at the ground, *i.e.*, the surface below the bus. (*iv*) Symmetry condition at the two side surfaces and the top surface of the domain.

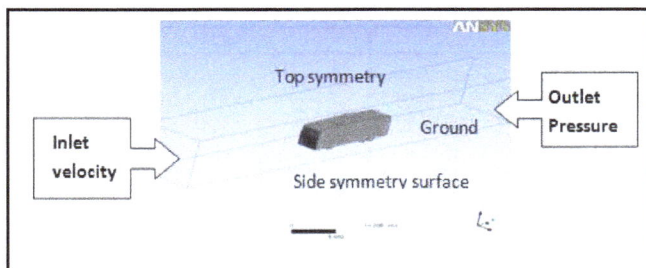

Figure 2. Computational domain and boundary conditions.

4.3. Computational Grid (Mesh)

The computational domain was discretized using unstructured grids. This type of grids usually guarantees the flexibility of generating enough computational points in locations of severe gradients. The computational domain was covered with tetrahedral elements (Figure 3). The grid is very fine next to the solid boundary. The dimensionless distance between the wall and first computational point is $y^+ \approx 1.8$, which is calculated as

$$y^+ = \frac{u_\tau y}{v} \qquad (9)$$

Where, y is the distance to the first point off the wall, v is the kinematic viscosity, u_τ is the friction velocity.

$u_\tau = \sqrt{\frac{\tau}{\rho}}$, τ is the wall shear stress and ρ is the flow density. The value of $y^+ \approx 1.8$ ensures good resolution of the complex turbulent flow.

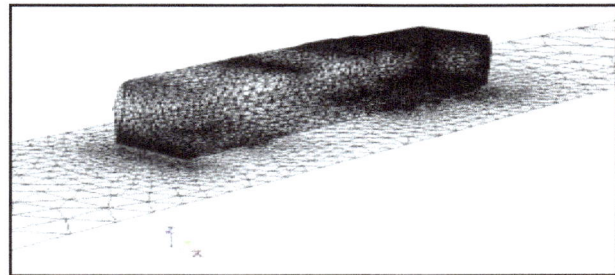

(a) Grid structure for "Original" shape.

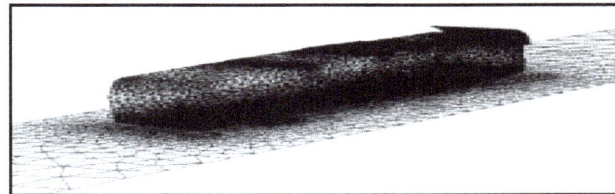

(b) Grid structure for MCOBS1.

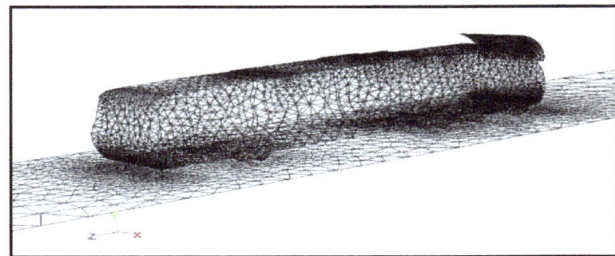

(c) Grid structure for MCOBS6.

Figure 3. Samples of grid structures.

4.4. Grid Independency

Careful consideration was paid to ensure the grid-independency of the computational results. Therefore, three grid sizes were used to test the grid-independency, namely: 50,000, 65,000 and 85,000 elements (cells).

Considering the flow characteristics, it was noticed that the difference between the results of the second and third grids is in the range of 2-3%. Thus, the second grid size (65,000) was used for all test cases.

4.5. Numerical Scheme

SIMPLE algorithm (Semi-Implicit Method for Pressure-Linked Equations) was used to solve the velocity and pressure fields. Each momentum equation was solved by the '' first-order upwind" scheme.

The ''standard wall function" was used as the near-wall technique in the turbulence model. The solution continues until the numerical error of all computed quantities gets below 10^{-5}.

4.6. Validation of the Present Computational Algorithm

The present numerical result of the total drag coefficient $C_{D_T} = 0.698$ for the ''original" compares very well to the range of 0.6-0.8 that was reported in [13]. The present value of C_{D_T} lies exactly in the middle of the range. This gives confidence in the present computational scheme.

5. Results and Discussions

This section shows the computational results of the different considered cases (Original and modified models) that were mentioned in Sec.3. The main objective is to find the modified model that gives maximum drag reduction. However, the flow field (pressure and velocity) around the bus model is illustrated.

5.1. Investigated Cases

As it is well-known, the pressure (form) drag represents the major part of the total drag on bus in comparison to the friction drag, the pressure distributions on the frontal and rear surfaces are considered. The pressure drag depends on the difference between the pressure distributions on the frontal and rear surfaces of the bus. Figures 4 and 5 illustrate the pressure contours on the frontal and rear surfaces of the bus, respectively, for all cases.

The velocity field in the zone adjacent to the rear surface affects the pressure distribution on rear surface. Thus, velocity vectors, in a vertical section, are shown in Figure 6 for all cases. The vertical section passes through the mid-section of the bus width.

The results of various cases are discussed as follows:

5.1.1. Original

As expected, Figures 4 and 5 show that the pressure is really high on the bus frontal surface due to flow stagnation. The pressure is very low on the rear surface due to wake formation behind the bus. Figure 6 shows that two main vortices are formed in the wake zone behind the bus. The two vortices have nearly equal size. For this case, at $Re = 23.9 \times 10^6$, the total drag coefficient (C_{D_T}) equals 0.698.

5.1.2. Mcobs1

As can be seen in Figure 5, the curved-surface device at the rear of the bus reduces slightly the pressure on the rear surface in comparison to the ''Original ''. However, this change of pressure does not considerably reflect on the value of C_{D_T} that becomes 0.649.

Another effect of the curved-surface device is seen in Figure 6. The upper vortex behind the bus becomes smaller then the lower vortex. The curved-surface device directed the flow from the top surface at the bus to the wake zone behind the bus.

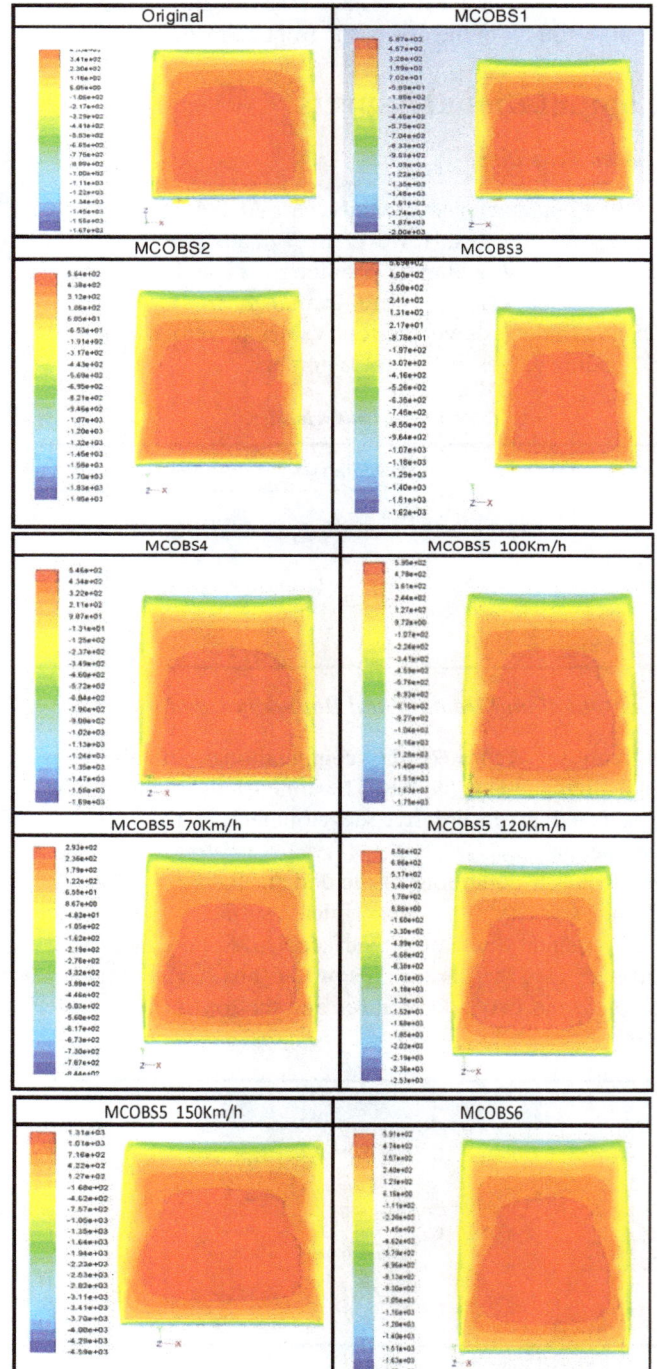

Figure 4. *Pressure contours on the frontal surface.*

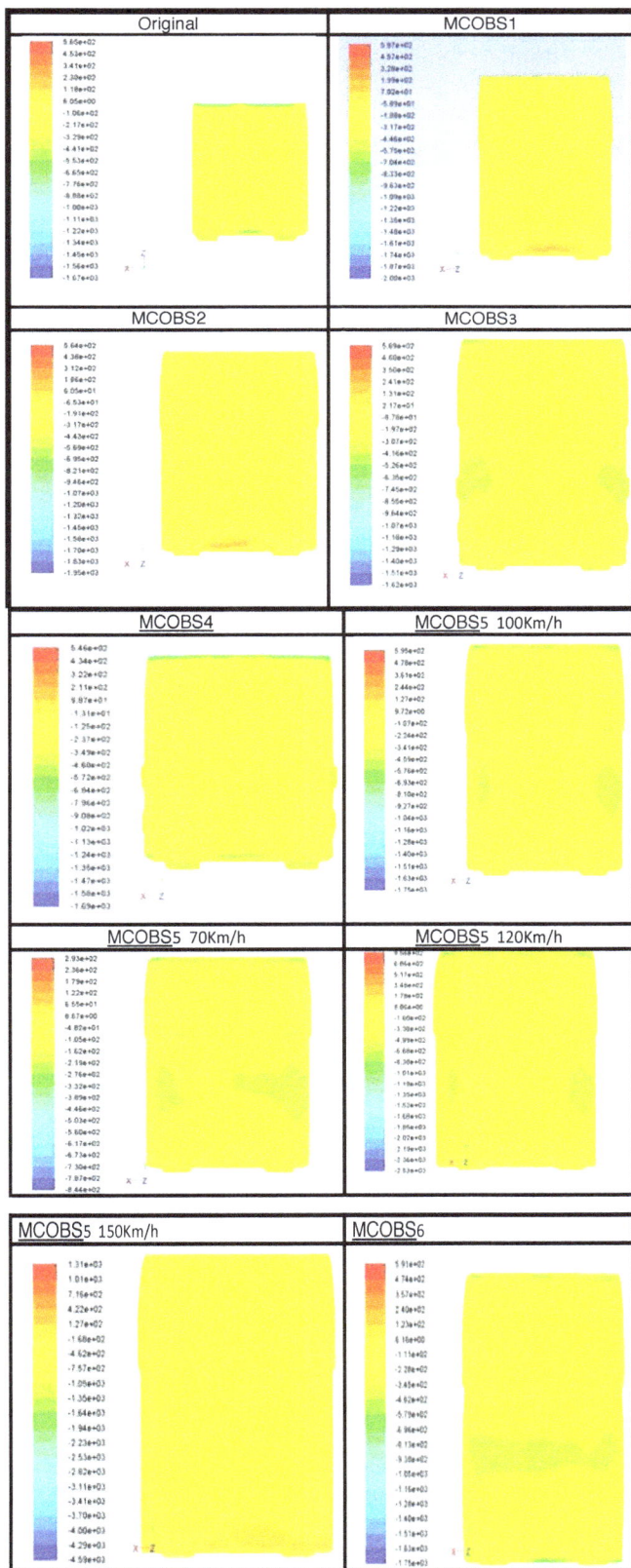

Figure 5. Pressure contours on the rear surface.

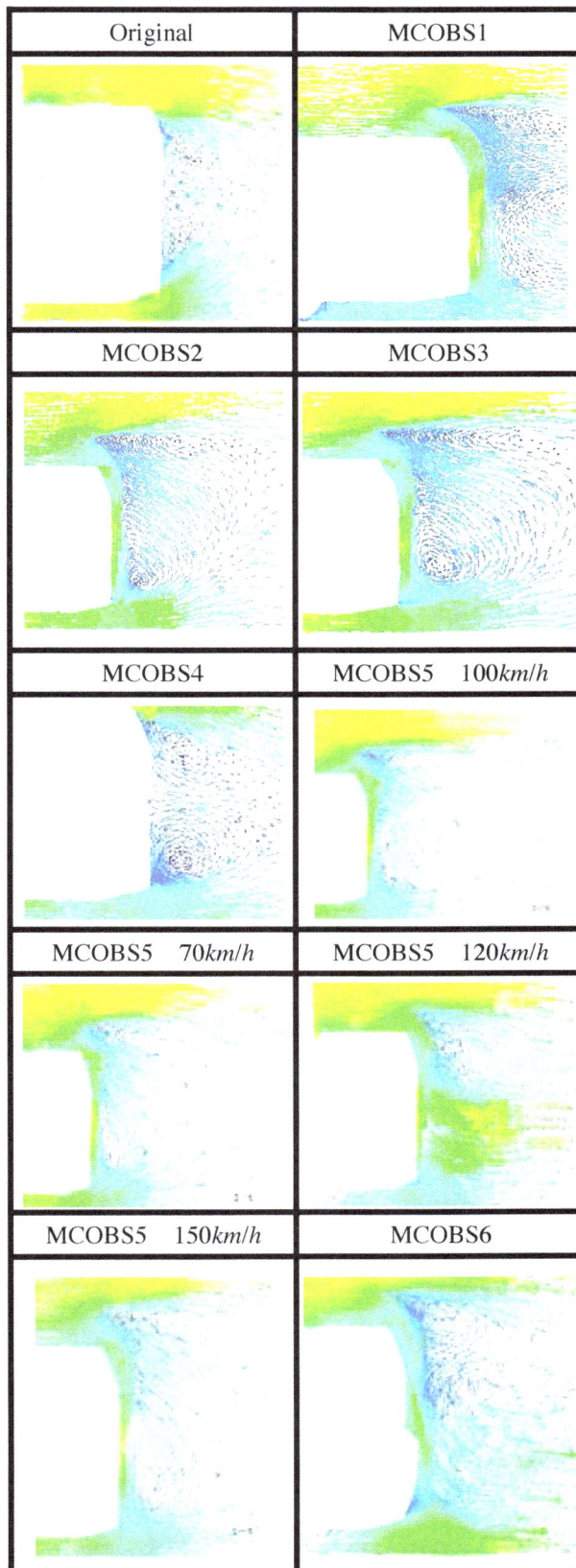

Figure 6. Velocity vectors behind the bus in a vertical section.

5.1.3. Mcobs2

It seems that the closed ends of the curved-surface device

have a minor effect on the pressure distributions, Figure 5. In Figure 6, the vortices behind the bus are altered in comparison to MCOBS1. The value of C_{D_T} is 0.652, which is very close to that of MCOBS 2. Thus, the curved-surface device reduces the value at C_{D_T} in comparison to the "Original".

5.1.4. Mcobs3

It seems that the two side ducts have a negative effect to the curved-surface device. They do not help in reducing the total drag on the bus. Unfortunately, they increase the value of C_{D_T} to reach 0.691, which is greater than the two values of MCOBS1 ($C_{D_T} = 0.649$) and MCOBS2 ($C_{D_T} = 0.652$). However, it is slightly lower than the value at the "original" ($C_{D_T} = 0.698$).

This may be attributed to the long path of the ducts, which causes big pressure drop inside them. Thus, air may be ever sucked inside them at their rear ends causing increase in the value of C_{D_T}. Generally, it is clear from Figure 6 that there is almost no change in comparison to MCOBS2.

5.1.5. Mcobs4

It is obvious that the two ducts alone have no effect in reducing the total drag on the bus. The value at C_{D_T} is 0.697, which is the same as the "Original" (C_{D_T}=0.698). However, the two ducts cause the lower vortex behind the bus to extend downstream, Figure 6.

5.1.6. Mcobs5

It is found that the curvature at the frontal surface causes a noticeable change on the total drag on the bus. Thus, the value of C_{D_T} becomes 0.632 for 100 km/h. This value is lower than that of the "Original" (C_{D_T}=0.698).

In Figure 6, as expected, the curvature at the frontal surface has no effect on two vortices in the wake region behind the bus.

5.1.7. Mcobs6

It is obvious that the curvatures of the frontal and rear surfaces cause a remarkable change on pressures of these two surfaces, Figures 4 and 5. Thus, the value of C_{D_T} reduces to 0.602.

This is the lowest value of C_{D_T} achieved in all test cases. Also, the wake zone behind the bus is favorably changed as can be seem in Figure 6.

5.2. Overall view of all cases

Based on the results of the previous section, an overall view of all cases can be demonstrated. Table 3 illustrates overall results of the total drag coefficient (C_{D_T}) for all cases. Also, Figure 7 shows the values of total drag coefficient (C_{D_T}) for all cases. Moreover, table 3 illustrates the percentage reduction in total drag coefficient for all cases of bus modifications.

It is clear from Table 3 and Figure 7 that the lowest value of C_{D_T} = 0.602, which corresponds a total drag reduction of 14%, is obtained for case MSCOBS6. This is the case of modifying the frontal and rear surfaces by slight curvature. This is the best case.

Whereas, the worst case is MCOBS4 with C_{D_T} = 0.697 and total drag reduction of 0.14%.

Thus, the idea of putting two side ducts seems useless. This is may be attributed to the relatively big length of the ducts. Internal friction at the walls of the ducts causes considerable pressure loss inside them. Thus, there is no pressure rise by the end of the ducts at the rear surface of the bus.

Table 3. Overall results of C_{D_T}.

No.	Case	Velocity (km/h)	Re × 10^6	C_{D_T}	C_{D_T} Reduction (%)	Fuel Reduction (%)
1	Original	100	23.9	0.698	----	----
2	MCOBS1	100	23.9	0.649	7	4.2
3	MCOBS2	100	23.9	0.652	6.6	4
4	MCOBS3	100	23.9	0.691	1	0.6
5	MCOBS4	100	23.9	0.697	0.14	0.1
6	MCOBS5	100	23.9	0.632	9.5	5.7
7	MCOBS5	70	16.73	1.03	47.6 increase	----
8	MCOBS5	120	28.68	0.558	20	12
9	MCOBS5	150	35.85	0.428	38.7	23.2
10	MCOBS6	100	23.9	0.602	14	8.4

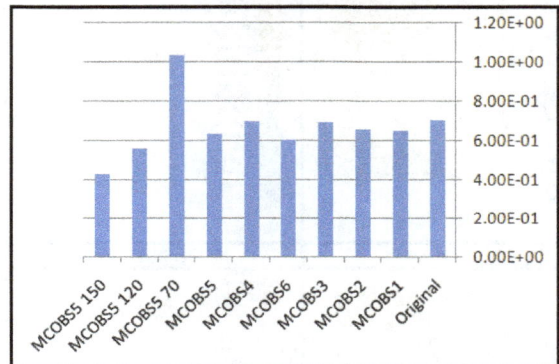

Figure 7. Values of total drag coefficient (C_{D_T}) for all cases.

Table 4. An example of fuel saving

Fuel consumption of "Original" (Liter/h)	Percentage fuel saving (%) MOCBS6	Fuel consumption of MOCBS6 (Liter/h)	Fuel saving (Liter/h) MOCBS6
66.88	8.4	61.26	5.618

Also, Table 3 indicates that a considerable drag reduction up to 38.7%, at 150 km/h, can be reached by increasing the bus velocity. This drag reduction may be attributed to the separation delay or even prevention due to the high momentum of the air surrounding the bus.

Unfortunately, high velocities are not always practical due to traffic and safely considerations. In table 3, the fuel reduction percentage was calculated based on the approximate relationship that was derived from [15]:

Fuel reduction %= (3/5) × Total drag reduction %   (10)

Thus, considerable fuel reduction can be obtained such as case MCOBS6 (8.4%). As an example of the saving of fuel consumption, based on the present findings, Table 4 shows the annual saving of an intercity bus that operates at an average velocity of 100 km/h.

6. Adaptive Neuro-Fuzzy Inference System (*ANFIS*)

6.1. Introduction

The Adaptive Neuro-Fuzzy Inference system (*ANFIS*) was used to predict the values of the pressure drag coefficient C_{D_T} based on the obtained data from the computational investigation. A well-trained nruro-fuzzy scheme can perform the prediction successfully based on enough computational data. After training, this technique enables the prediction of new values of C_{D_T} that were not predicted by the computational study. Application of *ANFIS* needs much less programming effort and computer run-time in comparison to traditional computational fluid dynamics (*CFD*) techniques.

Fuzzy logic (*FL*) is a form of many-valued logic, which deals with reasoning that is approximate rather than fixed and exact. Compared to traditional binary sets (where variables may take on true or false values), fuzzy logic variables may have a truth value that ranges in degree between 0 and 1.

ANFIS is a kind of neural network that is based on Takagi–Sugeno fuzzy inference system. Since it integrates both neural networks and fuzzy logic principles, it has potential to capture the benefits of both in a single framework. Its inference system corresponds to a set of fuzzy *IF–THEN* rules that have learning capability to approximate nonlinear functions. Hence, *ANFIS* is considered to be a universal estimator.

For more details about *ANFIS*, one may refer to Refs. [16] and [17].

6.2. Present ANFIS Model

Table 5. *Training data for all cases*

Case	Input		Output
	Code	Reynolds number	Pressure drag coefficient (C_{D_p})
Original	1	23.9×10^6	0.661
MCOBS1	2	23.9×10^6	0.611
MCOBS2	3	23.9×10^6	0.614
MCOBS3	4	23.9×10^6	0.654
MCOBS4	5	23.9×10^6	0.564
MCOBS6	6	23.9×10^6	0.659
MCOBS5	7	23.9×10^6	0.595
MCOBS5-70	8	16.73×10^6	0.994
MCOBS5-120	9	28.68×10^6	0.522
MCOBS5-150	10	35.85×10^6	0.392

ANFIS was constructed based on "Quadrangle prediction". The two input parameters to *ANFIS* are the bus case and the Reynolds number (*Re*). The bus cases are coded as numbers,

Table 5.

The output of *ANFIS* is the pressure drag coefficient (C_{D_p}). Table 5 shows the training data for all cases.

Figures 8 and 9 illustrate the structure and output performance surface of *ANFIS*, respectively. Figure 9 explains the variation of the Quadrangle predictions with the two inputs (bus case and *Re*), which end up with convergence.

After the success in training *ANFIS*, new set of data, which was not seen before by the *ANFIS*, is introduced to predict output (C_{D_p}).

Thus, the ability of ANFIS to predict correctly the values of C_{D_p} as output is confirmed.

Table 6 shows the new set of input data and the corresponding predictions of output.

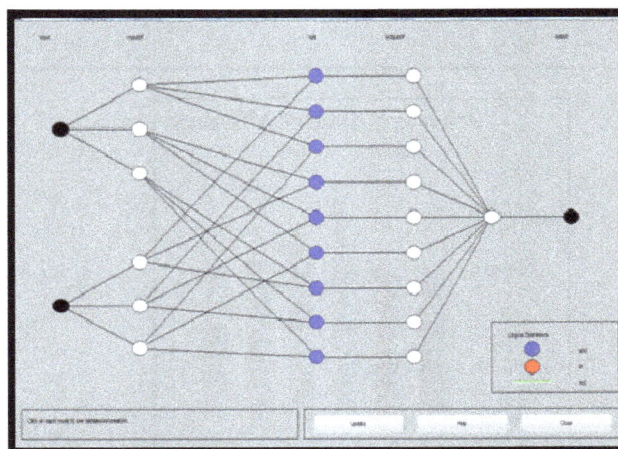

Figure 8. *Structure of ANFIS model (Quadrangle Prediction).*

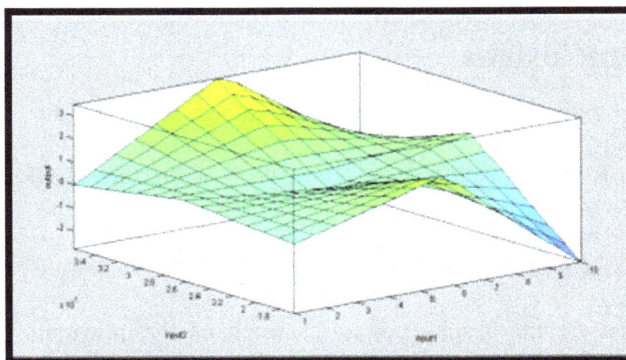

Figure 9. *Output surface of performance of ANFIS model (Quadrangle Prediction).*

Figure 10 represents validation of *ANFIS* new predictions. Comparison is made between the new predictions of *ANFIS* (*Red*) and the corresponding predictions of the present ANSYS-Fluent scheme (*Blue*) as *CFD* results.

Generally, it is clear that there is a fair agreement between *ANFIS* and *CFD* predictions. The differences between the two predictions may be attributed mainly to two reasons: (*i*) The small number of cases (10) that was used to train *ANFIS*. (*ii*) The generality of *ANFIS* that considered different cases of the bus with modifications. Better predictions are expected if

ANFIS is modeled for each bus case with different values of *Re*.

However, Figure 10 suggests that *ANFIS* technique is promising if the above two points are carefully considered.

Table 6. *Prediction of new data.*

Input		Output
Code	Reynolds number	Pressure drag coefficient (C_{D_p})
1	23.6×10^6	0.621
2	20.1×10^6	0.64
3	19.1×10^6	0.902
4	22.4×10^6	0.794
5	23.0×10^6	0.757
6	18.7×10^6	1.45
7	27.1×10^6	0.691

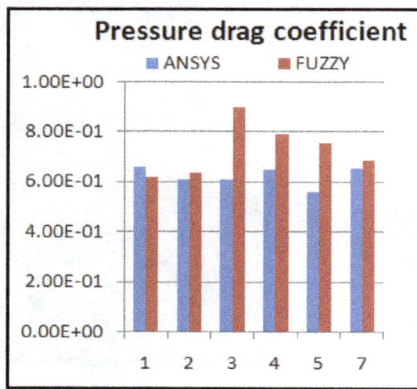

Figure 10. *Validation of ANFIS new data.*

7. Conclusions

A computational scheme was developed to study the possibility of drag reduction of buses. An actual bus was chosen to carry out the study. The values of the total drag coefficient C$_{(DT)}$, corresponding to suggested modifications to the bus, were computed. Seven case studies were investigated and the reductions in total drag and fuel were recorded.

Based on the results and discussions of the previous sections, the following concluding points can be stated:

1. The curvature modification of the frontal and rear surfaces, case MCOBS6, gives the best drag reduction of 14%. This gives a consequence fuel savings of about 8.4%.

2. The proposed curvature at the frontal and rear surfaces is accepted from the economic and manufacturing points of view. This modification is easy to be implemented as it does not affect the body/structure of the bus.

3. The idea of adding a rear curved–shape device seems interesting. The present curved–shape causes a maximum drag reduction of 7 %, which corresponds to a fuel saving of about 4.2 %.Perhaps introducing other profiles of the curved-shape is a good idea to find the optimum profile that gives the highest drag reduction.

4. The using of side ducts proved to be inefficient technique. The idea of transferring high pressure from the upwind zone of the bus to its downwind zone to increase the pressure at the rear surface of the bus did not succeed. This may be attributed to the relatively big length of the ducts. Thus, internal friction causes losses and overall pressure drops in the ducts. So, the pressure at the rear surface of the bus is nearly the same as that of the bus wake.

5. The total drag on the bus decreases with the bus speed. However, the considerations of traffic and safety limit this option of increasing the bus speed.

6. Although the maximum value of fuel saving of 8.4% at 100 *km/h* of the present study is not a big number, it represents a very good achievement when considering the huge amount of fuel consumption of buses allover the world, *e.g.*, intercity buses.

7. Fairly good predictions were obtained from *ANFIS*. Increasing the amount of training data will certainly improve its efficiency.

8. Other ideas may be considered in future investigations such as:(*i*) air jets, using a suitable pneumatic system, may be injected from the rear surface of the bus to increase the pressure in the wake zone. (*ii*) Minimizing flow separation by tapering or rounding the fore-body of the bus, and modifying the roof of the bus.

Nomenclature

A_F	: Frontal area.
A_S	: Side area = $2 \times L \times H$.
A_R	: Roof area $= L \times W$.
A_{RS}	$= A_R + A_S$.
C_{D_f}	: Friction drag coefficient.
C_{D_p}	: Pressure drag coefficient.
C_{D_T}	: Total drag coefficient.
C_p	: Pressure coefficient.
D_f	: Frictions drag force (*N*).
D_p	: Pressure drag force (*N*).
D_T	: Total drag force (*N*).
H	: Bus height.
L	: Bus length.
P	: Pressure (*kPa*).
P_b	: Outlet pressure (*kPa*).
Re	: Reynolds number.
Re_{Cr}	: Critical Reynolds number for flow on a flat plat.
U_∞	: Bus velocity.
W	: Bus width.
ΔP	: Pressure difference.

Abbreviations

ANFIS	: Adaptive Neuro-Fuzzy Inference system
CFD	: Computational fluid dynamics.
FL	: Fuzzy logic
GRV	: Ground research vehicle.
SIMPLE	: Semi-Implicit Method for Pressure-Linked Equations.

References

[1] L. E. Newland, "A Fuel Consumption Function for Bus Transit Operations and Energy Contingency Planning", Technical Report: UM-HSRI-80-53, Highway Safety Research Institute, The University of Michigan, July 1980.

[2] S. Roy, and P. Srinivasan, "External Flow Analysis of a Truck for Drag Reduction", International Truck and Bus Meeting & Exposition, Paper #: 2000-01-3500, 2000.

[3] C. Diebler and M. Smith, "A Ground-Based Research Vehicle for Base Drag Studies at Subsonic Speeds", Technical Report: NASA/TM-2002-210737, NASA Dryden Flight Research Center, Edwards, California, November 2002.

[4] A. K. M. Yamin, "Aerodynamic Study of a Coach Using Computational Fluid Dynamic (CFD) Technique", M.Sc. in Automotive Engineering, Faculty of Engineering and Computing, Coventry University, August 2006.

[5] A. A. Abdel Aziz, and A. F. Abdel Gawad, "Aerodynamic and Heat Transfer Characteristics around Vehicles with Different Front Shapes in Driving Tunnels," Proceedings of Eighth International Congress of Fluid Dynamics & Propulsion (ICFDP 8), December 14-17, 2006, Sharm El-Shiekh, Sinai, Egypt.

[6] D. G. François, J. S. Delnero, J. Colman, Di Leo J. Marañón and M. E. Camocardi, "Experimental determination of Stationary Aerodynamics loads on a double deck Bus", 11th Americas Conference on Wind Engineering, San Juan, Puerto Rico, June 22-26, 2009.

[7] M. M. Yelmule and S. R. Kale, "Aerodynamics of a Bus with Open Windows", Int. J. Heavy Vehicle Systems, Vol. 16, No. 4, 2009.

[8] Z. Mohamed-Kassim, A. Filippone, "Fuel savings on a heavy vehicle via aerodynamic drag reduction", Transportation Research Part D 15, 275-284, 2010.

[9] C. N. Patil, K.S. Shashishekar, A. K. Balasubramanian, and S. V. Subbaramaiah, "Aerodynamic Study and Drag Coefficient Optimization of Passenger Vehicle", International Journal of Engineering Research & Technology (IJERT), Vol. 1, Issue 7, September-2012.

[10] Bob Lloyd, "Dissecting the Basic Fuel Consumption Equation into Its Components to Improve Adaptability to Changing Vehicle Characteristics", 25th ARRB Conference-Shaping the Future: Linking Policy, Research and Outcomes, Perth, Australia, 2012.

[11] I. M. Berry, "The Effects of Driving Style and Vehicle Performance on the Real-World Fuel Consumption of U.S. Light-Duty Vehicles", M.Sc. Mechanical Engineering and M.Sc. Technology and Policy, Massachusetts Institute of Technology, February 2010.

[12] Fluent guide manual, (2011).

[13] F. M. White, "Fluid Mechanics", McGraw-Hill, 7th Edition, 2008.

[14] Mercedes-Benz, Reisebus/Coach Travego M, Mannual. http://ebookbrowsee.net/1843659-graphics-travego-pdf-d94749039

[15] J. Patten, B. McAuliffe, W. Mayda, and B. Tanguay, "Review of Aerodynamic Drag Reduction Devices for Heavy Trucks and Buses" NRC-CNRC Technical Report, Canada, 2012.

[16] Matlab guide manual, 2011.

[17] A. F. Abdel Gawad, "Computational and Neuro-Fuzzy Study of the Effect of Small Objects on the Flow and Thermal Fields of Bluff Bodies," Proceedings of 8th Biennial ASME conference on Engineering Systems Design and Analysis (ESDA2006), July 4-7, 2006, Torino, Italy.

13

Measurement of the Liquid Level in the Fuel Tank of Rocket

N. I. Klyuev

Department of Mathematical Modeling in Mechanics, Samara State Aerospace University, Samara, Russia

Email address:

nikolay_klyuev@mail.ru

Abstract: In this paper we consider the problem of mathematical modeling of fluid flow in the vertical channel of fuel management system. The system has a vertical measuring channel with sensors inside the channel for fixing the free surface level of fluid in the channel. By lowering the level of fuel in the tank fuel level decreases in the channel. When the level of fuel in the channel reaches the sensor, the sensor activation occurs. Thus, the level of fuel in the channel determining fuel level in the tank. Fluid flow in the vertical measuring channel describes by non-stationary equation of motion of parabolic type. For dynamic modeling used a viscous incompressible fluid. The fluid flow is called non-stationary parabolic equation of motion for the cylindrical channel. We got an approximate solution of the problem by averaging the terms of the equation of motion for the channel radius. The solution of the differential equation is satisfied in the package Mathcad applications. Graphs of displacement and velocity of the free surface of fuel in the measuring channel are represented over time. Measurement error of liquid level in the fuel tank has been determined. It is proposed engineering solution to eliminate error of the fuel level measurement.

Keywords: Viscous Liquid, Unsteady Flow, Liquid Level, Cylindrical Channel, Vibration, Measuring Error

1. Introduction

A fuel of the oxidizer tank and combustible tank flows into the combustion chamber of the rocket engine. Simultaneous supply of fuel at a predetermined proportion, ensures efficient operation of a rocket engine. Synchronous fuel delivery, at a predetermined proportion, enables efficient operation of the rocket engine. Efficient operation is dependent on accurate measurement of the fuel level in the tank. For this purpose, the fuel tank has a fuel management system. The system is a vertical measuring channel with sensors inside the channel for fixing the free surface level of fluid in the channel (Figure 1).

Vertical channel and fuel tank are communicating vessels. By lowering the level of fuel in the tank, fuel level decreases in the channel. When the level of fuel in the channel reaches the sensor, the sensor activation occurs. The signal goes to the fuel management system. As a result, the fuel consumption varies. Thus, the level of fuel in the channel determins fuel level in the tank. The problem is that the free surface of the fuel do not coincide in the channel and the tank. Error in measuring the level of fuel leads to inefficient fuel consumption. As a result,

the rocket engine is not operating optimally and tanks have the "extra" amount of fuel.

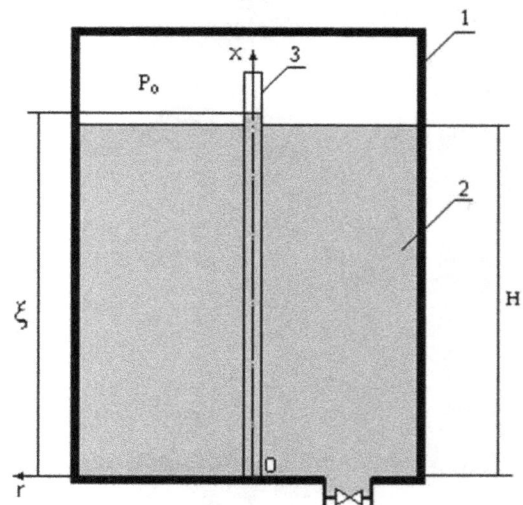

Figure 1. *Diagram of a fuel tank. 1- tank, 2- fuel, 3- measurement channel, p_0 - gas pressure, ξ - liquid level in the channel, H - liquid level in the tank, x, r - coordinate axis.*

We consider the problem of mathematical modeling of fluid flow in the vertical channel of fuel level management system of rocket. Fluid flow in the vertical measuring channel describes by non-stationary equation of motion of parabolic type. Unsteady motion equation was solved by many researchers. Problems discussed in [1-3] can be called a classic. The article [1] considers the laminar fluid flow in a cylindrical channel from standstill. The pulsating motion is presented in [2]. The results of calculation of a periodic motion are compared to the experimental data. The equations of motion are solved by operational calculus.

In [5] it is shown that in practical usage of rocket engines, the ratio of oxidizer and combustible is broken. What leads to a greater or lesser throttle opening. The paper [6] presents a non-linear mathematical model to simulate the pulsating flow. The governing equations are numerically solved by a composite scheme incorporating the two-step Lax-Wendroff method with the trapezoidal integration version of the method of characteristics at the pipe boundaries. The validity of the mathematical model and the methodology was verified by comparing the analytical predictions with experimental results.

Work [7] presents unsteady flow of a non-Newtonian fluid due to instantaneous valve closure. the appropriate governing equations are solved by a numerical approach. A fourth-order Runge-Kutta scheme is used for the time integration, and the central difference scheme is employed for the spatial derivatives discretization. To verify the proposed mathematical model and numerical solution, a comparison with corresponding experimental results from the literature are made.

In work [8] presents the method of characteristics (MOC) for simulation of unsteady flow in a pipeline. However, the relatively complex method of implicit (MOI) provides the advantages of unconditional convergence and mutual independence between time and space mesh parameters. This study combines the MOC and MOI to simulate pipeline unsteady flow and hydropower transient processes.

The paper [9] presents results of an investigation of the response of an incompressible fluid in a circular micro pipe to a sudden time-independent pressure drop. Solutions of the problem were obtained analytically using the Laplace transform technique and numerically using the lattice Boltzmann method.

The paper [10] presents steady and unsteady flows in a mildly curved pipe for a wide range of Reynolds numbers are examined with direct numerical simulation. In article [11] fully developed pulsed flow of an incompressible Newtonian fluid (crude oil) through a pipe line has been modeled and analyzed using the finite element method and compared with results based on an analytical solution. The flow is generated by a periodic pressure gradient superimposed on a constant Poiseuille flow. The results show good agreement between analytical and numerical solutions based on finite element method for the Newtonian fluid under unsteady regimes.

In article [12] the dynamic interaction between the pipe and unsteady flows is analyzed based on experiments and numerical models. The paper [13] presents a method of characteristics for solving one-dimensional model of fluid flow in pipe networks.

The goal of this research is to develop a mathematical model of the liquid flow in the vertical measuring channel of the fuel level measurement system of rocket.

2. The Mathematical Formulation of the Problem

At the initial time, $t = 0$, the fuel level in the tank and in the channel is the same (H_0). The free upper end of the cylindrical channel is located above the fuel level in the tank, so the flow of fuel from the tank into the channel at this location possible. The lower base of the cylindrical channel coincides with the bottom of the tank, and the fuel is communicated freely between the tank and the channel. Above the free surface of fuel ($t \geq 0$) in the tank and the channel maintains a constant pressure p_0 (boost pressure).

The fuel level in the tank varies by the law $H = H_0 - V \cdot t$. V - lowering speed of fuel in the tank. It is known that the liquid levels in the tank and the channel do not coincide. It is necessary to determine the level of liquid in the cylindrical channel at an arbitrary time. We introduce a coordinate system, the beginning of which was placed in the center of the lower base of the cylindrical channel.

As a model for the flow is used non-stationary motion of a viscous incompressible fluid in a cylindrical channel

$$\frac{\partial u}{\partial t} = -g - \frac{1}{\rho}\frac{\partial p}{\partial x} + \frac{\nu}{r}\frac{\partial}{\partial r}\left(r\frac{\partial u}{\partial r}\right) \qquad (1)$$

boundary conditions of the problem $r = 0$, $\frac{\partial u}{\partial r} = 0$, $r = R$, $u = 0$,

the initial condition of the problem $t = 0$, $x = H_0$, $u = 0$,

where $u = (r, t)$ - velocity of the liquid in the channel, p - pressure, ρ - density, t - time, ν - kinematic viscosity, g - acceleration of gravity.

3. Method for Solving Problem

The average speed of fuel in the cylindrical channel $<u> = \frac{2}{R^2}\int_0^R r u dr$. Multiplying the left and right sides of the equation (1) by r. We write down the individual terms of the equation of motion

$$\int_0^R r\frac{\partial u}{\partial t}dr = \frac{d}{dt}\int_0^R r u dr = \frac{R^2}{2}\frac{d<u>}{dt}, \quad \int_0^R rg dr = \left.\frac{gr^2}{2}\right|_0^R = \frac{gR^2}{2},$$

$$\int_0^R \frac{r}{\rho}\frac{\partial p}{\partial x}dr = \left.\frac{\partial p}{\partial x}\frac{r^2}{2\rho}\right|_0^R = \frac{R^2}{2\rho}\frac{\partial p}{\partial x},$$

$$\int_0^R v \frac{\partial}{\partial r}\left(r\frac{\partial u}{\partial r}\right)dr = \frac{1}{\rho}\int_0^R \partial\left(r\mu\frac{\partial u}{\partial r}\right) = \frac{1}{\rho}\int_0^R \partial(r\tau) = \frac{R\tau_w}{\rho} \quad (2)$$

where τ – friction, τ_w – wall friction, μ – dynamic viscosity, R – radius of the cylindrical channel.

We use formulas (2) and write the equation (1) in the form (slanting braces at medium speed hereinafter omitted)

$$\frac{R^2}{2}\frac{du}{dt} = -\frac{gR^2}{2} - \frac{R^2}{2\rho}\frac{dp}{dx} + \frac{R\tau_w}{\rho}$$

or

$$\frac{du}{dt} = -g - \frac{1}{\rho}\frac{dp}{dx} + \frac{2\tau_w}{R\rho} \quad (3)$$

Select in the cylindrical channel the volume of liquid by two cross sections at a distance ℓ. Write down the balance of pressure and friction

$$\Delta p \pi R^2 = 2\pi R\ell\tau.$$

We obtain the following expression

$$\Delta p = \frac{2\ell\tau_w}{R} \quad (4)$$

where λ – drag coefficient of friction.

Then, we use the equation of Darcy-Weisbach

$$\Delta p = \lambda\frac{\ell}{2R}\frac{\rho u^2}{2},$$

uniting the equation of Darcy-Weisbach and equation (4)

$$\lambda\frac{\ell}{2R}\frac{\rho u^2}{2} = \frac{2\ell\tau_w}{R},$$

where we find

$$\tau_w = \frac{\lambda\rho u^2}{8} \quad (5)$$

We substitute (5) into equation (3) and obtain the following expression

$$\frac{du}{dt} = -g - \frac{1}{\rho}\frac{dp}{dx} + \frac{\lambda u^2}{4R} \quad (6)$$

We calculate the pressure gradient of the following conditions: pressure decreases linearly from the boost pressure p_0 over the free surface of the fuel to the pressure $p_0 + \rho gH_0$. The pressure gradient is

$$\frac{dp}{dx} = -\frac{\rho g(H_0 - Vt)}{\xi} \quad (7)$$

where ξ – coordinate of free fuel surface in the measuring

channel.

In view of the formula (7) the equation of motion (6) takes the form

$$\frac{d^2\xi}{dt^2} = -g + \frac{g(H_0 - Vt)}{\xi} + \frac{\lambda}{4R}\left(\frac{d\xi}{dt}\right)^2 \quad (8)$$

we are adding initial conditions and obtain the Cauchy problem

$$t = 0, \xi = H_0, \frac{d\xi}{dt} = 0 \quad (9)$$

4. Results

Performed numerical solution of the problem: $2R = 0,078\,\text{m}$, $H_0 = 8,2\,\text{m}$, $g = 9,8\,\text{m}/\text{s}^2$, $\lambda = 4,83\cdot10^{-2}$, $V = 0,039\,\text{m}/\text{s}$. The solution of the boundary value problem is given numerically in package Mathcad applications. Results of the solution are shown in the graphs (Figure 2 - Figure 4)

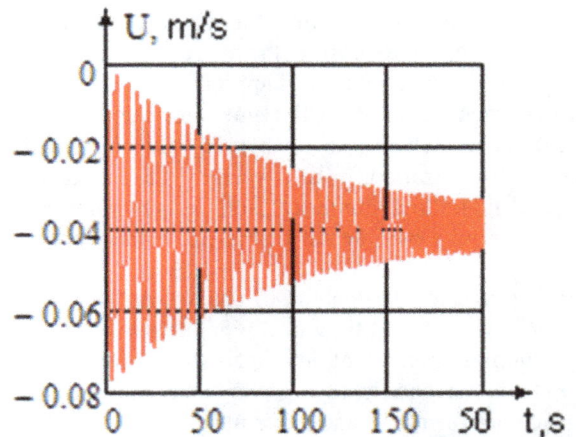

Figure 2. Changing the average fluid velocity in the channel.

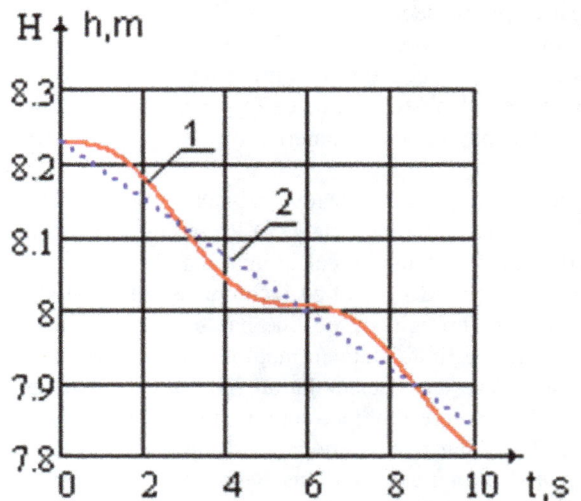

Figure 3. Level of liquid: 1- in the channel, 2- in the tank.

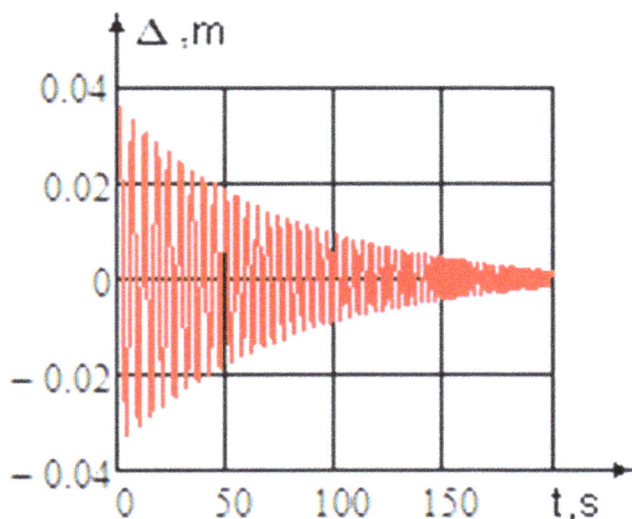

Figure 4. *Error in the measurement the liquid level in the range* $0 \leq t \leq 200\,s$, $\Delta(t) = \xi(t) - H(t)$.

5. Discussion

We can see (Figure 2), that the average velocity of the fluid in the channel has a decreasing amplitude oscillations. The liquid level in the cylindrical channel fluctuates relative to the liquid level in the fuel tank (Figure 3). The cause of the oscillation is the inertia of the liquid column in the channel. At the initial time the liquid has a very great acceleration, that causes a great inertia force of the. Consequently, the force of inertia is the cause fluctuations of the liquid in the channel. Summand in the equation of motion, which determines the oscillation process, is a pressure gradient.

Fluctuations have a variable period. Mathematical experiment shows that the period of oscillation decreases with time: in the time interval $0 \leq t \leq 10\,s$ period of oscillations $T \approx 5,712\,s$; in the time interval $50 \leq t \leq 60\,s$ period of oscillation $T \approx 4,986\,s$. On the flow characteristics significantly affected by acceleration, so for $g = 20\,m/s^2$ and $\lambda = 4,83 \cdot 10^{-2}$ in the range $0 \leq t \leq 10s$ the oscillation amplitude is reduced to $a \approx 0,026\,m$, and the period is reduced to $T \approx 3,97\,s$. In comparison with the acceleration channel resistance effect was insignificant, so for $\lambda = 9,7 \cdot 10^{-2}$ and $g = 9,8\,m/s^2$ in the range $0 \leq t \leq 10s$ oscillation amplitude is practically unchanged $a \approx 0,038\,m$ and the period of oscillations is reduced to $T \approx 5,56\,s$.

The graph (Figure 4) shows the error in measuring the liquid level versus time. We can see that the magnitude of the error is a periodic function, in which the amplitude and period of oscillations decreases with time. The maximum error in determining the level of fuel in the tank is observed in the beginning of the flight of a rocket and reaches the value $\Delta = 0,035\,m$. Error in measuring the liquid level is not regular, making it difficult to adjustment errors in software.

Note that the sensor may operate as in the positive and as in the negative deviation from the actual position of the fuel level in the tank. In this case, the error is doubled. Given the size of the diameter of the rocket, we are talking about hundreds of fuel kilograms.

6. Conclusion

The results suggest that the level sensors may provide false information, depending on the site of attachment. The error is not systematic. Experimental research is time-consuming and expensive. Therefore, the mathematical experiment is the most efficient method of solving the problem. The task is to signal from the sensor corresponded to the actual level of the liquid. For this purpose, the sensors must be installed at the points of intersection of the functions $H(t)$ and $\xi(t)$.

It should be noted that the mathematical model of fluid flow in the vertical channel of management systems do not take into account fluctuations of liquid in the fuel tank rocket and some rocket design features. Such a study will be done in the next article.

References

[1]   Slezkin N. A. Dynamics of viscous incompressible fluid / A. T. Slezkin. M.: Gostekhizdat, 1955. 520 pp. (in Russ.)

[2]   Loytsyansky L.G. Fluid Mechanics. M. Science, 1970. 904 pp. (in Russ.)

[3]   Popov D. N. Unsteady hydromechanical processes. M. G. Machinery, 1982. 240 pp. (in Russ.)

[4]   Babe G. D., Bondarev E. A., Voevodin A. F., Kanibolotsky M. A. Identification of hydraulic models. Novosibirsk. Science. 1980, 160 pp. (in Russ.)

[5]   Ozawa, K., Shimada, T. Flight performance simulations of vertical launched sounding rockets using altering-intensity swirling-oxidizer-flow-type hybrid motors. 51st AIAA/SAE/ASEE Joint Propulsion Conference. 2015. Orlando, United States. 2015, 21 p.

[6]   Liu, Z., Feng, Q. Numerical analysis of gas pulsation attenuation characteristics of a perforated tube in a reciprocating compressor piping system. Proceedings of the Institution of Mechanical Engineers, Part A: Journal of Power and Energy. V. 230, Issue 1, 1 February 2016, pp. 99-111.

[7]   Majd, A., Ahmadi, A. Keramat, A. Investigation of non-Newtonian fluid effects during transient flows in a pipeline. Strojniski Vestnik/Journal of Mechanical Engineering. V. 62, Issue 2, 2016, pp. 105-115.

[8]   Wang, C., Yang, J. D. Water hammer simulation using explicit-implicit coupling methods. Journal of Hydraulic Engineering. V. 141, Issue 4. Article number 04014086.

[9]   Avramenko, A. A., Tyrinov, A. I., Shevchuk, I. V. An analytical and numerical study on the start-up flow of slightly rarefied gases in a parallel-plate channel and a pipe. Physics of Fluids. V. 27, Issue 4, 2015, Article number 1.4916621.

[10]   Noorani, A., Schlatter, P. Evidence of sublaminar drag naturally occurring in a curved pipe. Physics of Fluids V. 27, Issue 3. 2015, Article number 1.4913850.

[11] Jafari Behbahani, T., Dahaghin, A., Behbahani, Z. J. Modeling of Flow of Crude Oil in a Circular Pipe Driven by Periodic Pressure Variations. Energy Sources, Part A: Recovery, Utilization and Environmental Effects. V. 37, Issue 13, 2015, pp. 1406-1414.

[12] Simão, M., Mora-Rodriguez, J., Ramos, H. M. Mechanical interaction in pressurized pipe systems: Experiments and numerical models. Water (Switzerland). V.7, Issue 11, 2015, pp. 6321-6350.

[13] Korade, I , Virag, Z., Šavar, M. Numerical simulation of one-dimensional flow in elastic and viscoelastic branching tube. 11th World Congress on Computational Mechanics, WCCM 5th European Conference on Computational Mechanics, ECCM 2014 and 6th European Conference on Computational Fluid Dynamics, ECFD 2014. 2014, pp. 7124-7131.

14

CubeSail Displaced Orbit Design for Near Earth Object Observation

Yang Yang[1], Xiaokui Yue[1], Yong Li[3], Andrew G. Dempster[2], Chris. Rizos[3]

[1]School of Astronautics, Northwestern Polytechnical University, Xi'an, China
[2]School of Electrical Engineering & Telecommunications, University of New South Wales, Sydney, Australia
[3]School of Civil & Environmental Engineering, University of New South Wales, Sydney, Australia

Email address:
yiyinfeixiong@gmail.com (Yang Yang)

Abstract: Microsatellites known as "CubeSats" have recently been developed to enable comparatively inexpensive and timely access to space for small payloads. As a new standard for small satellites, the CubeSat has shown great promise for space applications such as earth observation, planetary science and space physics mission. In this paper a "CubeSail" mission – a CubeSat deployed with a solar sail –for near earth object (NEO) observation is introduced. It is important to observe a NEO which may intersect or pass close to earth space before instigating any procedure for hazard avoidance. Furthermore, close observation of NEO may also be important for exploiting the new resources and exploring new living environment in outer space. This paper describes the concept of a large numbers of CubeSails deployed in the vicinity of the NEO for observation purposes. The dynamic model of the NEO-centreddis placed orbit in space is analysed. The solar radiation pressure on the sail can be utilised as propulsion to compensate for third body gravitational perturbation. To maintain the relative motion/position between a CubeSail and the NEO, periodic initial conditions are searched, which also must satisfy some observation mission constraints. A simulation study is carried out using the near earth asteroid Apophis 99942, discovered in recent years.

Keywords: CubeSail, Multi-Object Global Optimisation, Relative Motion, Apophis 99942

1. Introduction

The original concept of a "CubeSat" was proposed by Bob Twiggs of Stanford University, California, USA (currently at Morehead State University, Kentucky, USA) and Jordi Puig-Suari of California Polytechnic University at San Luis Obispo, California, USA in 1999[1]. The CubeSat's low cost, short development period and global user community, combined with its value as a teaching tool, has made this a new standard for small satellites. CubeSat, as the name implies, is a cube-shaped satellite with a 10 cm×10 cm×10 cm volume having a mass no greater than 1 kg. This standard size is termed 1 unit (U). CubeSats are often configured in 1.5U, 2U, 3U and 6U sizes, maintaining the 1U shape[2]. In recent years CubeSats have attracted great interest for space applications such as Earth observations, planetary science and space physics missions. QB50 is a proposed international space network of 50 CubeSats for multi-point, in-situ measurements in the lower thermosphere and for re-entry research[3]. And the Surrey Space Centre is currently developing a nano-solar sail mission based on the 3U CubeSat standard[4]. One of the primary objectives of the "CubeSail" mission is to demonstrate deployment of a 25 m² solar sail for one year deployment in a low Earth orbit (LEO). NanoSail-D, a recent NASA mission that demonstrated successful deployment of a solar sail in LEO, is similar to the CubeSail project[5]. Both make use of the 3U CubeSat platform, and both have four-quadrant square sails supported by metallic booms. What's more, ideas of launching CubeSats into deep space and "beyond low earth orbit" orbits have been proposed more recently. Pergola introduced an innovative space mission devoted to the survey of the small Earth companion asteroid by means of nano platforms[6].

A near Earth object (NEO) is an asteroid or comet with an orbit that intersects or passes near that of the Earth. The growing interest in these objects has led to an increasing number of missions to study NEOs, such as the sample return mission Hayabusa of JAXA[7], impact or mission Deep Impact of NASA[8] and possible deflector demonstrator mission Don Quixote of ESA[9]. Also many research work has been focusing on redirecting or deflecting the orbit of

NEOs to avoid their impact on our Earth[10-12].

The concept of multiple CubeSats with very large solar sails orbiting in the vicinity of NEOs is presented in this paper. A NEO-centred displaced orbit[13, 14] chosen for the CubeSats in deep space is analysed. As periodicity of motion is a prerequisite to minimise the use of propulsion for the relative motion maintenance and control, the initial conditions are searched using a multi-object goal optimisation algorithm. The perturbation due to solar radiation pressure and third body gravity effects on a CubeSail relative to the NEO is compared. Simulations were conducted using the Apophis 99942 near Earth asteroid as an example NEO.

2. Displaced Orbit for CubeSail

The non-Keplerian orbit displaced near a planet was researched by McInnes and co-workers, and their results are summarised in the paper [13]. In our case, a NEO-centred displaced orbit is chosen for the CubeSail observation mission, as shown in Figure 1.Such a displaced orbit can be maintained by suitably orienting the thrust direction so as to balance the centrifugal component of spacecraft acceleration.

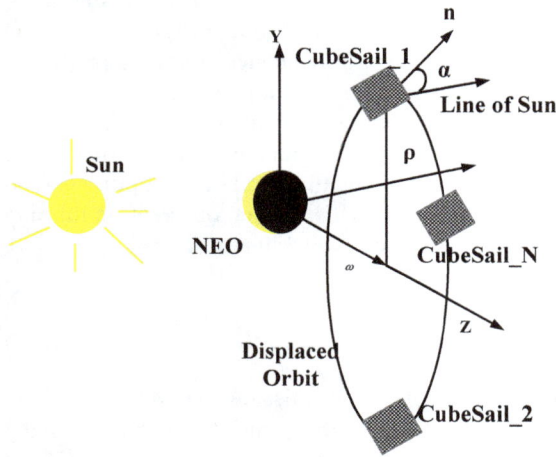

Fig. 1. NEO-centred displaced orbit[13].

Firstly, the motion of the displaced orbit should be modelled for the CubeSail. In order to investigate the relative motion between the CubeSail and the NEO, the Hill frame is used: originating on the NEO, x-axis directed towards the radius vector of the NEO, z-axis pointing along the angular velocity of the NEO, and the y-axis completing the right hand orthogonal triad (see Figure 2). In Figure 2, the r_A r_{sc} are the absolute radius of the NEO and the spacecraft from the centre of the Sun respectively. ρ is the relative radius vector.

In the Hill frame, the orbital mechanics for the CubeSail with respect to the NEO can be described (see Eq. (1)). The multi-body gravity effects and the solar radiation pressure perturbations are neglected in the initial study for simplicity, assuming that the CubeSail's motion is only influenced by the Sun. In the following section, however, the idea of

utilising the negative solar radiation pressure perturbation as positive thrust to compensate for the third body gravity effect of the NEO will be discussed.

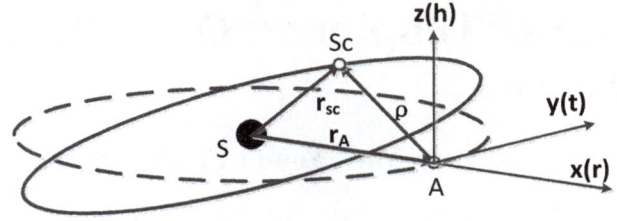

Fig. 2. The Hill coordinate system for relative motion.

$$\ddot{x} = 2\dot{f}_A\left(\dot{y} - y\frac{\dot{r}_A}{r_A}\right) + x\dot{f}_A^2 + \frac{\mu_S}{r_A^2} - \frac{\mu_S}{r_{sc}^3}(r_A + x) + u_x$$

$$\ddot{y} = -2\dot{f}_A\left(\dot{x} - x\frac{\dot{r}_A}{r_A}\right) + y\dot{f}_A^2 - \frac{\mu_S}{r_{sc}^3}y + u_y \qquad (1)$$

$$\ddot{z} = -\frac{\mu_S}{r_{sc}^3}z + u_z$$

Where f_A is the true anomaly; r_A is the radius of the NEO in the sun-centred coordinate system; μ_S is the gravity coefficient of the sun; r_{sc} is the distance between NEO and the CubeSail; $[\mu_x;\mu_y;\mu_z]$ is the thrust acting on the CubeSail.

3. Initial Conditions for Periodic Relative Motion

Since the CubeSats are of small size and relatively simple structure, there is limited capacity for complex payloads. Hence multiple CubeSats are proposed, working together to fulfil the observation mission. Several considerations should be taken into account in the mission design. Firstly, the distance between the CubeSail and the NEO should be as short as possible for better imaging or sensing. Secondly, the radius of the displaced orbit should be large enough in order to avoid possible collisions between multiple CubeSails. Finally, the CubeSails should probably fly out of the Hill sphere of the NEO to escape the gravity effect of the NEO. The limitation conditions are formulated as shown in Eq. (2) – Eq. (4):

$$\min J_{D1} = \rho_T \qquad (2)$$

$$\min J_{D2} = -\sqrt{x_T^2 + z_T^2} \qquad (3)$$

$$C_{ineq} = \rho_T - r_{\lim} > 0, \quad r_{\lim} = a_A(1 - e_A)\left(\frac{m_A}{m_S}\right)^{2/5} \qquad (4)$$

Where $r_{\lim}$ is the radius of the NEO's Hill sphere, inside of which the NEO dominates the attraction of the satellites; a_A, e_A are the semi-major axis and the eccentricity of the NEO orbit; m_A, m_S are the mass of the NEO and the Sun, respectively. For the Apophis 99942 near Earth asteroid, $r_{lim} \approx 1.5592 \times 10^3 m$. Three equations mentioned above are necessary to be considered in the initial value search for the dynamic equations (Eq. (1)) of the CubeSail.

Above all, the initial conditions should satisfy the periodicity so as to ensure that fuel for the formation maintenance is conserved, hence extending the life of the CubeSat. The formula for the periodicity limitation is:

$$\min J_{D0} = \sqrt{(x_T - x_0)^2 + (y_T - y_0)^2 + (z_T - z_0)^2} \quad (5)$$

In this paper, the Matlab® Optimization Tool is used to execute the above multi-objective optimisation, in which the solver adopts the Genetic Algorithm. A family of optimal results can be obtained as the Pareto results. One set of initial values is chosen according to Eq. (5) for one CubeSail in the displaced orbit:

$$X^{opt} = \begin{bmatrix} x_0 = -365.95m \\ y_0 = 663.20m \\ z_0 = 1511.52m \\ v_{x0} = 1.9257 \times 10^{-4} m/s \\ v_{y0} = 1.7214 \times 10^{-4} m/s \\ v_{z0} = -1.3970 \times 10^{-4} m/s \end{bmatrix} \quad (6)$$

4. Solar Radiation Pressure for Thrust Compensation

When considering the third body gravity effect from the NEO, extra propulsion should be added to maintain the CubeSat orbit. The solar radiation pressure on the solar sail of the CubeSat can provide some valid thrust. Assume a spherical and homogenous gravity field for the asteroid, a more precise relative dynamics equation is:

$$u_{pert} = \mu_A \left(\frac{\rho}{\rho^3} - \frac{T_I^H r_A^I}{r_A^3} \right) + \frac{F_{SRP}}{m_A} \quad (7)$$

Here $\rho = \sqrt{x^2 + y^2 + z^2}$ is the relative distance between the CubeSail and the NEO; μ_A is the gravity constant of the NEO; and $F_s = \begin{bmatrix} F_{SRPx} & F_{SRPy} & F_{SRPz} \end{bmatrix}^T$ is the force vector due to the solar radiation pressure on the sail. The transformation matrix between the initial and Hill frames is:

$$T_I^H = R_3(\omega_A + f_A) R_1(I_A) R_3(\Omega_A) \quad (8)$$

where the R_1, R_3 denote the rotation matrix along the x-axis and z-axis, respectively. Some artificial equilibrium points exist where both the acceleration and the velocity are zero, namely $\begin{bmatrix} \dot{x} & \dot{y} & \dot{z} \end{bmatrix}^T = 0_{3\times1}, \begin{bmatrix} \ddot{x} & \ddot{y} & \ddot{z} \end{bmatrix}^T = 0_{3\times1}$, and obtained by a search in the solution domain[15]. An important

parameter of the CubeSail design is the area of the solar sail. Currently a $5m \times 5m$ CubeSail has been developed by the Surrey Space Centre and has completed testing for ground deployment[3]. For the simulations described in this paper, this value for the sail area is used. The force due to the solar radiation pressure is given by the following model. In order to observe the NEO, the CubeSail is assumed to point towards the NEO, which leads to an angle α between the Sun light vector los and the solar radiation force direction n (see Figure 3). The force is expressed in Eq. (9).

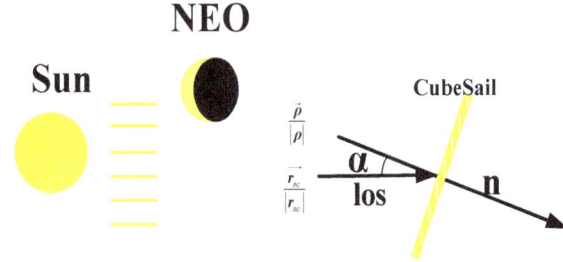

Fig. 3. The sun light vector on the solar sail.

$$F_{SRP} = \sigma_S P_{SR} A \cos^2 \alpha n$$

$$P_{SR} = \frac{P_0}{c} \left(\frac{A_u}{\rho} \right)^2$$

$$\alpha = \pi - \arcsin \frac{\rho \cdot r_{sc}}{|\rho| \cdot |r_{sc}|} \quad (9)$$

$$n = \frac{\rho}{|\rho|}$$

where σ_S is the efficiency of the solar sail; P_{SR} is the radiation pressure at the CubeSail surface (N/m^2); A is the Sun-facing cross-sectional area (m^2); A_u is the astronomical unit ($1.4959787 \times 10^{11} m$); P_0 is the intensity of solar radiation (w/m^2); c is the speed of light (m/s).

5. Simulation and Results

Apophis 99942 is a near Earth asteroid discovered in 2004, which will pass close to the Earth in 2029, may pose subsequent impact threats in 2036, 2037 or later[16]. This paper will discuss the challenge of Apophis observation using CubeSail concept. The orbital parameters of the Apophis are listed in the Table 1.

Table 1. Estimated and observed orbital parameters of Apophis 99942[17].

Orbital Elements		Measurements
Semi-major axis	a	0.922300 AU
Eccentricity	e	0.191076
Inclination	i	3.33196°
Right Argument of Ascending	Ω	204.430410°
Argument of periapsis	ω	126.424477°
Period	T	323.596917d
Mean motion	n	1.112745°/d

According to the initial conditions in Eq. (6), the CubeSail orbit is propagated to verify the periodicity of the relative motion. The initial epoch for the Appophis is 13:37:00 on 13 April, 2031, which is 5 years ahead of the potential impact. The simulation time lasts for 5 orbital periods of the Apophis NEO. The relative motion is illustrated in Figure 4. It is periodic, closed and bounded.

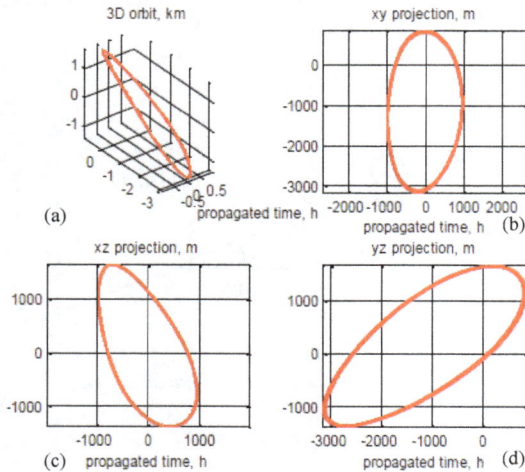

Fig. 4. *Relative orbit between the CubeSail and the NEO.*

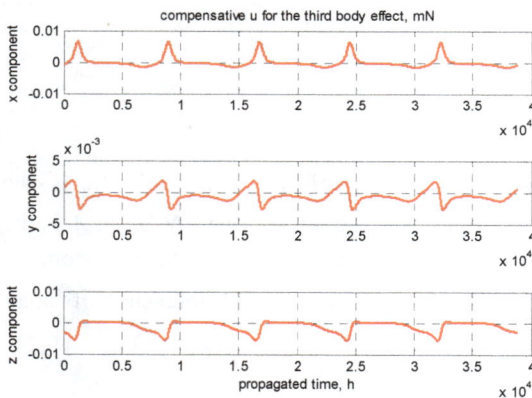

Fig. 5. *Third body gravity perturbations due to the NEO.*

As discussed above, third body gravity perturbations on the CubeSails cannot be ignored, as they will damage the periodicity of the relative motion and cause the orbit to drift. Figure 5 shows the extra thrust needed for the third body gravity perturbations from the NEO for the three axes with the order of 1mN. For the CubeSails, the solar radiation pressure can be used as positive continuous thrust to compensate for the third body gravity perturbations. Using Eq. (9), the accelerations induced by the solar radiation pressure can be calculated. The components of solar radiation pressure acceleration are shown in Figure 6, from which one can see that the solar radiation pressure is one order of magnitude greater than the third body gravity perturbations, indicating that it is possible to compensate for these perturbations if the spacecraft attitude (relative to the Sun)

are appropriately selected and controlled. The total disturbing forces in the three directions are shown in Figure 6.

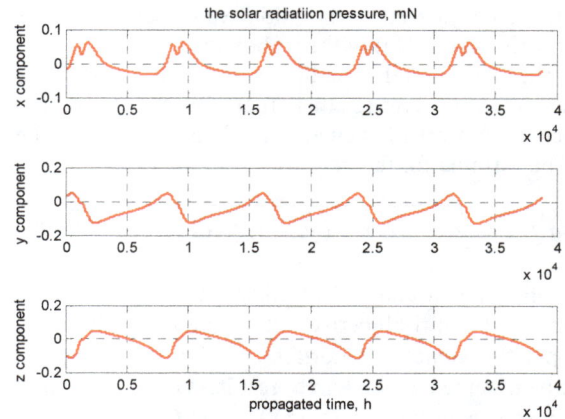

Fig. 6. *The solar radiation pressure perturbation.*

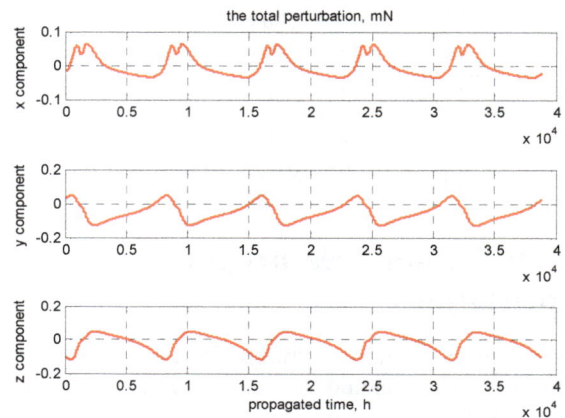

Fig. 7. *The total perturbation.*

6. Concluding Remarks

This paper presented a concept of NEO observation using a proposed Cube Sail spacecraft flying near the NEO. The NEO-centred displaced orbit mechanics was analysed. The initial conditions were searched using a multi-object goal genetic optimisation algorithm. A precise dynamic model for the Cube Sail relative to the NEO was defined, and the solar radiation pressure and the third body gravity perturbations were discussed. Utilising the solar radiation pressure as a form of propulsion for CubeSats shows great potential.

Acknowledgement

China Scholarship Council (CSC) is gratefully acknowledged for supporting the author's visiting period at the School of Surveying and Geospatial Engineering, the University of New South Wales. Also the author appreciates the help from the researchers in the Australian Centre for Space Engineering Research.

References

[1] Pang, A.S.-K. and B. Twiggs, Citizen Satellites. Scientific American, 2011. 304(2): p. 48-53.

[2] Shiroma, W., et al., CubeSats: A bright future for nanosatellites. Central European Journal of Engineering, 2011. 1(1): p. 9-15.

[3] Gill, E., et al. Formation Flying to Enhance the QB50 Space Network. 2010.

[4] Lappas, V., et al., CubeSail: A low cost CubeSat based solar sail demonstration mission. Advances in Space Research, 2011. 48(11): p. 1890-1901.

[5] Johnson, L., et al., Status of solar sail technology within NASA. Advances in Space Research, 2011. 48(11): p. 1687-1694.

[6] Pergola, P., Small satellite survey mission to the second Earth moon. Advances in Space Research, 2013. 52(9): p. 1622-1633.

[7] Kawaguchi, J.i., A. Fujiwara, and T. Uesugi, Hayabusa—Its technology and science accomplishment summary and Hayabusa-2. Acta Astronautica, 2008. 62(10–11): p. 639-647.

[8] A'Hearn, M., et al., Deep impact: excavating comet Tempel 1. Science, 2005. 310(5746): p. 258-264.

[9] Gálvez, A. and I. Carnelli, ESA'S DON QUIJOTE MISSION: AN OPPORTUNITY FOR THE INVESTIGATION OF AN ARTIFICIAL IMPACT CRATER ON AN ASTEROID. 2006.

[10] Eneev, T.M., R.Z. Akhmetshin, and G.B. Efimov, On the asteroid hazard. Cosmic Research, 2012. 50(2): p. 93-102.

[11] Gates, M. and L. Johnson, NASA's Asteroid Redirect Mission, in Handbook of Cosmic Hazards and Planetary Defense, F. Allahdadi and J.N. Pelton, Editors. 2014, Springer International Publishing. p. 1-7.

[12] Sugimoto, Y., et al., Hazardous near Earth asteroid mitigation campaign planning based on uncertain information on fundamental asteroid characteristics. Acta Astronautica, 2014. 103(0): p. 333-357.

[13] McInnes, C.R., Solar sail mission applications for non-Keplerian orbits. Acta Astronautica, 1999. 45(4-9): p. 567-575.

[14] McInnes, C.R., Passive Control of Displaced Solar Sail Orbit. Journal of Guidance, Control, and Dynamics, 1998. 21(6).

[15] Vasile, M., A multi-mirror solution for the deflection of dangerous NEOS. Communications in Nonlinear Science and Numerical Simulation, 2009. 14(12): p. 4139-4152.

[16] Ferraz-Mello, S. and J. Fernández, Potential impact detection for near-Earth asteroids: The case of 99942 Apophis (2004 MN4). 2005.

[17] Vasile, M., C.A. Maddock, and G. Radice. Mirror formation control in the vicinity of an asteroid. in AIAA/AAS Astrodynamics Specialist Conference and Exhibit. 2008. Honolulu, Hawaii.

15

Hands-on engineering education by construction and testing of models of sailing boats

Ahmed Farouk Abdel Gawad

Professor of Computational Fluid Mechanics, Mech. Eng. Dept., Umm Al-Qura Univ., Makkah, Saudi Arabia

Email address:

afaroukg@yahoo.com

Abstract: This paper introduces involvement of the hands-on learning method. According to the modern environment of technology, engineering students have to realize the multidisciplinary nature of engineering systems. This learning technique is essential to offer students the necessary skills to master practical, organizational and work-group cleverness. The work is concerned with the redesign, construction and operation of two models of sailing boats. The approach of the work and final outputs are illustrated.

Keywords: Hands-On Learning, Multidisciplinary Engineering, Sailing Boats, Laboratory Investigations

1. Present Project

The scheme of hands-on learning technique is established in this work though supervising B.Sc. graduation project. Students had to redesign, construct and test a sailing boat model. Sailing boats represent a very rich multidisciplinary teaching field. Investigation of sailing boats covers the areas of sail aerodynamics, boat hydrodynamics, control systems, boat stability, material choice, manufacturing techniques, design aspects, etc. Also, this type of projects increases the knowledge of students about marine activities and the different types of sailing boats.

The students were divided into two groups. The first group (four students) concerned the case of a mono-hull sailing boat. This type of boats is the widely known and used allover the world. The other group (six students) concerned a multi-hull (catamaran) boat. This type of boats has many advantages and practical applications.

The technique of dividing the students into two groups has some objectives, namely: (*i*) Inspiring competition between the two groups for better achievement. (*ii*) Motivating the cooperation between the two groups in the common issues of the work. Thus, students learn how to organize activities between working groups in the same field. (*iii*) Increasing the knowledge and experience about different types of sailing boats instead of concentration on one type only. (*iv*) Reducing the overall effort and time-needs of every student by relatively increasing the students' number.

2. Background

2.1. Sailing

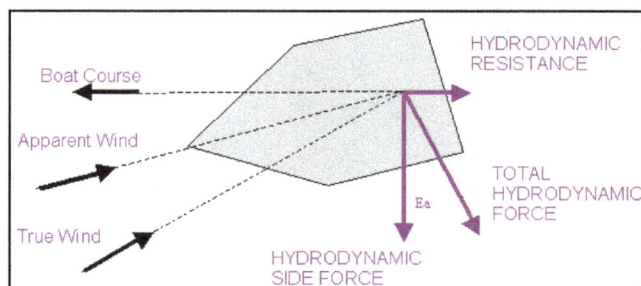

Fig 1. Forces on a sailing boat [1].

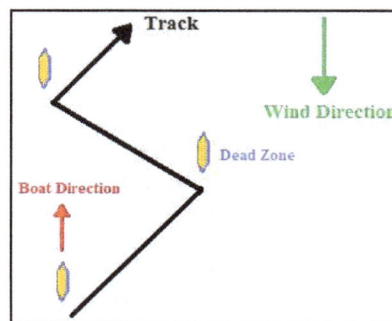

Fig 2. Sailing upwind.

Sailing is the skillful art of controlling the motion of a sailing ship or smaller boat, across a body of water. A boat moves as wind pushes on its sails. This is obvious when the boat is sailing downwind. The keel of a boat keeps it from strafing to the sides. This allows a boat to sail downwind but at an angle.

The force of the wind is used to create motion by using one or more sails, Fig. 1.

When sailing downwind (away from the wind source) the vessel's motion is derived from the simple force of the wind pushing the sail. When sailing upwind (towards the wind source), the movement of air over the sails acts in the same way as air moving over an aircraft's wing. Air flowing over the sail generates lift. This pulls the sail (and the boat) ahead, but also pushes it downwind rather strongly. A basic rule of sailing is that it is not possible to sail directly into the wind. Generally speaking, a boat can sail 45 degrees off the wind, Fig. 2. Since a boat cannot sail directly into the wind, but the destination is often upwind, one can only get there by sailing close-hauled with the wind coming.

2.2. Balance of Hull and Sails

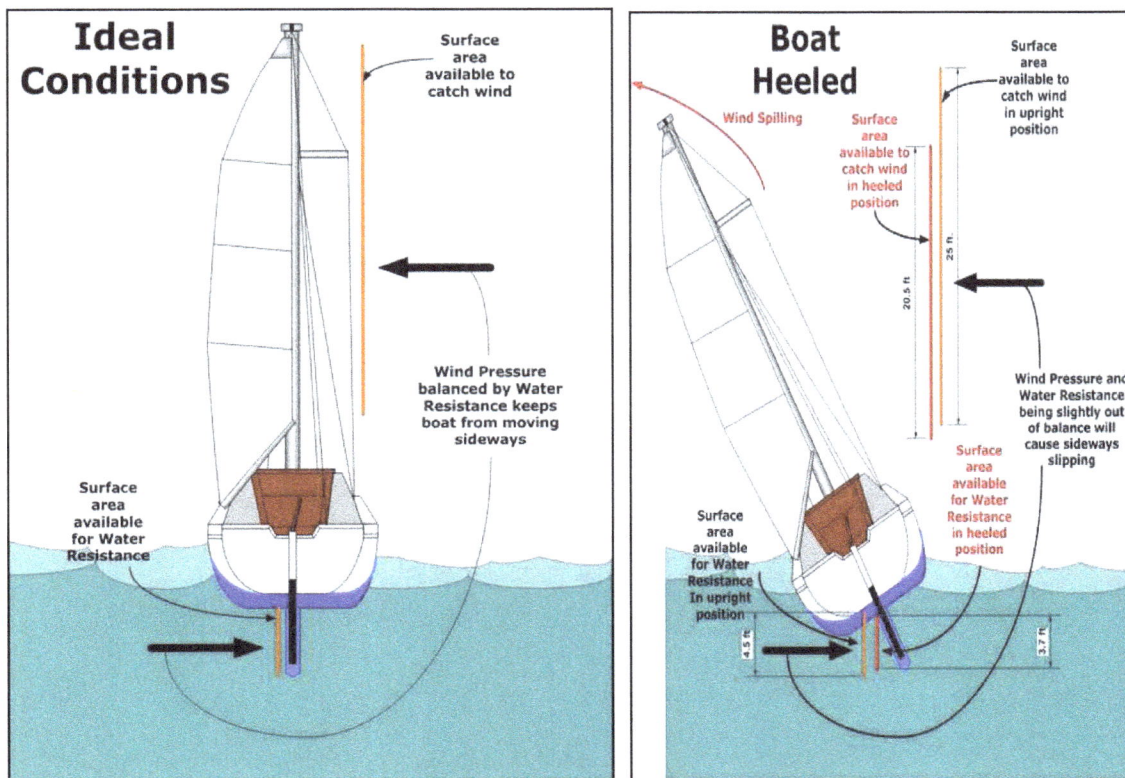

Fig 3. Balance of hull and sails [2].

Due to the pressure of the wind in the sails, a sailboat side-slips a little as it goes forward. This is called "making leeway." Since the water has to travel a greater distance on the windward side of the keel, an area of reduced pressure produces "lift" to windward. The more lift from the underwater surfaces, the less leeway the boat makes, Fig. 3.

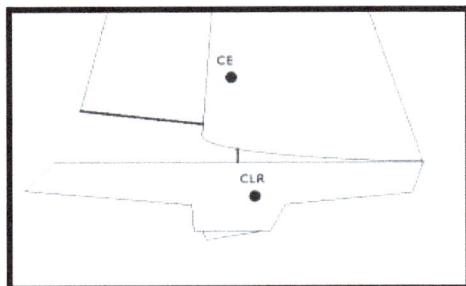

Fig 4. Center of forces (CE) and center of lateral resistance (CLR) [3].

The *CE* of the boat is the "Center" of all the forces acting to push the boat sideways against the center of all the forces resisting that push. The *CLR* is the "Center of Lateral Resistance" of the hull shape, Fig. 4.

3. Types of Boats

The boats can be classified according to rigs, meaning the way they set their sails, as in the following sections.

3.1. Single Rigs

3.1.1. Sloop

A sloop has one mast and two sails, a headsail (jib) and a mainsail. The sloop rig is the most popular rig for small and medium-size sailing craft because of its efficiency and simplicity [4], Fig. 5a.

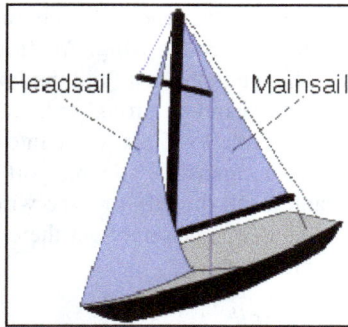

Fig 5a. Sloop boat [4].

3.1.2. Catboat

A catboat has one mast and one sail, with the mast usually stepped forward. Since there is no second sail on a catboat, it is a good choice for sailing shorthanded or with children. Cruising catboats have cabins and normally range in overall length from 5-10 meters. Others are fully or partially decked and suitable for day sailing or camp cruising [5], Fig. 5b.

Fig 5b. Catboats [5].

3.1.3. Sunfish (Lateen Rig)

Fig 5c. Sunfish (lateen rig) boat [6].

The Sunfish sailboat is a personal size, beach launched sailing dinghy utilizing a pontoon type hull carrying a lateen sail mounted to an un-stayed mast. Having a lateen sail with its simple two line rigging makes a Sunfish simple to learn sailing on and to set up. Upgrades can be added to enhance sail control for competitive sailing, making the boat attractive

to novice and experienced sailors [6], Fig. 5c.

3.1.4. Catamaran

A catamaran is a multi-hulled vessel consisting of two parallel hulls of equal size. A catamaran is geometry-stabilized, that is, it derives its stability from its wide beam, rather than having a ballasted keel like a mono-hull. Being ballast-free and lighter than a mono-hull, a catamaran can have a very shallow draught. The two hulls are much finer than a mono-hull's, the reduced drag allowing faster speeds. A sailing multi-hull heels much less than a sailing mono-hull, so its sails spill less wind and are more efficient. The limited heeling means the ride may be more comfortable for passengers and crew although catamarans can exhibit an unsettling "hobby-horse" motion. A catamaran's two hulls are joined by some structure, the most basic being a frame. More sophisticated catamarans combine accommodation into the bridging superstructure [7], Fig. 5d.

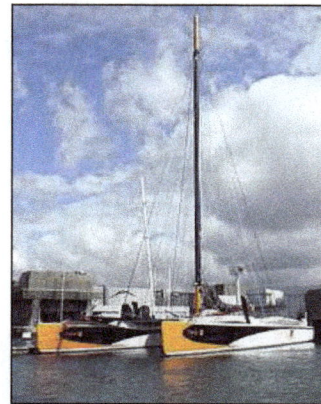

Fig 5d. Catamaran boat [7].

Fig 5. Single-rig boats.

3.2. Divided Rigs

3.2.1. Schooner

Fig 6a. Schooner boat [8].

A schooner is a type of sailing vessel with fore-and-aft sails on two or more masts, the foremast being no taller than the rear mast(s). Such vessels were first used by the Dutch in the 16th or 17th century. The most common type of schooners, with two-masts, were popular in trades that required speed and windward ability, such as slaving, blockade running, and fishing. Schooners were popular on both sides of the Atlantic in the late nineteenth and early twentieth centuries [8], Fig. 6a.

3.2.2. Ketch

Fig 6b. Ketch boat [9].

A ketch is a sailing craft with two masts, both rigged fore-and-aft: a mainmast and a shorter mizzen mast abaft the mainmast but forward of the rudder post. To assist going to windward, a ketch may carry one or more jibs or foresails. If a ketch has no jibs, it is called a cat ketch. The large fore-and-aft sail on the mainmast is the mainsail, while the sail on the mizzen mast is the mizzen. These sails may be any type of fore-and-aft sail, in any combination. Most modern ketches are Bermuda rigged, but other possible rigs on a ketch include gunter rigs and gaff rigs [9], Fig. 6b.

3.2.3. Yawl

A yawl is a two-mast sailing craft similar to a sloop or cutter but with an additional mast (mizzenmast or mizzen mast) located well aft of the main mast, often right on the transom, specifically aft of the rudder post. The yawl was originally developed as a rig for commercial fishing boats. In the 1950s and 60s, yawls were developed for ocean racing, to take advantage of the handicapping rule that did not penalize them for flying a mizzen staysail, which on long ocean races, often downwind, were a great advantage [10], Fig. 6c.

Fig 6c. Ketch boat [10].

Fig 6. Divided-rigs boats.

4. Present First Model (Mono-Hull Boat)

4.1. Description of the Boat Model

The first model represents a mono-hull boat. Its type is "PILGRIM 590" [11]. The overall shape and dimensions of the prototype (full-scale) are shown in Fig. 7.

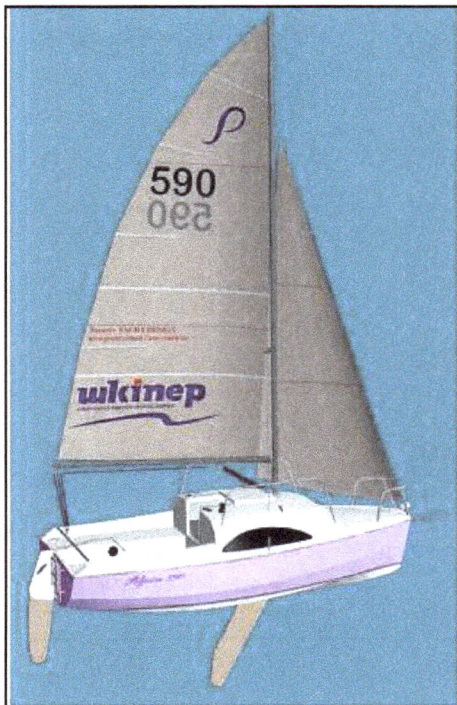

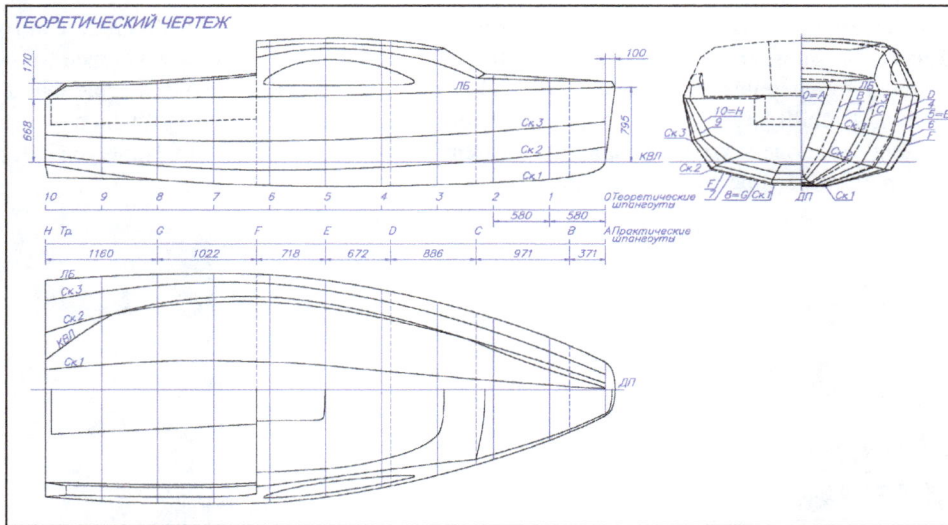

Fig 7. Overall shape and dimensions of the boat prototype [11].

The layout of the prototype is shown in Fig. 8.

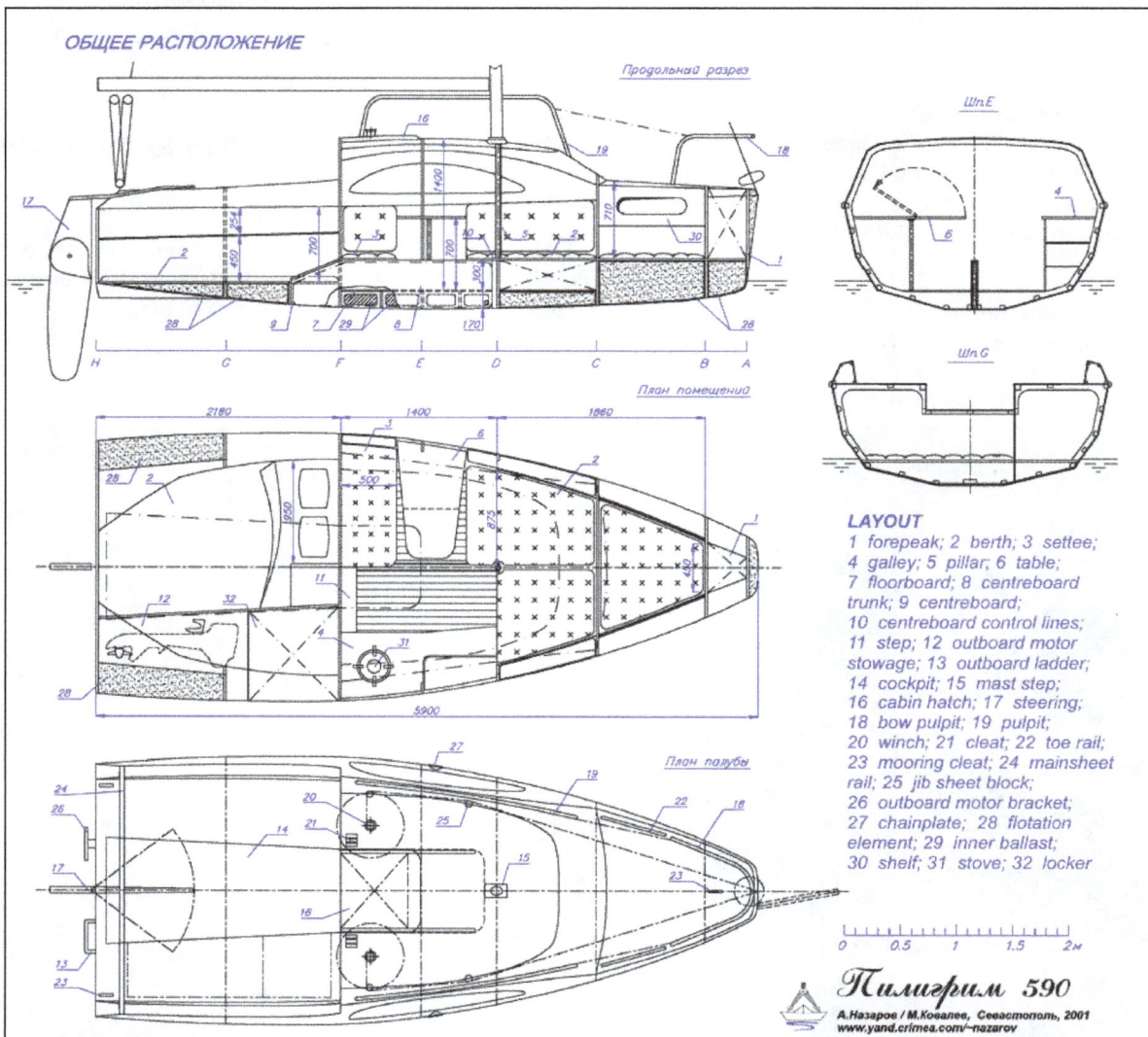

LAYOUT

1 forepeak; 2 berth; 3 settee;
4 galley; 5 pillar; 6 table;
7 floorboard; 8 centreboard
trunk; 9 centreboard;
10 centreboard control lines;
11 step; 12 outboard motor
stowage; 13 outboard ladder;
14 cockpit; 15 mast step;
16 cabin hatch; 17 steering;
18 bow pulpit; 19 pulpit;
20 winch; 21 cleat; 22 toe rail;
23 mooring cleat; 24 mainsheet
rail; 25 jib sheet block;
26 outboard motor bracket;
27 chainplate; 28 flotation
element; 29 inner ballast;
30 shelf; 31 stove; 32 locker

Fig 8. Layout of the boat prototype [11].

Table 1. List of the layout items of Fig. 8, [11].

Item No.	Description	Item No.	Description	Item No.	Description
1	Forepeak	12	Outboard	23	mooring
2	Berth	13	Outboard	24	mainsheet
3	Settee	14	Cockpit	25	Jib
4	Galley	15	Mast	26	outboard
5	Pillar	16	Cabin	27	Chainplate
6	Table	17	Steering	28	flotation
7	Floorboard	18	Bow	29	Inner
8	Centreboard	19	Pulpit	30	Shelf
9	Centreboard	20	Winch	31	Stove
10	Centreboard	21	Cleat	32	locker
11	Step	22	Toe		

The items that appear in Fig. 8 are listed in Table 1. The construction details and dimensions of the model are shown in Fig. 9. The items that appear in Fig. 9 are listed in Table 2.

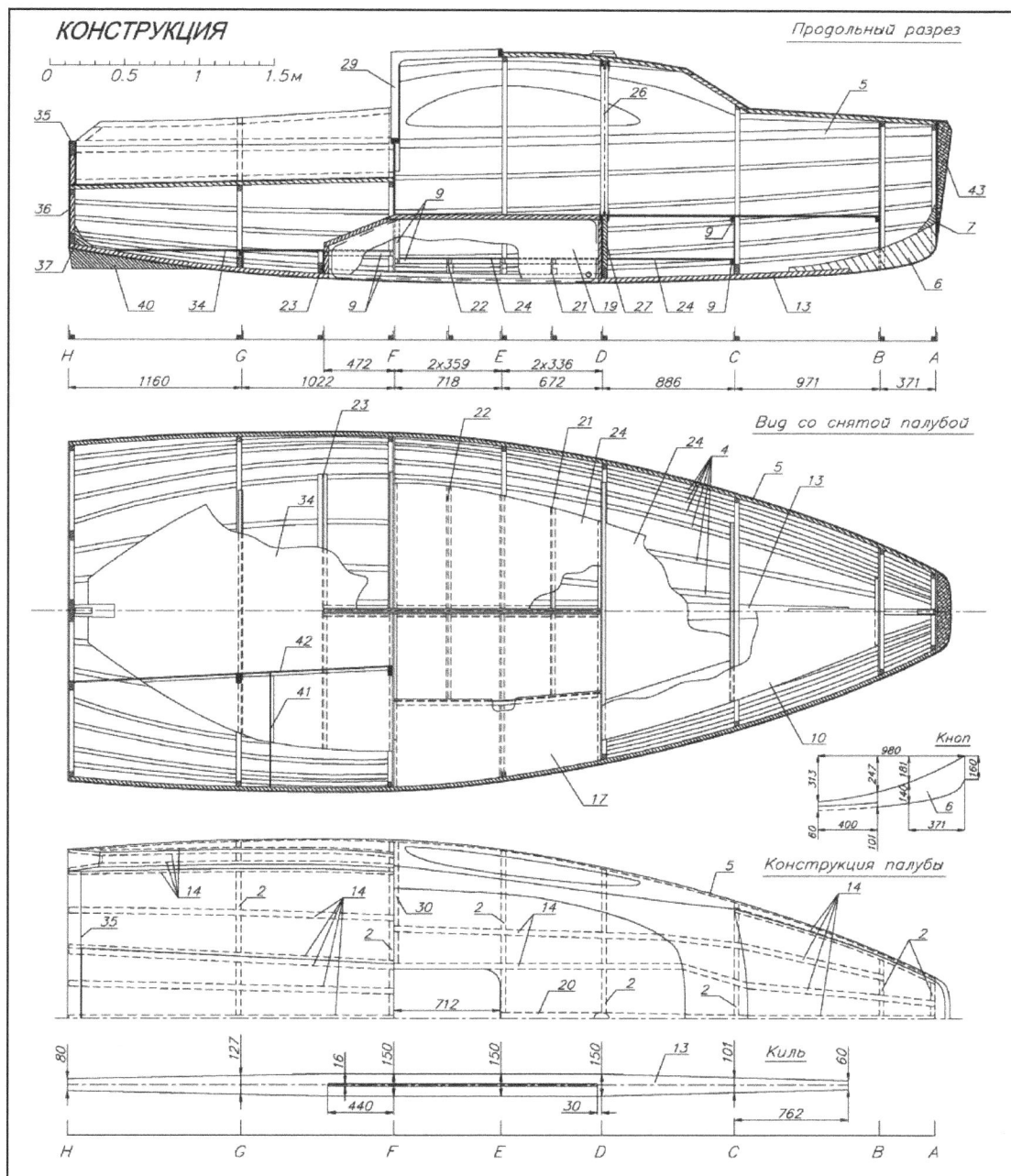

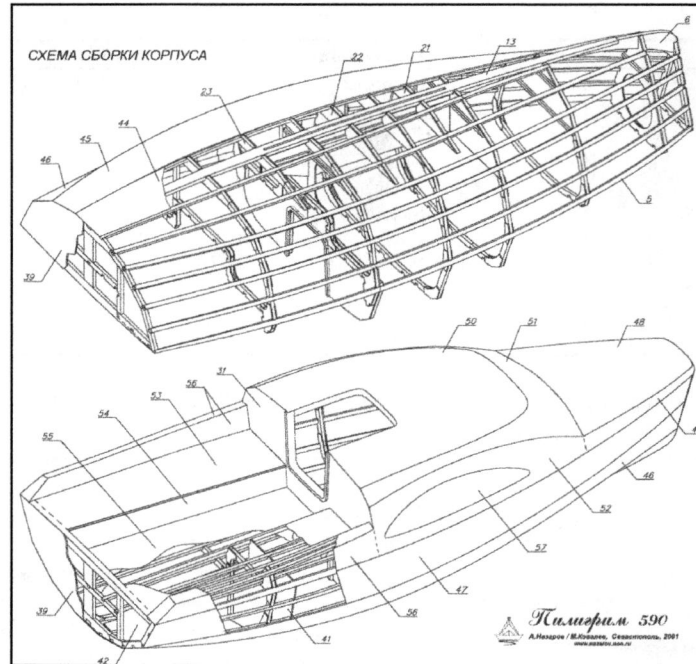

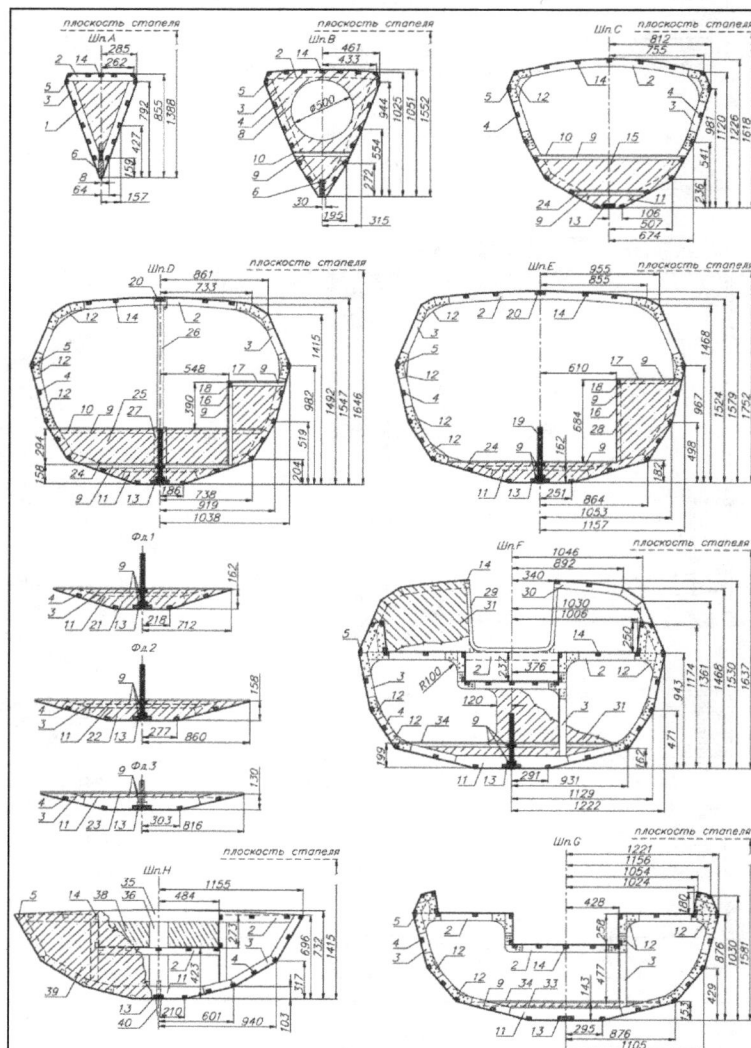

Fig 9. *Construction details and dimensions of the boat model [11].*

Table 2. *List of the layout items of Fig. 9, [11].*

Item No.	Description
1	bow transom (A), plywood, =10 mm
2	beam, pine, 30×60 mm
3	frame, pine, 30×60 mm
4	stringer, pine laminate, 25×40 mm
5	sheer clamp, oak, 25×60 mm
6	forefoot, oak laminate, =40 mm
7	bow transom knee, oak, =30 mm
8	bulkhead (B), plywood, =10 mm
9	plank, pine, 20×30 mm
10	berth plating, plywood, =8 mm
11	Floor timber, pine, 30×100 mm
12	knee (both sides), plywood, =6 mm
13	keel, oak laminate, =30 mm
14	carling, pine, 25×40 mm
15	bulkhead (C), plywood, =10 mm
16	galley side, plywood, =6 mm
17	galley table, plywood, =6 mm
18	plank, pine, 25×40 mm
19	centreboard trunk side, plywood, =8 mm
20	carling, pine 25×80 mm
21	floor, plywood, =10 mm
22	floor, plywood, =10 mm
23	floor, plywood, =10 mm
24	floorboard, plywood, =10 mm
25	bulkhead (D), plywood, =10 mm
26	pillar, steel tube, d=32 mm
27	pillar, pine, 30×100 mm
28	bulkhead (E), plywood, =10 mm
29	hatch framing, 30×40 mm
30	Half-beam, pine,30×60 mm
31	bulkhead (F), plywood, =10 mm
32	pillar, pine, 30×60 mm
33	bulkhead (G), plywood, =8 mm
34	berth plating, plywood, =8 mm
35	beam, =30 mm
36	pillar, pine, 30×150 mm
37	stern knee, oak, 30×120×120 mm
38	transom plating, plywood, =8 mm
39	transom (H), plywood, =10 mm
40	skeg, oak, =40 mm
41	bulkhead of outboard motor compartment, plywood, =8 mm
42	longitudinal bulkhead, plywood, =8 mm
43	bow piece, foam
44	bottom plating, plywood, =8 mm
45	bottom plating, plywood, =8 mm
46	chine plating, plywood, =6 mm
47	board plating, plywood, =6 mm
48	deck plating, plywood, =8 mm
49	deck chamfer, plywood, =6 mm
50	roof superstructure, plywood, =10 mm
51	superstructure coaming, plywood, =6 mm
52	superstructure coaming, plywood, =6 mm
53	cockpit seat, plywood, =8 mm
54	cockpit side, plywood, =6 mm
55	cockpit plating, plywood, =8 mm
56	cockpit coaming, plywood, =4 mm
57	window, plexiglass, =10 mm
58	centerboard, alloy or steel, =10 mm

4.2. Software Construction of the Boat Model

The students used a web software (FREE!ship, *Ver.* 2.6) that can be downloaded freely to reconstruct the boat model based on its construction details and dimensions [12].

FREE!ship was developed to offer an alternative to hull-form definition programs based on *NURB* (Non-Uniform Rational Basis Spline) surface modeling. Most hull modeling packages use these parametric spline surfaces, which can be very tricky to use. FREE!ship uses subdivision surfaces instead, which offer many advantages over NURB surfaces, such as: no need for a rectangular control grid divided into rows and columns; more freedom in modeling knuckle lines; surfaces can contain holes; even the most complex shapes can be created with just one surface; and the possibility to insert just one single control point [13].

The drawings of the students of the model boat can be seen in Fig. 10 (wire drawings) and Fig. 11 (solid drawings). The printouts of these drawings were used by the students to construct the real model of the boat. Figure 12 shows the students' assembly drawings of the boat model.

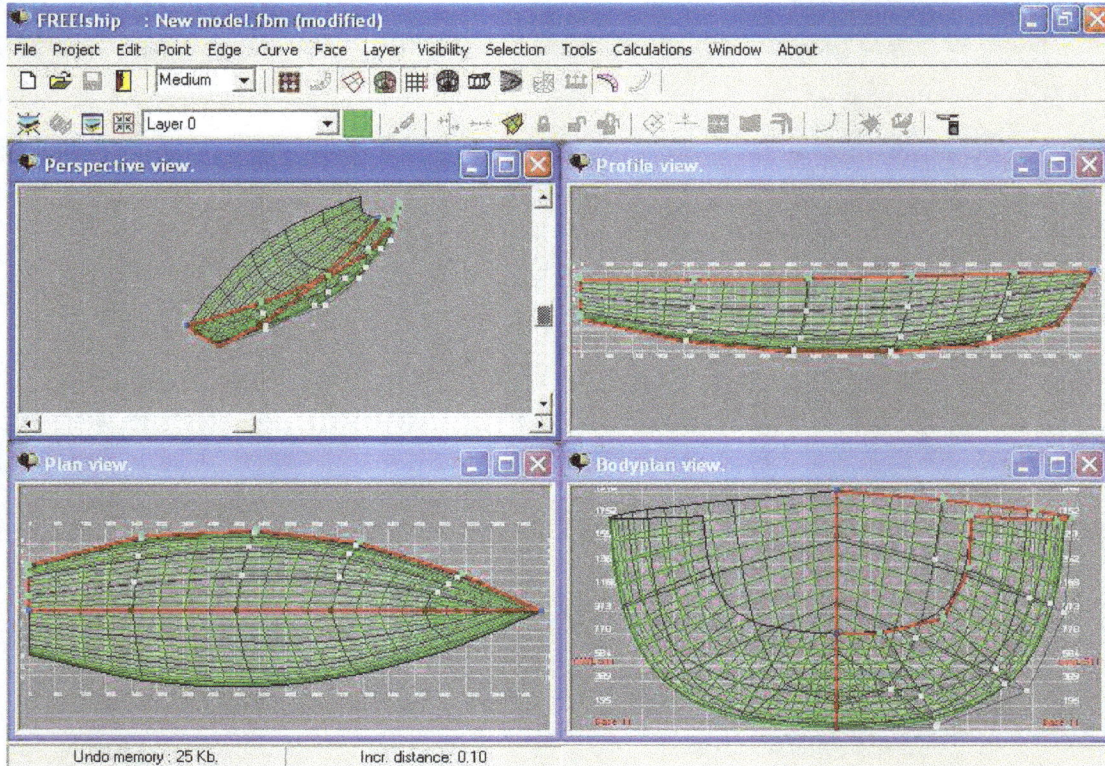

Fig 10. Students' wire drawings of the boat model.

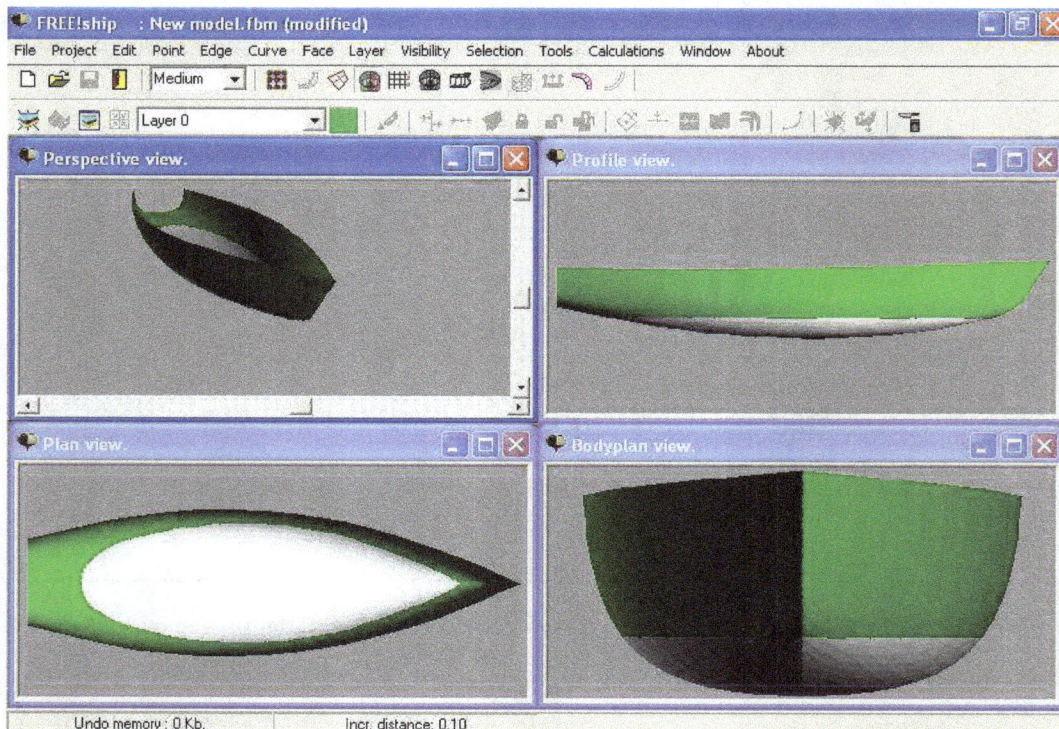

Fig 11. Students' solid drawings of the boat model.

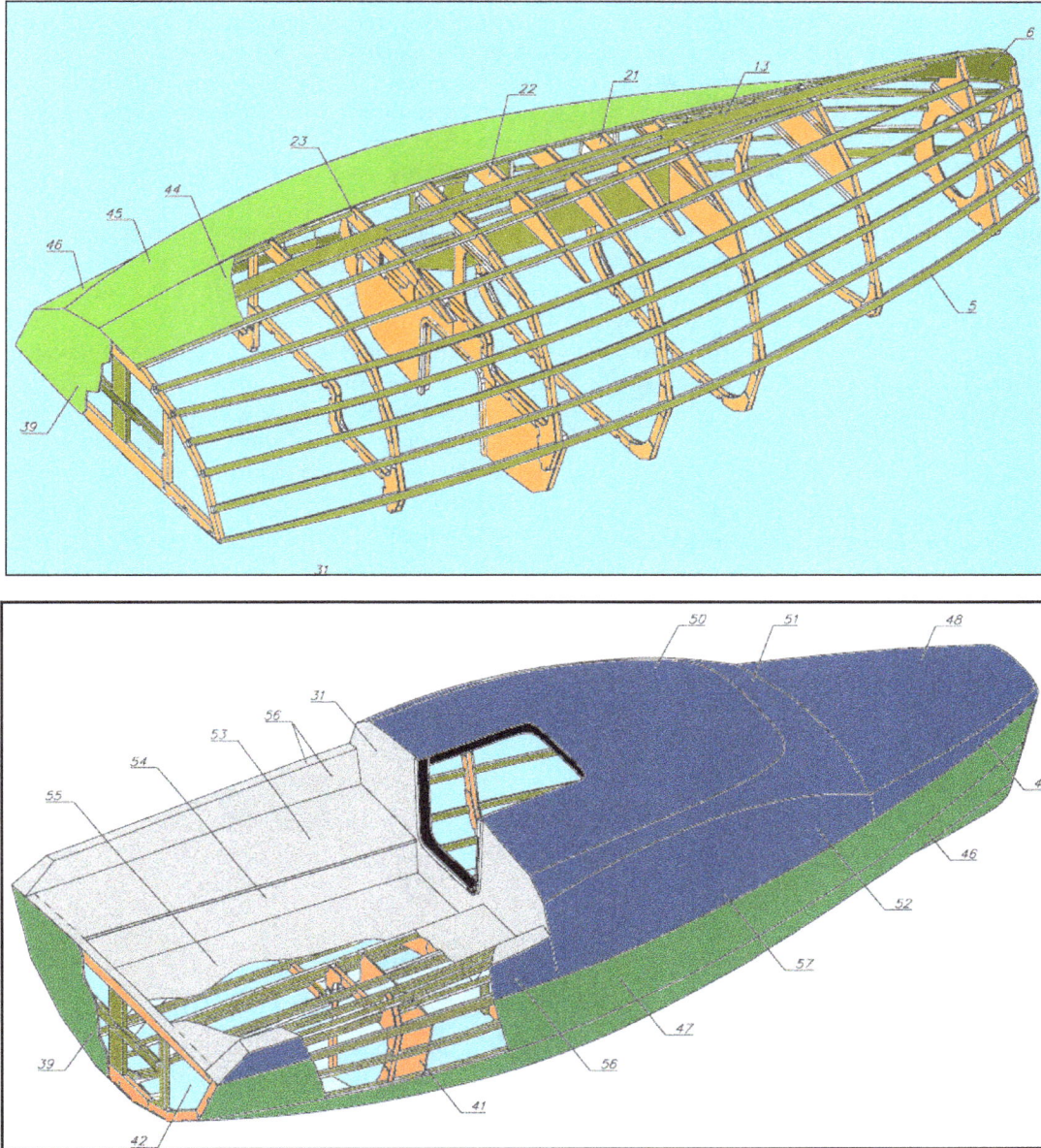

Fig 12. *Students' assembly drawings of the boat model.*

4.3. Construction of the Boat Model

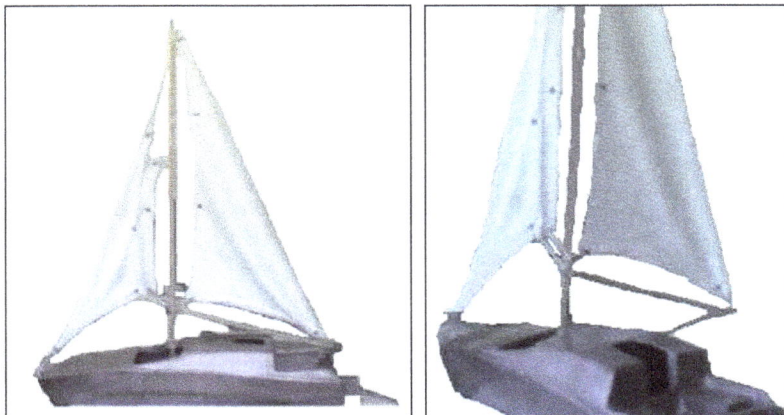

Fig 13. *Real model after fabrication.*

The dimensions of the model were taken with a scale of approximately 1:10 relative to the dimensions of the boat prototype. Thus, the overall length and maximum width of the model are 600 *mm* and 200 *mm*, respectively. The big sail has a height of 600 *mm* and a base of 320 *mm*. The small sail has a height of 400 *mm* and a base of 255 *mm*.

The model was totally constructed by the students from wood. Wooden strips were used to construct the main frame (Skelton) of the model with necessary accessories. Suitable pieces of plywood were used to cover the frame to complete the model body. Then, the model surface was cover by a water-resistant coating. Finally, the model was carefully painted. The two sails were made from fabric. Fig. 13 shows the real model after fabrication.

4.4. Control

A control system was used to direct the sail model by controlling the model rudder. The control system consists of a control circuit and a stepper motor.

4.4.1. Control Circuit

The control circuit is manually operated by the operator of the model through electrical wires. Figure 14 shows the main components of the control circuit. Theses main components can be summarized as:

1. Microcontroller:

A microcontroller is a computer-on-a-chip used to control electronic devices. It is a type of microprocessor emphasizing self-sufficiency and cost-effectiveness, in contrast to a general-purpose microprocessor such as the kind used in a PC. The microcontroller is programmed to guide the stepper motor to rotate to the right or left by a certain angle according to the signal of the right-left switch.

2. Right-left switch:

The switch controls the motion of the sail model by giving appropriate signals to the microcontroller. The duration of pressing the switch controls the angle of rotation of the stepper motor. The longer the pressing duration, the bigger is the rotation angle of the stepper motor.

3. Battery:

A set of *DC* batteries is used to supply the necessary current/voltage of the control circuit.

4. Screen:

A small *LED* screen is used to show the direction of rotation (right/left) as well as the magnitude of the rotation angle of the stepper motor.

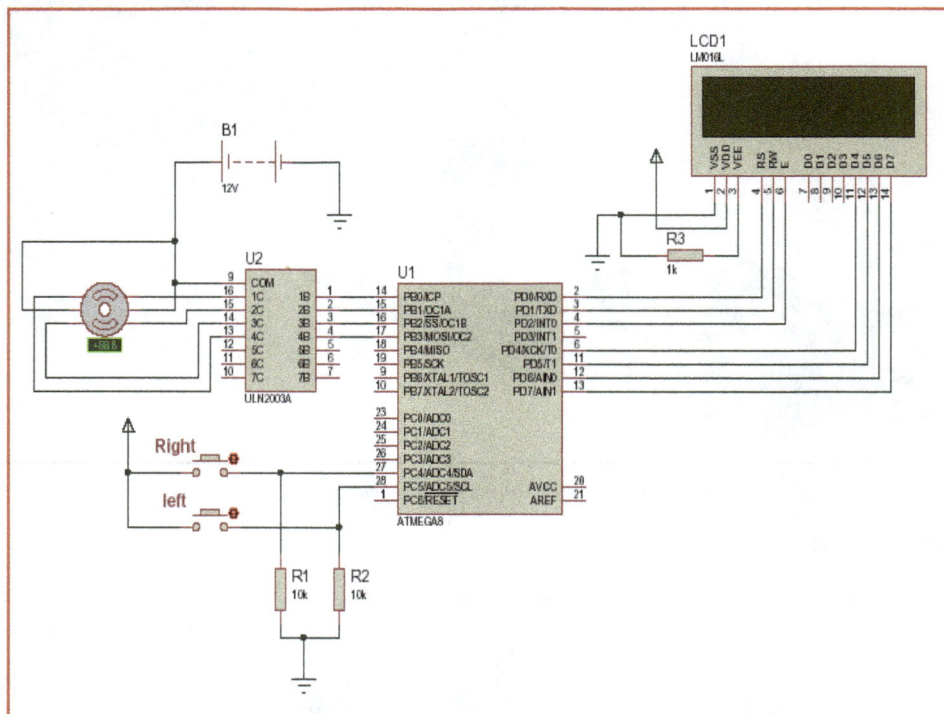

Fig 14. *Control circuit of the mono-hull model.*

4.4.2. Stepper Motor

Stepper motor is a brushless, synchronous electric motor that can divide a full rotation into a large number of steps, for example, 200 steps. This is achieved by increasing the numbers of poles on both rotor and stator, Fig. 15.

Computer-controlled stepper motors are one of the most versatile forms of positioning systems, particularly when digitally controlled as part of a servo system. Stepper motors are used in flatbed scanners, printers, plotters and many more devices.

In the present work, the stepper motor is used to direct the sail model to the right/left direction according to the signal of the microcontroller based on the corresponding signal of the right-left switch.

Fig 15. Stepper motor [14].

5. Present Second Model (Multi-hull Boat)

5.1. Description of the Boat Model

The second model represents a multi-hull boat that is known as catamaran. A catamaran is a multi-hulled vessel consisting of two parallel hulls of equal size. A catamaran is geometry-stabilized. It derives its stability from its wide beam, rather than having a ballasted keel like a mono-hull. A catamaran can have a very shallow draught. The two hulls will be much finer than a mono-hull's. Thus, the reduced drag allows faster speeds. Having no ballast, an upturned catamaran will be unlikely to sink [15].

A catamaran's two hulls are joined by some structure (frame). More sophisticated catamarans combine accommodation into the bridging superstructure. Catamarans may be driven by sail and/or engine. Originally catamarans were small yachts, but now some ships and ferries have adopted this hull layout because it allows increase speed, stability and comfort [15]. Figure 16 shows modern engine-powered ferry catamaran.

Fig 16. Engine-Powered Ferry Catamaran [15].

The second model resembles a class of boats that is known as "Tektron 50" [16]. Fig. 17 shows the overall shape of the model. The overall dimensions of the prototype (Tektron 50) and the second model are listed in Table 3. The reduction scale was intended to be 1:38. This scale was kept for the overall length and width of the model. However, for constructional, stability and floating reasons, other dimensions were taken according to another scale of 1:25.

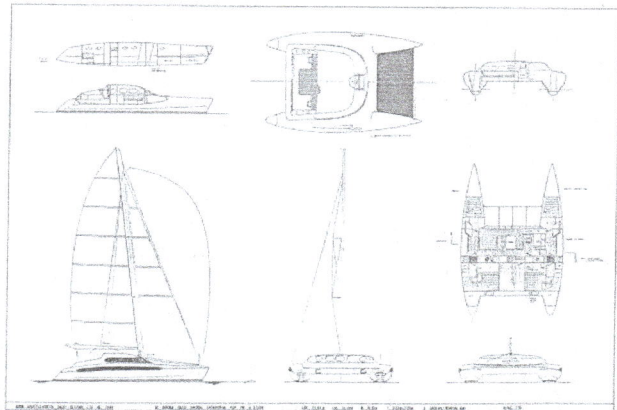

Fig 17. Overall shape of the second model [16].

Table 3. Overall dimensions of the prototype and second model.

No.	Quantity	Prototype (Tektron 50) [16]	Present Model
1	Length Overall (LOA)	15.24 m	40 cm
2	Loaded Waterline Length (LWL)	14.33 m	37.5 cm
3	Model maximum width	10 m	26 cm
4	Maximum beam at waterline (BWL)	1.02 m	4 cm
5	Width of Hull (B-hull)	1.08 m	4.25 cm
6	Height of Hull (H-hull)	0.58 m	2.5 cm
7	Draft	0.424 m	1.8 cm
8	Mid-Sec. Area	0.315 m2	5.3 cm2
9	Water Plane Area	10.288 m2	105.5 cm2
10	Displacement	2.5 m3	111.7 cm3

5.2. Design of Important Parts of Model

5.2.1. Sail Design

Figure 18 shows the main components of the sail.

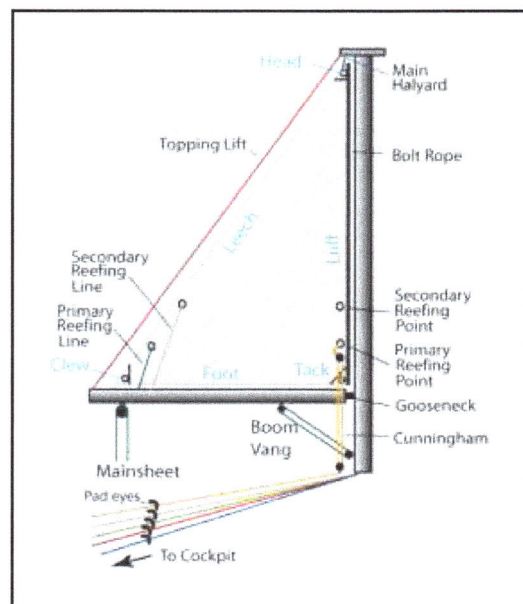

Fig 18. Main components of the sail [17].

5.2.1.1. Sail Forces and Moments

The heeling moment is caused on one hand by the sail heeling force acting in the center of pressure of the sails, and on the other hand the side force developed by the keel, the rudder and the underwater hull. This couple trying to overturn the boat is balanced by another couple, the righting moment, caused by the buoyancy of the boat and the weight of the keel and the hiking crew (these forces are not shown), Fig. 19 [18].

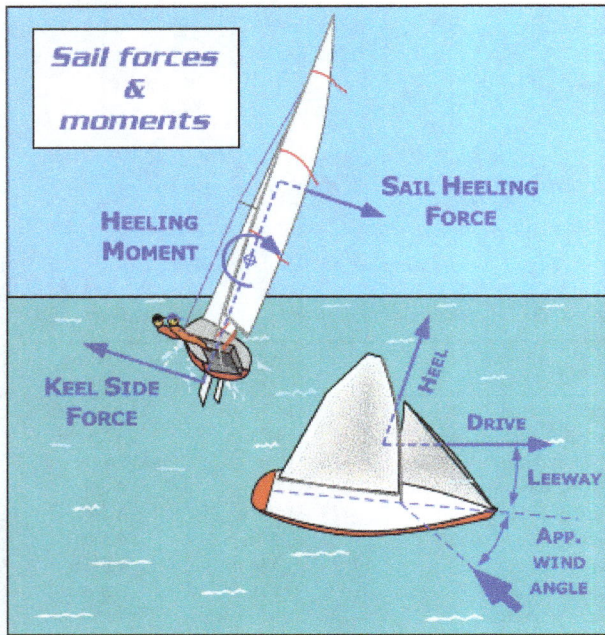

Fig 19. Sail forces and moments [18].

While the heeling force grows in a quadratic manner with wind speed, the heeling is best controlled by feathering the sails (twisting the head-off) and flattening them especially in the upper part. This lowers the aerodynamic center of effort, making it possible to keep the boat upright.

5.2.1.2. Defining a Sail

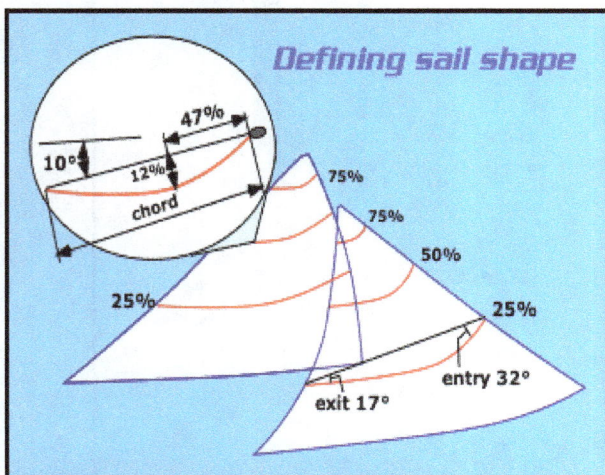

Fig 20. Definition of sail shape [18].

The shape of a sail section is defined with sufficient accuracy by two percentages and three angles: the camber, expressed in percentage of the local sail chord (width, 12%), the position of the maximum camber, similarly expressed in percentage of the local sail chord (47%), the twist expressed in degrees relative to the sail foot chord (10 degrees), the entry angle (32 degrees) and the exit angle (17 degrees), as defined in the illustration.

To define the geometry of a complete sail, we usually take three sections, at 25% - 50% - 75% heights, and the foot section plus the headboard. We also need to know the sheeting angle between the centerline of the boat and the foot chord of the sail, and the mast bend or forestay sag, to be able to fully describe one setting of the sail, Fig. 20 [18].

5.2.1.3. Parameters of Sail Design

Firstly, students had to obtain some data that will help them to make a proper design for the sail. The key word for designing a sail is the Main Sail Area, which is determined from empirical formulae. Most of these formulae give a range of possible values. So, the average value is usually considered.

The design procedure marches as [19-29]:
Formula (1):
Sail area/Cubic root of (displacement)^2 = 15 - 17

$$\text{For ratio } =16, \text{ the sail area } =371 \ cm^2 \qquad (1)$$

Formula (2):
$LWL \times BWL \times 2.75$ = approximately sail area

$$\text{Sail area } =412 \ cm^2 \qquad (2)$$

Formula (3):
Water plane area $\times 3.75$ = sail area

$$\text{Sail area } = 398 \ cm^2 \qquad (3)$$

Formula (4):
(Sail Area)^2/(Displacement)^2 = 3.8 - 4

$$\text{Sail area } = 371 \ cm^2 \qquad (4)$$

Then, from (1)-(4), the average sail area can be taken as 400 cm^2.

5.2.1.4. Software Design of the Boat Sail

The sail was designed using software called "Sailcut". It is free software [30]. This software simplifies the design process as it contains the fundamentals of design, which allows designing a proper sail and jib for the sailing boat. The students designed about 14 alternative models for the sails of the Tektron-boat. Some of these sail models are accompanied by jib and some depend on main sail only. The model "sail 2012-jib" was chosen as it is the most familiar to the "Catamaran Sailing Boats".

5.2.1.4.(a). Sail 2012-jib (Main Sail)

Based on the design results of the "Sailcut" software, the following dimensions of the main sail and jib are obtained. Table (4) shows the main dimensions of the main sail.

Table 4. *Main dimensions of the main sail.*

No.	Quantity	Value
1	Luff Length	350 *mm*
2	Foot Length	200 *mm*
3	Diagonal Length	400 *mm*
4	Leech Length	402 *mm*
5	Sail Area	0.04 *m²*
6	Luff Round	10 *mm*
7	Luff round Position	50 %
8	Foot Round	10 *mm*
9	Leech Round	30 *mm*
10	Leech Round Position	60 %

5.2.1.4.(b). Sail 2012-jib (Jib Sail)

To obtain some information about the jib, students got some relations from other models and made sure that these relations are right by testing them on other detailed models [19-29]. Table (5) shows the relations between main sail and jib sail.

Table 5. *Relations between main sail and jib sail.*

No.	Relation	Ratio
1	Jib luff/Sail luff	0.8
2	Jib foot/Sail foot	0.54
3	Jib area/Sail area	0.36
4	Boom/Foot	1.02

Thus, the jib main dimensions are listed in the following table (6).

Table 6. *Main dimensions of the jib sail.*

No.	Relation	Value
1	Jib Area	0.0144 *m²*
2	Jib Hoist (Luff)	295 *mm*
3	Jib Base (foot)	108 *mm*
4	Boom/Foot	1.02

Fig. 21 shows the results of the design of the main and jib sails based on the above values of tables (4-6) and using "Sailcut" software.

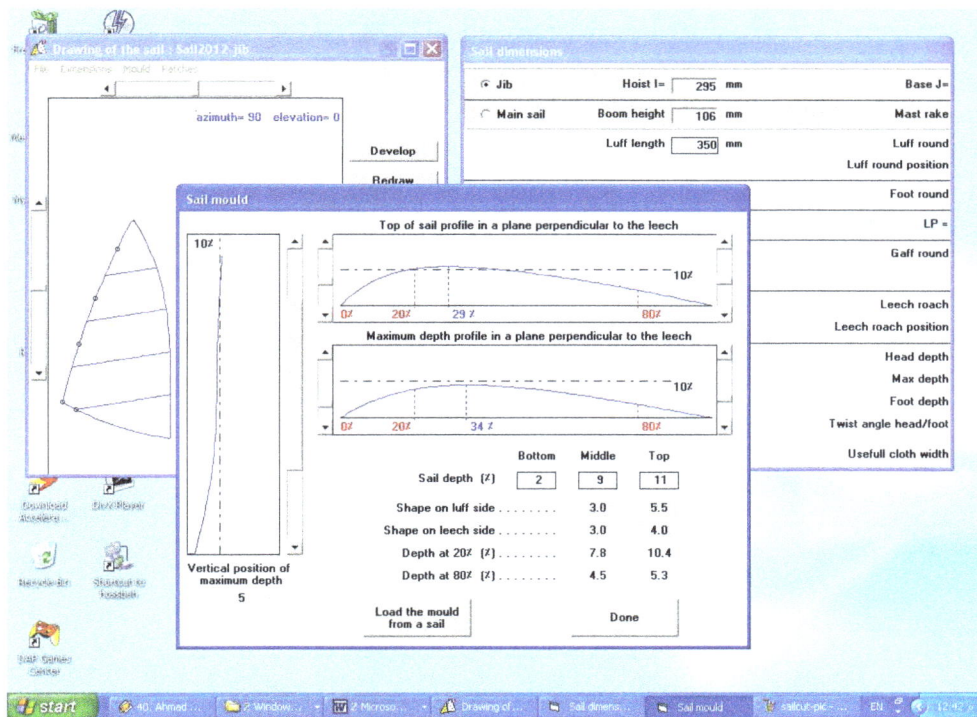

Fig 21. *Results of the design of the main and jib sails using "Sailcut" software.*

5.2.2. Fin Keel Design

5.2.2.1. Definition and Advantages/Disadvantages

(i) Definition

The keel is basically a flat blade sticking down into the water from a sailboat's bottom. It has two functions: it prevents the boat from being blown sideways by the wind, and it holds the ballast that keeps the boat right-side up.

Keels come in many styles. Traditional boats have graceful keels built into the shape of the hull; the ballast is either bolted to the bottom of the keel or placed inside it. The keel is built of whatever the boat is built of, usually fiberglass, aluminum or wood, and the ballast is lead. This is a sturdy, time-proven design, especially good for a cruising boat, which might run aground on an uncharted reef or require hauling out in a remote part of the world [31].

A fin keel is much shorter (fore-and-aft) than a full keel, Fig. 22. A fin keel is often deeper, in order to move the ballast weight as low as possible in the water [32].

(ii) Advantages of fin keel sailboats [32]

With less wetted surface and drag, fin keel boats are usually faster than their full-keel counterparts. With less keel length to resist the turning action of the rudder, a fin-keel boat turns more quickly and usually tacks easily. Most racing sailboats have fin keels (or a centerboard that is similarly shaped).

(iii) Disadvantages of fin keel sailboats [32]

Because the shorter keel provides less resistance to forces that act to throw a sailboat off course, such as wind gusts and waves, a fin-keel sailboat does not track as well as a full-keel boat and requires more attention to the helm. Its motion may not be as sea-kindly.

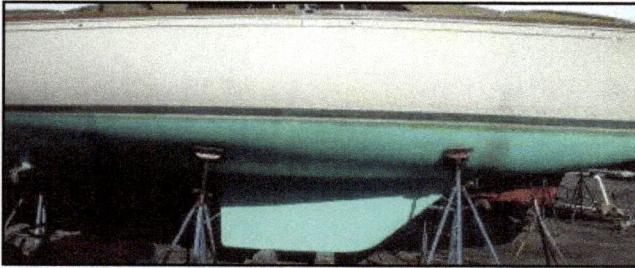

Fig 22. Shape of boat fin keel [32].

5.2.2.2. Keel of the Present Model

The keel of the present model takes the shape of an airfoil section. At first, the symmetrical airfoil section *NACA 0010* was chosen due its simplicity and easy-manufacturability, Fig. 23a. Then, the airfoil section was changed to *NACA 0010-66* as it gives better stability to the model, Fig. 23b. Two similar keels were manufactured. A keel was fixed to each of the two bodies of the boat model, Fig. 24. The keel has a length of 8 cm and a height of 5 *cm*.

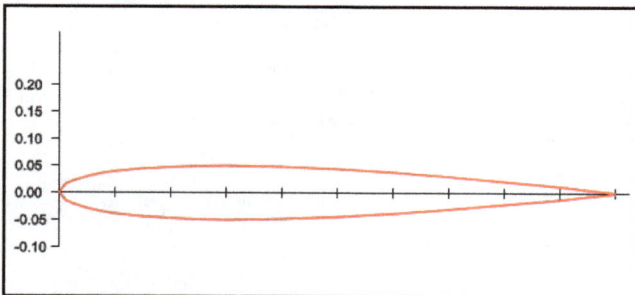

(a) NACA 0010.

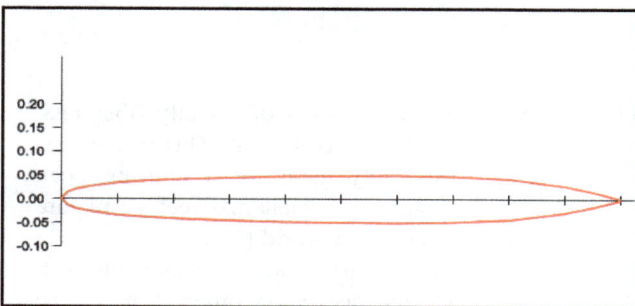

(b) NACA 0010-66.

Fig 23. Airfoil section of the present model [33].

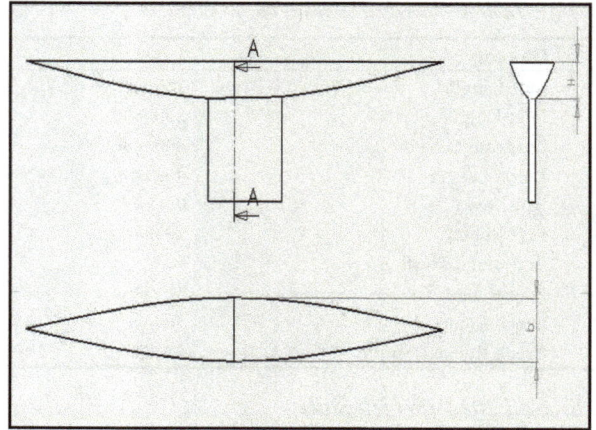

Fig 24. Shape of the keel of the present model, not to scale.

5.2.2.3. Numerical Simulation of Keel

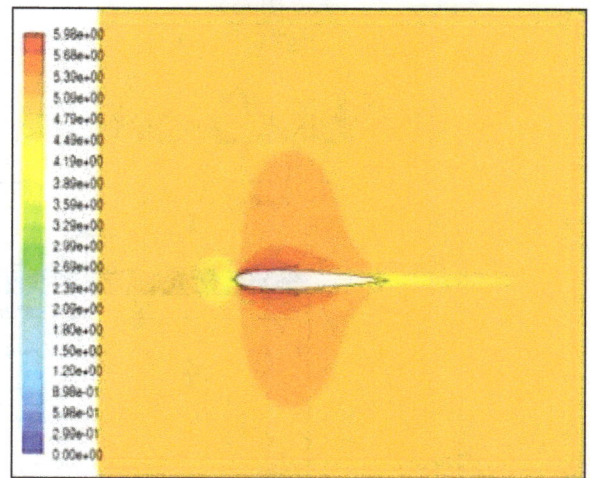

α = 0°

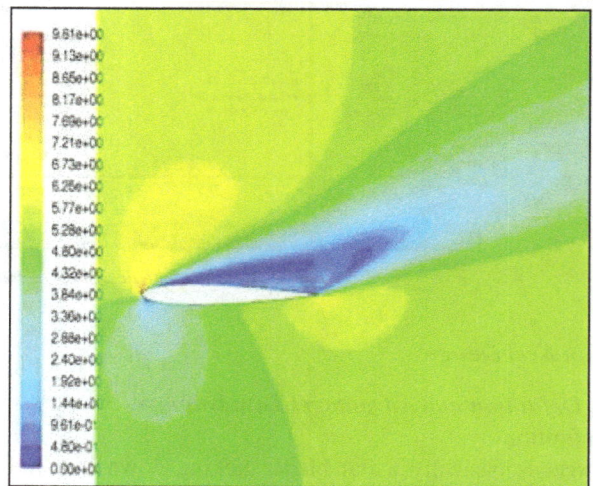

α = 20°
NACA 0010

Fig 25. Continued.

As an attempt to teach students the aspects of computational fluid dynamics (*CFD*), the students carried out numerical simulation of the flow around the two airfoil sections *NACA 0010* and *NACA 0010-66* at different angles of attack. The commercial software "Fluent 6.2" [34] was used to carry out the *2-D* simulations. Fig. 25 shows the results of their simulations at two angles of attack (α).

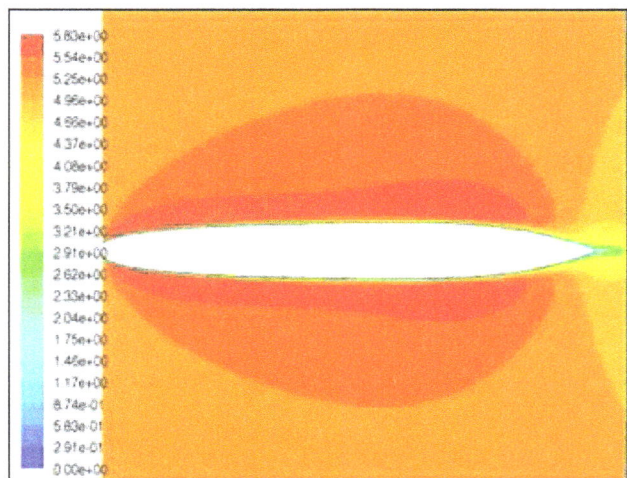

$\alpha = 0°$

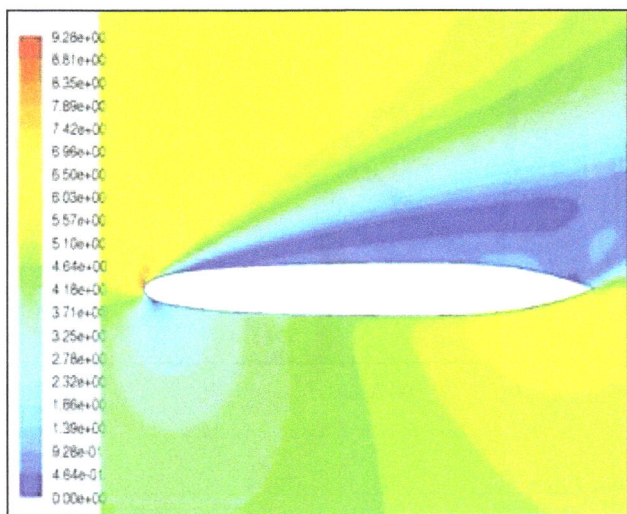

$\alpha = 20°$
NACA 0010-66

Fig 25. *Computational predictions of the velocity contours of the keel sections.*

5.3. Construction of the Second Model

The model was totally constructed by a professional craftsman from wood. The two hulls were made from two solid pieces of wood. The main part of the boat rests on a plywood piece that takes a rectangular shape. This rectangular wooden piece connects the two similar hulls of the model. Then, the model surface was cover by a water-resistant coating. Finally, the model was carefully painted. The two sails were made from fabric. Fig. 26 shows 3-D drawings of the model. Fig. 27 shows the real model after fabrication.

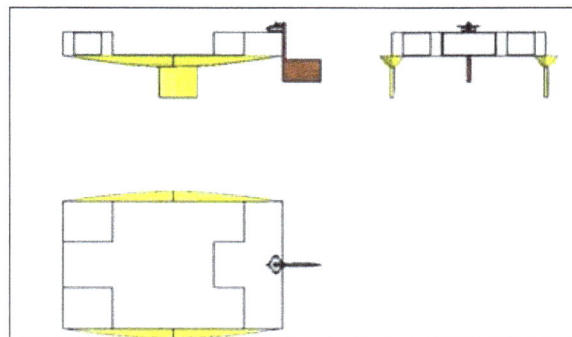

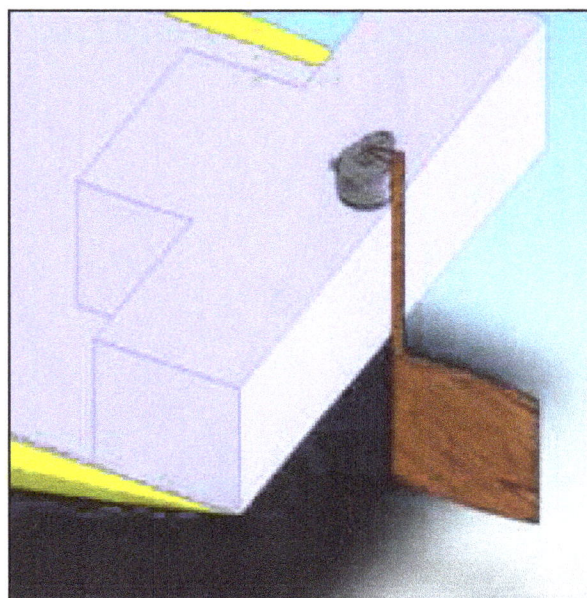

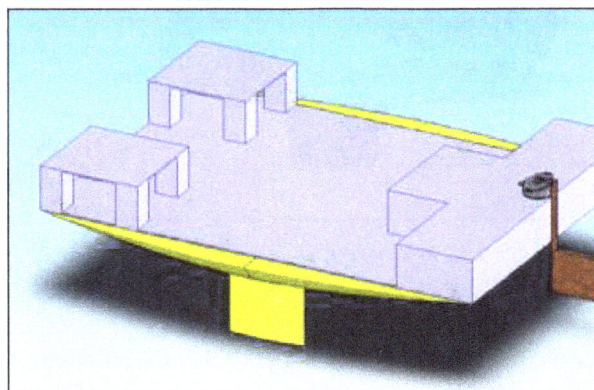

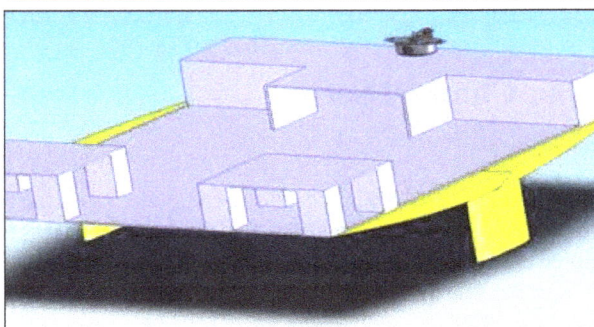

Fig 26. *3-D drawings of the model.*

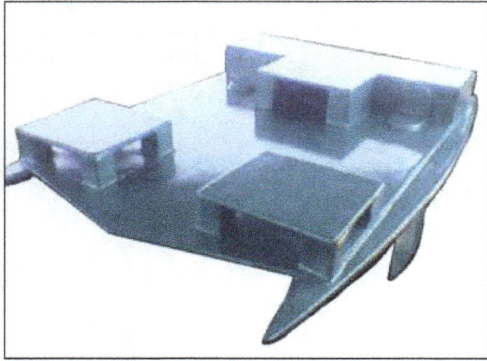

Fig 27. Real model after fabrication.

5.4. Control

A control circuit similar to that of Sec. 4.4 of the mono-hull model was used to control the rudder rotation of the catamaran model.

6. Testing Channel Set

6.1. Description

A testing channel was designed and fabricated to test the two boat models. A closed circuit of water circulation was used to given enough water stream to test the two models. The channel length was designed to be at least 7 times the length of the sailing boat. The channel is designed to have maximum height of water equals 15 cm. This height gives maximum

volume of water equals $15 \times 75 \times 300 = 337500 \ cm^3$, Fig. 28.

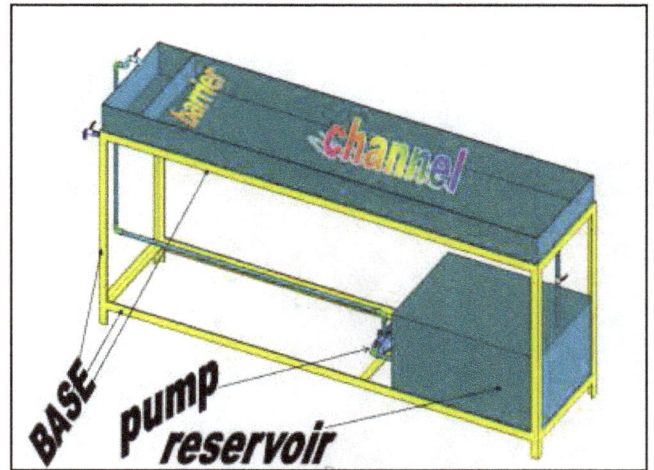

(a) Drawing.

(b) Actual.

Fig 28. General view of the channel set.

6.2. Main Parts

6.2.1. Channel

Its main purpose is to accomplish a place that has uniform flow of water for testing the motion of the sailing boat. The pump draws water from the tank and elevates it to the main channel to have enough quantity of water to move the sailing boats.

6.2.2. Barrier

Its purpose is to avoid flow turbulence and avoid waves. Thus, the flow becomes uniform. It is located 25 cm form beginning of the channel.

6.2.3. Base

It is the fixation of the channel, tank and pump. The base is designed to withstand the heavy weight of water in channel and tank, which reaches approximately half a tone.

6.2.4. Tank

It stores a volume of water that is sufficient to supply the channel with the necessary amount of water to carry out the experiments.

6.2.5. Pump

It supplies the channel with the necessary amount of water to carry out the experiments. Also, it grantees the continuous circulation of the water stream during the experiment.

6.3. Dimensions and Specifications of the Main Parts

Table (7) shows the dimensions and specifications of the main parts of the channel set.

Table 7. Dimensions and specifications of the main parts of the channel set.

Part	Quantity	Value
Channel	Length	300 cm
	Width	75 cm
	Height	20 cm
	Volume	$300 \times 75 \times 20 = 450000 \ cm^3$
	Wall thickness	0.15 cm
	Material	Iron Sheets
Tank	Length	100 cm
	Width	75 cm
	Height	50 cm
	Material	Iron Sheets
	Door	$20 \times 20 \ cm^2$
Base	Length	300 cm
	Width	75 cm
	Height	120 cm
	Thickness	3.0 mm
	Material	Iron
Pump	Type	Centrifugal
	Maximum Head	40 m
	Volume flow rate	5-40 L/min
	Power	0.5 hp
	Frequency	50 Hz
	Voltage	220 V

7. Experimental Tests

The objective of the experimental tests is to confirm the proper operation of the two sailing boats. The tests were performed in the water channel. These tests demonstrated the proper floating and cursing of the sailing boats as well as confirmed the operation of control circuits and the steering operation. Air blowing was generated by a suitable blower. All the tests were recorded by a suitable video camera. Students were guided to solve the uprising technical problems that appeared during the tests, Fig. 29.

As expected, the mono-hull model faced some instability problem. This problem was solved by carefully adding additional two small barrels; one on each side of the model. On the other, the multi-hull (catamaran) model did not face any stability problems at all.

Tests showed that control circuits operate well. The objective of proper steering and maneuver of the two models is well-performed. Moreover, the designed sails gave the models a suitable thrust to move and accelerate them. Sometimes, the wire connection causes shift in the boat direction while cursing.

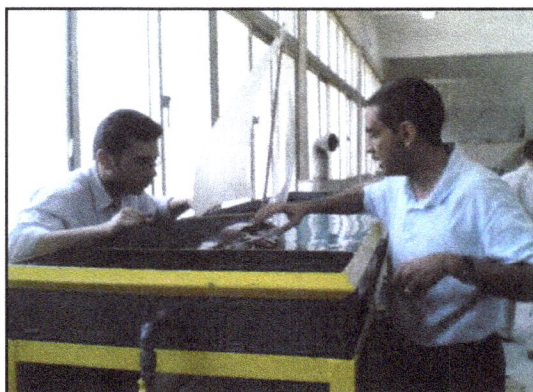

Fig 29. Experiments in the channel set.

8. Internet Dissemination

As part of the activities of the students and to teach them how to disseminate for their work, the author advised the students to initiate an internet group for their project.

The multi-hull (catamaran) group responded and constructed a Yahoo-group. Fig. 30 shows some shoots of that group. It was easy for interested persons to join their group. Thus, communication as well as exchange of knowledge and experience was available in a worldwide scale.

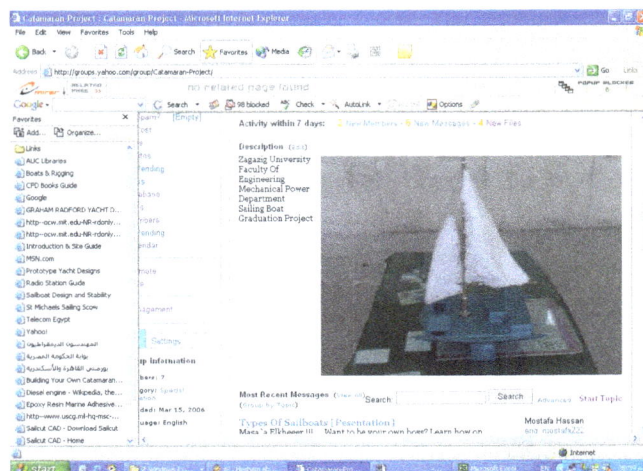

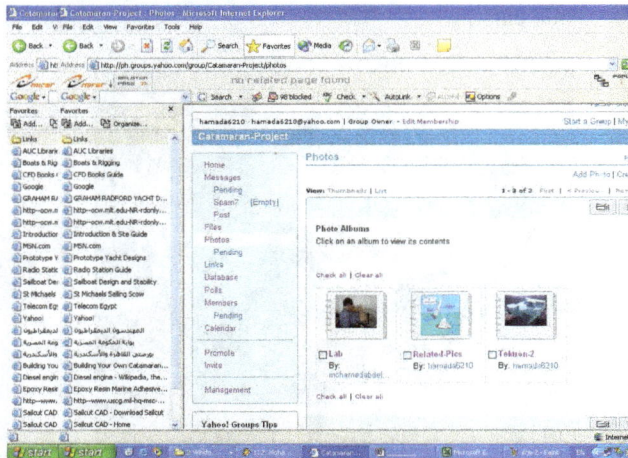

Fig 30. *Shoots of the Yahoo-group.*

9. Conclusions

Based on the above illustrations and test observations, the following points can be stated:

1. The hands-on learning method confirmed to be very effective technique for the understanding and construction of multidisciplinary engineering systems.
2. Students sufficiently learned that optimization of the design, material, manufacturing and construction gives a notable result in reducing the total cost of the engineering product.
3. The wire-connection of the control circuits causes shift in the boat direction while cursing. Thus, a wireless control method is recommended.
4. In spite of their simplicity, the control circuits proved to be suitable for the operation of the two model boats
5. As expected, the mono-hull model faced some instability problems. On the other, the multi-hull (catamaran) model did not face any stability problems at all.
6. The designed sails proved to be quite successful in gaining the required thrust to push the two models with a suitable speed.

Recommendations for Future Work

Based on the above discussions, the following recommendations can be listed:

1. The two boat models are to be used as demonstration tools for the students of: (*i*) "Fluid Mechanics Course"; especially the topic of "Stability of Floating Bodies" and "Wind Aerodynamics". (*ii*) Graduation projects as a real example of multidisciplinary engineering.
2. Distance sensors, depth sounder, and wireless digital camera are to be installed on the boat model with a suitable control circuit for safer operation.
3. The boat model may be supplied with Global Positioning System (*GPS*) to determine exact position and direction of the boat model.
4. Infra-Red (*IR*) or Radio-frequency (*RF*) control system is recommended to avoid the shift of the direction of the boat model due to the wire-connection of the control circuit. RF system has the advantage of the longer range and more flexibility of control.
5. Other designs of the sails of the two boat models may be applied for better thrust force.

Acknowledgements

The author would like to acknowledge Engs. Ali M. Elkoshnea, Haitham A. Baz, Mohamed F. Elnagar, and Mahmod S. Elden of Mono-hull group as well as Mohamed Gaber and his colleagues of Multi-hull group as being members of the team of the B.Sc. Graduation project of the present work under the author's supervision. Also, acknowledgement is extended to Eng. Wael Elwan for his help in guiding the students.

Nomenclature

2-D	Two-dimensional
3-D	Three-dimensional
A	Bow transom (Mono-hull)
B, C, D, E, F, G	Sections of bulkhead (Mono-hull)
H	Transom (Mono-hull)
Hp	Horse power
α	Angle of attack

Abbreviations

NACA	National Advisory Committee for Aeronautics
B-hull	Width of Hull
BWL	Maximum beam at waterline
CE	Center of forces
CFD	Computational fluid dynamics
CLR	Center of lateral resistance
DC	Direct current
foot	Sail Base
GPS	Global Positioning System
H-hull	Height of Hull
IR	Infra-Red
LED	light-emitting diode
LOA	Length Overall
Luff	Sail Hoist
LWL	Loaded Waterline Length
NURB	Non-Uniform Rational Basis Spline
PC	Personal computer
RF	Radio-frequency

References

[1]   http://web.mit.edu/2.972/www/reports/sail_boat/sail_boat.html

[2]   https://sites.google.com/site/catalina22experiment/home/basics-for-advanced-sailors-501

[3]   http://www.jordanyachts.com/archives/4023

[4] http://en.wikipedia.org/wiki/Sloop

[5] http://www.catboats.org/index.php

[6] http://en.wikipedia.org/wiki/Sunfish_(sailboat)

[7] https://en.wikipedia.org/wiki/Catamaran

[8] http://en.wikipedia.org/wiki/Schooner

[9] http://en.wikipedia.org/wiki/Ketch

[10] http://en.wikipedia.org/wiki/Yawl

[11] http://mateykatsarski.blog.bg/hobi/2010/01/04/proekt-na-malk ata-iahta-quot-pilgrim-590-quot.466300

[12] http://sourceforge.net/projects/freeship/

[13] http://download.cnet.com/Freeship/3000-6677_4-10558861.ht ml

[14] http://simple.wikipedia.org/wiki/Stepper_motor

[15] http://en.wikipedia.org/wiki/Catamaran

[16] http://tek-composites.com/images/50ft_cat_cruiser/tek50c.php

[17] http://sail-lbs.com/?p=54

[18] http://www.wb-sails.fi/Portals/209338/news/98_11_PerfectSh ape/Main.htm

[19] Brewer, Ted (1994) Understanding Boat Design 4th Ed, International Marine

[20] Bruce, Peter (1999) Adlard Coles' Heavy Weather Sailing 30th Anniversary Ed, International Marine

[21] Chapelle, Howard I. (1967) The Search for Speed under Sail 1700-1855, WW Norton and Company

[22] Gerr David (1992) The Nature of Boats, International Marine

[23] Garrett, Ross (1996) The Symmetry of Sailing: The Physics of Sailing for Yachtsmen, Sheridan House

[24] Johnson, Peter (1971) Yachtsman's Guide to the Rating Rule, Nautical Publishing Co.

[25] Larsson, Lars and Rolf Eliasson (1994) Principles of Yacht Design, McGraw Hill

[26] Marchaj, C.A. (1964) Sailing Theory and Practice, Dodd, Mead & Company

[27] Marchaj, C.A. (1996) Seaworthiness: The Forgotten Factor, Adlard Coles Nautical / Tiller

[28] Marshall, Roger (1986) A Sailors Guide to Production Sailboats, Hearst Marine Books

[29] Technical Committee of the Cruising Club of America, John Rousmaniere Ed (1987) Desirable and Undesirable Characteristics of Offshore Yachts, W.W. Norton & Company

[30] www.sailcut.com.

[31] http://www.discoverboating.com/resources/article.aspx?id=25 1

[32] http://sailing.about.com/od/typesofsailboats/ss/Keelshapes_2.h tm

[33] http://www.airfoildb.com/

[34] http://cdlab2.fluid.tuwien.ac.at/LEHRE/TURB/Fluent.Inc/flue nt6.2/help/pdf/ug/pdf.htm

16

Relativity and Aeroelasticity Effects on the Supersonic Objects

Arezu Jahanshir

Buein Zahra Technical University, Department of Eng. Physic, Qazvin, Iran

Email address:

jahanshir@bzte.ac.ir

Abstract: Flutter is one of the aerodynamic problems; it mainly occurs on the moving object, especially with wide wings, blade or aerospace vehicles when they cruise at ultra-high speeds. Development and applications of flutter and its related issues in usual speed such as structural design, material section and aerodynamic frame study by many authors like Baurmgart, Jureczko, Guo, Baxevanou and Larsen (see ref. [1-5]). But at ultra-high speeds where the Galilean space and time invariant change to the Lorentz spacetime invariant, the flutter phenomenon will be important to describe the stability of the moving objects at ultra-high speeds. In this limit the torsional stiffness of the wings or the body of the object is very large, so the self-variation causes the instability motion on aerospace-crafts. Therefore, the moving body displacement against the flow field plays an important role in dynamic stability studies. It is the main source of instability in an ultrasonic airplane, which is subjected to aerodynamic forces and velocity of a moving object. Instability and self-oscillation are one of the important reasons of studying the characteristics of an airplane and velocity conditions at the ultra-high speeds, which we can see the relativistic effect of motion, as predicated many years ago by Einstein's theory, i.e. the general theory of relativity. Nowadays, prediction of flutter in the field of aerospace science plays a fundamental role because the aviation safety of ultra-high objects in military and high technology equipment growth day by day. In this article in order to determine the aeroelasticity effects of ultrasonic aerospace-crafts, the theoretical methods based upon physical characteristics of four dimensional spacetime at high velocity (relativity theory) were selected.

Keywords: Lorentz Invariant, Relativity and Aeroelasticity Effects, Supersonic Aerospace-Craft, Relativistic Energy

1. Introduction

Aeroelasticity or flutter phenomenon in the supersonic aerospace-crafts involved in the study on the interaction between aerodynamic physics and relativity velocity at ultra-high speed. The flexibility of the modern supersonic aerospace-craft's structures make aeroelastic study an important aspect of modern re-entry equipment and aerospace-craft design and stability verification procedures. The flutter is the major aeroelastic phenomena considered in supersonic moving objects and dynamic aeroelastic instability characterized by sustained self-oscillation of moving object at high speed that arise from interaction between inertial, elastic and aerodynamic forces acting on the moving object with ultra-high speed [6-9]. The values for predication of flutter are taken by the theoretical and experimental results of aerospace-craft, wing, and blade, etc. model. This appears when the aerospace-crafts are subjected to aeroelastic forces at

the ultra-high speed. At acute speeds that known as the flutter speed; the supersonic structure sustains oscillations following some initial disturbance. Below this flutter speed the oscillations are damped [11-12], whereas above it any one of the modes becomes negatively unstable oscillations. Aeroelasticity can take various forms involving different pairs of interacting modes and at the high velocity of moving objects. Various approaches to this problem are to describe and determine flutter velocity and moment that will be occurred flutter. As we know in the moving object at ultra-high speed flutter phenomena occurs at a critical speed that is defined as the lowest air velocity at which a moving supersonic object will oscillate with sustained simple harmonic motion [12-14]. Supersonic flight at velocities below and above the flutter speed represents conditions of stable and unstable structural oscillation. It is found most frequently in aerospace-crafts subjected to large wings, tail units and control surfaces and also with ultra-high speed. In engineering studies the forces can interact, causing divergent oscillations for given period differences. The flutter of a supersonic moving object is important example instability that we can study it based on relativistic characteristics of motion. Supersonic-crafts compared to other airplanes have a very thin and wide surface (Fig. 1).

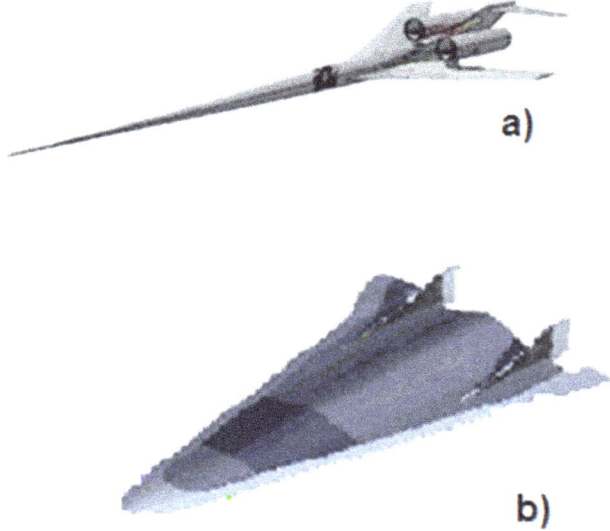

a)

b)

Fig. 1. Modern supersonic-crafts: a) The X-54A Low Boom Experimental Vehicle (LBEV), b) HTV3 [13].

Therefore, supersonic crafts under the influence of aerodynamic forces, especially at ultra-high velocities have specific type of aerodynamic forces that can change its properties and structure constants such as stiffness coefficient and natural frequencies. As a result, the aerospace-craft is faced with strong instabilities that cannot be prevented even by increasing the reliability during the design. This destruction has been created due to a fixed force at ultra-high velocity, and this value of force is created because of a specific relative velocity of flow that presented as flutter phenomenon and the fluid speed destruction presented as flutter speed. Predication the flutter velocity, we can change flight characteristics in order to get the safety of supersonic moving objects. In supersonic-crafts, flutter velocity is considered as the limiting velocity. Limiting velocity is the velocity which must not be reached by an aerospace-craft that depends on the main aerospace-craft's structure, surface, wings, blade and speed. Therefore, based on physical invariant's effect of the moving supersonic objects at relativistic spacetime coordinate system, preventing flutter close to fluid velocity can be predicated, if suppose that supersonic-craft move with ultra-high velocity, which is close to the speed of light. Flutter instability is characterized by critical velocity and pulsation in the moving object. To describe flutter we should introduce to physical principles and dynamic model of motion at ultra-high speeds. Mathematical and theoretical model describing aeroelasticity phenomenon is obtained from the Lagrange formalism which consists of the following equation:

$$\frac{d}{dt}\frac{\partial}{\partial \dot{q}}(T - \tilde{V}) - \frac{\partial}{\partial q}(T - \tilde{V}) = F_{\ddot{q}} \qquad (1)$$

where T is the total kinetic energy, $\tilde{V}$ is the potential energy, $\dot{q}$ is speed, $F_{\ddot{q}}$ is a force, and q is coordinate parameters expressed in terms of generalized coordinates of external forces on the supersonic moving aerospace-crafts. Flutter analysis of supersonic-crafts was often performed using a simple model such as the one shown in Fig. 2. The points P, C, Q, and T, which refers to the reference point, the center of mass, the aerodynamic center, and the three-quarter-of supersonic-crafts. The dimensionless parameters and determine the location of the point C and P, when these parameters are zero, the point lies on the mid of craft length.

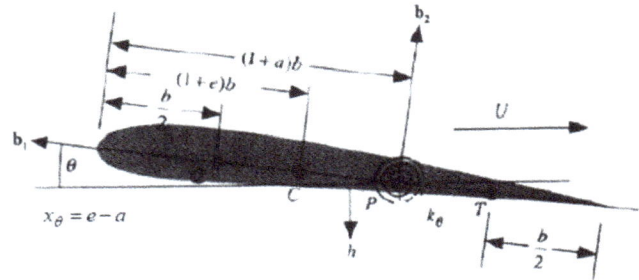

Fig. 2. Fig. Aeroelastic modeling of supersonic-craft.

In the aerodynamic subjects, the total kinetic energy of the supersonic-craft can be deduced based on the kinetic energy and potential of the mass element, as a result of the flap-wise h and torsional velocity and also the flap-wise and torsional stiffness, σ and ε respectively. Therefore, the total energy for the supersonic- craft is determined as follows [16]:

$$T - V = \frac{m}{2}(\dot{h} + x\dot{\theta}) - V =$$
$$\frac{1}{2}\left(m(\dot{h}^2 + 2x\dot{h}\dot{\theta} + x^2\dot{\theta}^2) - (\sigma h^2 + \varepsilon\theta^2)\right) \qquad (2)$$

Now, we present the uncoupled frequency and natural frequencies at zero airspeed for detail see [16]:

$$\omega = \sqrt{\frac{\sigma}{m}}, \Omega = \sqrt{\frac{\varepsilon}{m(\mu^2 + \rho^2)}} \qquad (3)$$

μ is the mass moment of inertia and ρ is the perpendicular distance between an axis through the center of mass. Here, we consider the dimensionless free stream speed (V) of the air that is sometimes called the reduced velocity:

$$U = V\omega \qquad (4)$$

Based on the Lagrange formalism we determine the system of equations and determine the complex roots as functions of V and find smallest value of v to determine flutter speed as follows [16]:

$$U_{flutter} = \omega V_{flutter} \qquad (5)$$

2. Kinetic Energy at Supersonic Speed

In some part of theoretical physics subjects, we have an object that moves with ultra-high speed close to the speed of light, the physical parameters like energy, mass, time will be changed based on the relativistic effect of motion at the ultra-high speed, which described by Einstein and relativistic equations and called ultra-relativistic limit. Description of aeroelasticity at the ultra-high sonic speed should be presented by relativistic theory. In this case, we must present 4D spacetime coordinate system. The theory of 4D spacetime is quite different from the theory of space and time, and also the Euclidean geometry of ordinary three dimensional spaces. The theory of spacetime is relativity theory. It is seen to be a theory of the geometry of the single entity, 'spacetime', rather than a theory of space and time [17-21]. The main idea of spacetime is the coordinates of an event have transformation properties analogous to space and time for ordinary three coordinate vectors that involve with a time coordinate. We will consider each event E occurring in spacetime coordinates: *(x,y,z,t)* in the frame of reference K and *(x',y',z',t')* in the new other inertial frame coordinate K' [15-20]. Relativity theory is the major theory on which modern technology and modern physics are based. This theory is general principles in moving which all specialized theories are required to satisfy and apply to all physical systems that move at ultra-high speeds. The role of relativity appears to be that of specifying the properties of spacetime that all physical processes take place. The principles of relativity are: a) all the laws of physics are the same in every inertial frame of reference. This postulate implies that there is no experiment whether based on the laws of mechanics or the laws of electromagnetism from which it is possible to determine whether a frame of reference is in a state of uniform motion, b) the speed of light is independent of the motion of its source. On the mathematical meaning we can say that the laws of physics are expressed in terms of equations, and the form that these equations take in different reference frames moving with constant velocity with respect to one another can be calculated by using of transformation equations, which require that the transformed equations have exactly the same form in all frames of reference, in other words that the physical laws are the same in all frames of reference. In order to describe the moving object at ultra-high velocity, determination the geometrical properties of spacetime four coordinate, is probably useful. Therefore, we should make use of the constancy of the speed of light and derive the general form that the transformation law must take place. In theoretical field of physics all objects that move with speed near the speed of light have a specific's motion equation and their dynamical description of motion should be based on the relativistic theory. Thus, hypersonic airplane include in this category. To predicate the occurring moment of flutter in the hypersonic objects we study relativistic combination of spacetime at ultra-high speeds. In the theoretically method, relativistic theory is employed to study the occurring moment of the flutter phenomena. Now, we describe spacetime coordinate relation in relativistic limit [30-31]. As we know, Newtonian mechanics gives an excellent description of motion, but when the object travels very fast, we should use new motion equations, i.e., the Einstein theory that is called special relativity [14, 15]. The effect appears when the speed of moving body or object becomes comparable to the speed of light in the vacuum. In the inertial frames K' and K coordinate systems that move with relative speed v to each other for simplicity, the direction x is parallel to the direction of motion. Both inertial frames come with Cartesian coordinates: (x', t') for K' and (x, t) for K, relation between space and time base on Lorentz invariant:

$$x' = \frac{x - vt}{\sqrt{1 - v^2/c^2}}, t' = \frac{t - xv/c^2}{\sqrt{1 - v^2/c^2}} \qquad (6)$$

Kinetic energy of motion for the moving object with very small speed compared with the speed of light is Kinetic energy is $K = \frac{m\dot{q}}{2}$ i.e. in the Newtonian frame work, but in the 4D framework that involved with velocity value of the moving objects it is completely different. Relativistic kinetic energy equation shows that the energy of an object approaches infinity as the velocity approaches the speed of light. Thus, it is impossible to accelerate an object across this boundary. Relativistic Kinetic energy can be expressed as:

$$K_R = \frac{m_0 c^2}{\sqrt{1 - v^2/c^2}} \qquad (7)$$

where m_0 is rest mass, v is velocity, c is speed of light. At an ultra-high sonic speed for the supersonic-crafts, the relativistic kinetic energy should be approximated by equation (7). Thus, the total energy can be partitioned into the energy of the rest mass ($E = m_0 c^2$) plus the relativistic energy at ultra-high speeds:

$$E = mc^2 = \frac{m_0 c^2}{\sqrt{1 - v^2/c^2}} + m_0 c^2 \qquad (8)$$

This formula can be blended with the relativistic momentum expression

$$p = \frac{m_0 v}{\sqrt{1 - v^2/c^2}} = mv \qquad (9)$$

Thus, to give an alternative expression for energy, the combination momentum and energy can be manipulated as follows:

$$E = \sqrt{p^2 c^2 + m_0^2 c^4} \qquad (10)$$

As we see total energy that presented in equation (8) directly depends on the mass and length. Therefore, at the relativistic limit energy, mass, length, time, etc., dependent on the velocity value of the moving object. Thus, the main formula of determination flutter and other aeroelasticity characteristics in the supersonic-crafts involved with relativity effects in spacetime coordinate system.

3. Spacetime Characteristics

In order to understand that how ultra-high velocity effects on spacetime, we try shortly shortly describe relativistic limit of motion and Einstein's general theory in the four dimensional (4D) spacetime system. The most important achievement of the Lorentz transformation (equation (6)) is relation of time in two coordinate systems. Consider an object placed at rest in a frame of reference K at some point x on the x axis and suppose that this frame is moving with a velocity v relative to some other frame of reference K'. The time interval between two events at K' and the exact event at K can be determined. Thus, the time interval between events in K is longer than in K'. It means that in the frame of reference K the time is running slow. It appears from K that time is passing more slowly in K'. Hence, after a series of mathematical transformations using relations in (4) we can find time duration in moving spacetime coordinate [17-24]:

$$\Delta t' = \frac{\Delta t}{\sqrt{1 - v^2/c^2}} \qquad (11)$$

or

$$\Delta t > \Delta t' \qquad (12)$$

This is the phenomenon of time dilation that we will use to describe how one can predicate events at ultra-high moving spacetime. Now based on relativistic Lorentz invariant we present the length contraction in supersonic moving object. The first of the interesting consequences of the Lorentz transformation is that length no longer has an absolute meaning: the length of an object depends on its motion relative to the frame of reference in which its length is being measured. Let us consider an object moving with a velocity u relative to a frame of reference K', and lying along the x axis. The object is then stationary relative to a frame of reference K, which is also moving with a velocity v relative to K'. We measure the x coordinates of the two ends of the object at the same time t, as measured by the clocks in K and K'. Turning now to the Lorentz invariant transformation equations (5), we have

$$\Delta l' = \Delta l \sqrt{1 - v^2/c^2} \qquad (13)$$

or

$$\Delta l' < \Delta l \qquad (14)$$

Thus, the length of the moving object as measured in the frame of reference K' with respect to which the length of moving object is shorter than the length as measured from a frame of reference K relative to which the object is stationary. A moving object will have the maximum length when it is stationary in a frame of reference. The length so-measured, l is known as proper length. This phenomenon is known as the Lorentz contraction. It is not the consequence of some force 'squeezing' the object, but it is a really physical phenomenon with observable physical effects. Note, however, that someone who actually looks at this moving object as it passes by will not see a shorter object. If the time that is required for the light from each point on the moving object to reach the observer's eye is taken into account, the overall effect is that of making the object appear as if it is rotated in spacetime. Now, we suppose that an object has a velocity v relative to a frame K'

and velocity of this object that measure in the frame of reference relative to a frame K moving with a velocity u relative to a frame K'. The moving object has coordinate x at time t in K' and it has coordinate x at time t in K. On the other words, if the observer in K measures an object moving along the x axis at speed u, then the observer in the K' coordinate system that moving at speed v in the x direction with respect to K, will measure the object moving with speed u' where from the Lorentz transformations (5) [20-23]:

$$u' = \frac{u - v\Delta t}{1 - uv/c^2} \qquad (15)$$

and

$$u = \frac{u' + v}{1 + u'v/c^2} \qquad (16)$$

In low speed both u and v are small with respect to the speed of light and we will recover

$$u' \cong u - v \qquad (17)$$

the intuitive Galilean transformation of velocities. If objects moving with ultrasonic speed βc, we can determine delay time as follows:

$$\Delta t = \Delta t' \sqrt{1 - \beta^2} \qquad (18)$$

and one may predicate the exact delay time in spacetime coordinate when the flutter will be happened in moving spacetime coordinate.

4. Hypersonic Flutter

Dynamic effects in mechanism of the moving supersonic objects due to aerodynamic characteristics during normal moving or rotation are significant [25-30]. As we know the velocity of moving supersonic-crafts and its stiffness of a structure need to find the various periods at which it will naturally resonate [16, 31]. The goal of relativistic analysis in structural mechanics is to determine the time dilation of flutter and natural mode shapes and frequencies of an object or structure during free vibration in second spacetime where supersonic-craft was observed. In order to do relativistic and aeroelastic analysis and ensure the non-occurrence of flutter phenomenon, time dilation in second spacetime coordinate are studied. For hypersonic moving objects using equation (18), we can determine delay time when flutter will be happened in the K' coordinate system. If objects moving with ultrasonic speed βc, so we can determine delay time.

5. Conclusion

In this study, the relativistic effects of aerodynamic loading at ultra-high velocity were considered. Based on the Lorentz invariant equations the time dilation for flutter was determined. In theoretical method the flutter velocity and natural frequency of the system was studied, and flutter speed related to the coupling of bending, and torsion were defined. We exactly determined that flutter involves with aerodynamic-elastic and velocity at the ultra-high limit. Equations were shown that some critical speed that known as the flutter speed, could be predicated by an observer in the initial frame reference. Therefore, the body displacement against the flow field plays a fundamental role in dynamic solidity analysis. In order to maintain the supersonic object stability in ultrahigh speed, predication's time in the occurring flutter can be determined in the relativistic limit. The main focus on this study is to merge relativistic equation of motion in mechanical engineering science and determination of predication flutter's moment.

References

[1]   A. A. Baumgart, mathematical model for wind turbine blades, J. Sound Vib., 251(1), 1-12, 2002.

[2]   M. E. Jureczko, M., Pawlak, and A., Mezyk, Optimisation of wind turbine blades, J. Mater. Process. Tech., 167(2), 463-471, 2005.

[3]   S. Guo, Aeroelastic optimization of an aerobatic aircraft wing structure, Aerospace Sci. Technol., 11(5), 396-404, 2007.

[4]   C. A. Baxevanou, P. K. Chaviaropoulos, S. G. Voutsinas, and N. S. Vlachos, Evaluation study of a Navier–Stokes CFD aeroelastic model of wind turbine airfoils in classical flutter, J. Wind Eng. Ind. Aerod., 96(8), 1425-1443, 2008.

[5]   J. W. Larsen, and Nielsen, S. R., Nonlinear parametric instability of wind turbine wings, J. Sound Vib., 299(1), 64-82, 2007.

[6]   A. Andronov, A. A. Vitt and S. E. Khakin, Theory of Oscillator, Mineola, NY: Dover, 1987.

[7]   L. Mirovitch, Elements of Vibration analysis, Dover Publications, 2011.

[8]   L. Raymond et al., Principles of aeroelasticity, Dover Publications, 2002.

[9]   T. H. G. Megson, Introduction to Aerospace Structural Analysis, Butterworth-Heinemann, 2013.

[10] W. C. Hurty and M. F. Rubinstein, Dynamics of Structures, Prantice Hall of India Pvt. Ltd., 1967.

[11] H. Babinksy, How do wings work? Phys. Educ. 38, 2003.

[12] E. H. Dowell, D. A. Peters, R. H. Scanlan and F. Sisto, A Modern course in Aeroelasticity, III edition, Kluwer Academic Publishers, 1995.

[13] a)http://www.uasvision.com/2012/04/04/nasa-tests-supersonic -aircraft-without-boom/, b)http://www.globalsecurity.org/space/systems/x-41-htv-3.htm

[14] R. Fitzpatrick, Oscillations and Waves: An Introduction, CRC Press, 2013.

[15] P. Scheerbart, The perpetual Motion Machines, Wakefield Press, 2011.

[16] A. H. Gasemi, A. Jahanshir, Numerical and analytical study of aero elastic characteristics of wind turbine composite blades, International WIND and Structure Journal, 18(2), 103-116, 2014.

[17] H. Goldstein, C. P. Poole and J. L. Safko, Classical Mechanics, 3rd ed., San Francisco: Addison Wesley, 2002.

[18] C. Moller, The Theory of Relativity, Oxford, Clarendon Press, 1972.

[19] A. Einstein, Relativity: The Special and General Theory (Translation 1920), New York: H. Holt and Company, 1916.

[20] S. Carroll, From Eternity to Here: The Quest for the Ultimate Theory of Time, New York: Dutton, 2010.

[21] E. J. Saletan, Classical Dynamics: A Contemporary Approach, Cambridge: Cambridge University Press, 1998.

[22] S. H. Strogatz et al., Theoretical mechanics: Crowd synchrony on the Millennium Bridge, Nature 438, 43-44, 2005.

[23] H. Goldstein, C. P. Poole and J. L. Safko, Classical Mechanics, 3rd ed., Boston: Addison Wesley, 2001.

[24] L. D. Landau and E. M. Lifshitz, Mechanics, 3rd ed., Oxford: Elsevier, 1976.

[25] A. B. Pippard, The Physics of Vibration, omnibus ed., Cambridge: Cambridge University Press, 1989.

[26] R. Altman, Sound theory, Sound practice, Routledge, 1992.

[27] A. Jenkins, Self-oscillation, Phy. Reports 525 (2), 167–222. arXiv:1109.6640, 2013.

[28] E. H. Dowell, Theoretical and experimental panel flutter study AIAA J., 3(12), 1995.

[29] H. J. Pain, The Physics of Vibrations and Waves, 6th ed., Chichester: John Wiley & Sons, 2005.

[30] J. Dugundji, Theoretical consideration of Panel flutters at high supersonic Mach No., AIAA J. 4(7), 1966.

[31] M. P. Paldoussis, S. Price and E.de Langre, Fluid Structure Interactions: Cross-Flow-Induced Instabilities, Cambridge University Press, 2011.

17

Performance evaluation of the tandem C4 blades for axial-flow compressors

Atef Mohamed Alm-Eldien[1], Ahmed Farouk Abdel Gawad[2], Gamal Hafaz[1], Mohamed Gaber Abd El Kreim[3, *]

[1]Mech. Power Eng. Dept., Faculty of Eng., Port Said University, Port Said, Egypt
[2]Professor of Computational Fluid Mechanics, Mech. Eng. Dept., Umm Al-Qura Univ., Makkah, Saudi Arabia
[3]Gas Turbine Maintenance Engineer, East Delta Electricity Production Company, Ismailia, Egypt

Email address:

atef_alameldin@yahoo.com (A. M. Alm-Eldien), afaroukg@yahoo.com (A. F. A. Gawad), mgaber1_eg@hotmail.com (M. G. A. E. Kreim)

Abstract: The purpose of this work is to study the aerodynamic performance of a tandem C4 base-profile compressor blade using numerical tools. In this paper, the flow along the tandem blade is studied for various relative blade positions. In all the studied cases, the front blade is fixed and the position of the rear blade is varied as a function of the axial and tangential displacements. A computer code was developed in "Visual Basic" using linear strength vortex-panel method to predict the aerodynamic performance of the tandem blade.

Keywords: Tandem Blades, Linear Strength Vortex Panel Method, Rotor, Stator, Axial-Compressor

1. Introduction

Looking at the Euler Turbomachinery Equation (1), it obviously indicates that there are two ways of achieving high loading at compressor blades. First one is the increase of the rotational speed. Second one is the increase of the tangential-velocity difference between inlet and exit of a compressor rotor. That is increasing the turning angle which is the tandem blade concept.

$$\frac{-P_{shaft}}{\dot{m}} = U_2 V_{\theta 2} - U_1 V_{\theta 1} \qquad (1)$$

Where, U_2 is blade speed at radius r_2, U_1 is blade speed at radius r_1, $V_{\theta 1}$ is tangential velocity at radius r_1, $V_{\theta 2}$ is tangential velocity at radius r_2, $\dot{m}$ is mass flow rate (kg/s), P_{shaft} is shaft power (kW).

A major limitation on the pressure rise in a subsonic axial flow compressor stage is boundary layer separation on the blade suction surface. One method of mitigating the suction surface separation is to employ tandem blades. Tandem blading is a method of increasing the flow deflection by delaying the separation in diffusing cascade arrangements. The two parts of a blade, the front and the rear, are arranged so that the air from the pressure surface of the forward blade is injected on the suction surface of rear blade. Due to this injection of air the boundary layer on the suction surface gets more momentum to follow the rest of the suction surface. Thus, the separation of the boundary layer is delayed. According to the axial and tangential displacements, the interference area between the blades plays the role of a convergent or a convergent-divergent nozzle-type of passage, through which the air from the pressure surface of the foreword blade blows on the suction surface of the rear blade. The configuration of the front and rear blade is arranged in such a way that the front blade is truncated approximately at the middle and then the rear blade is configured so that the suction surface keeps its continuity of shape to provide the required camber. By induction of extra momentum, the tandem blade increases turning or loading capability of a given blade row. It also gives a wider stall-free operating range and hence efficiency of the compressor can be increased at off-design conditions. All these benefits of tandem blade gives the solution to one of the oldest challenges faced by axial-flow compressor designers; that is to use as few stages as possible to achieve the desired pressure rise without compromising efficiency. The obvious benefits of using fewer stages are the improvement of engine power-to-weight ratio and the reduction in manufacturing parts.

2. Literature Review

Many attempts have been made to understand the flow in a tandem cascade. Experimental as well as analytical approaches are reported for many flow conditions and different blade parameters.

Wennerstrom [1] studied the potential of tandem-airfoil blading for improving the efficiency and stable operating range of compressor stages. The investigation included testing of one single-airfoil blading, one tandem-blade with a 20-80% loading split and one tandem-blade with 50-50% loading split (the loading was estimated using the diffusion factor criterion).The results of his effort showed that tandem configuration, with the majority of loading on the rear airfoil should have a larger operation range than conventional blades, was not substantiated. But with 50-50% loading split between the airfoils, the tandem arrangement demonstrated higher pressure rise and efficiency than the corresponding single one.

Saha and Roy [2],[3] conducted various aerodynamic performance evaluations of a single and a tandem cascade for a wide range of inlet angles. The purpose of the investigation was to determine the high deflection capabilities of the tandem blade and to compare the results at off-design with an equivalent single one. The results of the study showed that the diffusion capabilities of the tandem blade are higher compared to a single airfoil.

Other investigations in tandem airfoils [4],[5] indicated that the flow deflection capabilities shown by the tandem configuration at design point, is higher compared with a single one. This fact has been explained by the formation of a new boundary layer at the rear blade.

3. Applications of Tandem Blade Rows

In the practice, the tandem cascades are applied to compressors in the subsonic, transonic and even supersonic range for rotors and stators [5]. But the main use of this arrangement is in the stator of the final stage in axial compressor, where the flow enters with high swirl velocity and it has to be turned to the radial direction. Therefore, the flow-turning angle is rather high and the last stator row is heavily loaded with the danger of flow separation. Figure 1 shows a General Electric heavy-duty gas turbine that uses a triple tandem blade row in the stator of the last stage in the axial compressor [6]. Characteristics of the compressor are shown in table 1.

Table 1. *Heavy Duty tandem compressor characteristics*

Manufacturer	General Electric
Type	MS 7001 EA
Power	86.2 *MW*
Thermal efficiency (η_{th})	33.0 %
Pressure ratio (π)	11.9
Mass flow	299 [*kg/s*]
Turbine outlet temperature	537
Revolutions	3600 *rev./min*

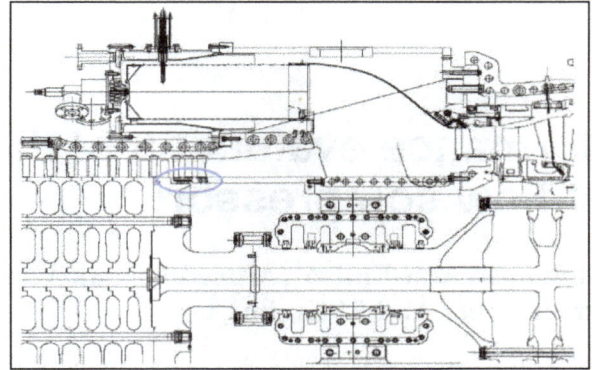

Figure 1. *Heavy-duty gas turbine with tandem compressor blades in the last stator row [6].*

Till now, tandem rotor concept has applications only for experimental purposes. Figure 2 shows a photo of tandem rotor built at The Institute of Turbomachinery of The Hannover University [7].

Figure 2. *Tandem rotor [7].*

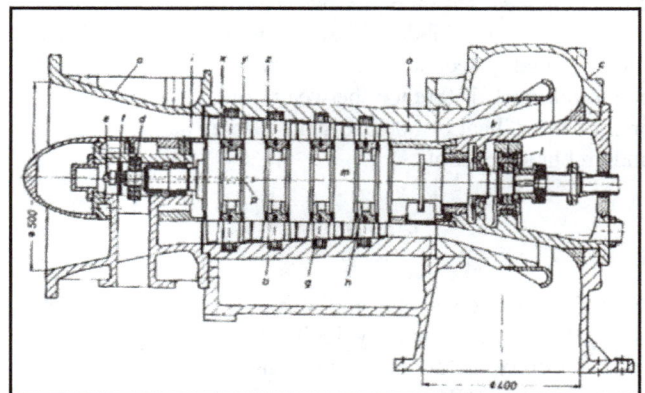

Figure 3. *Tandem rotor compressor [7].*

One of the first tandem blade studies was performed in Germany [7],[8], Figure 3. This was a 4-stage axial compressor where 3 stages were built as tandem rotors and having a design pressure ratio of 2.5. Detailed characteristics of the experimental axial compressor are listed in Table 2.

Other interesting studies of tandem blades can be found in [9]-[11].

Table 2. *Experimental set-up characteristic*

Name	Value	Units
Mass flow rate	6.58	*kg/s*
Speed	11000	*rpm*
Number of stages	4	-
Design pressure ratio	2.5	-
Hub to tip ratio	0.64	-
Blade tip speed	195.82	*m/s*
Internal power	672	*kW*
Inlet total pressure	1	*Bar*
Inlet total temperature	288.15	*K*
Outlet total pressure	2.51	*Bar*
Outlet total temperature	389.9	*K*

4. Geometry Description

The geometry and aerodynamic parameters for a tandem blade row are almost the same as those used for single airfoils, but two additional variables appear in the arrangement. These variables are the axial displacement and the tangential displacement. Figure 4 shows the geometry and standard nomenclature related to tandem airfoils in cascade [12].

The tandem configuration has two key physical effects that are of interest to compressor designers. The first is the circulation effect. Whenever an object, even a blunt one such as a cylinder, is placed downstream of an airfoil, the effect of that object is to increase the circulation around the airfoil. The Kutta- Joukouski law, expressed as, indicates that the lift force is proportional to the circulation. In terms of a tandem compressor rotor, this means that the aft blade will increase the loading on the forward blade, resulting in a greater combined pressure rise between both blades.

$$F_L = \rho U \Gamma \qquad (2)$$

The second effect is the fresh boundary layer that is formed on the aft blade. Ideally, the aft blade would be placed such that it relieves the forward blade just prior to the point of separation, as shown in Figure 4. This allows for greater overall turning of the airflow, hence more work (*i.e.*, pressure rise) while not incurring substantially higher losses.

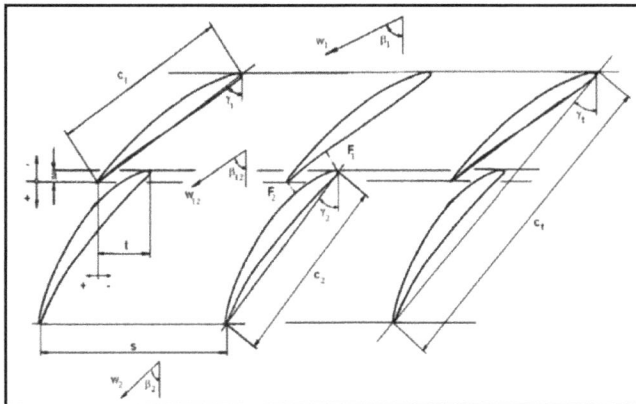

a	Axial displacement	w_{12}	Inlet velocity at rear blade
F_1	Inlet gap distance	w_2	Outlet velocity
F_2	Outlet gap distance	β_1	Inlet flow angle
c_1	Chord length front blade	β_{12}	Inlet flow angle at rear blade
c_2	Chord length rear blade	β_2	Outlet flow angle
c_t	Overall chord length	γ_1	Stagger angle front blade
s	Blade spacing	γ_2	Stagger angle rear blade
t	Tangential displacement	γ_t	Overall stagger angle
w_1	Inlet velocity		

Figure 4. *Tandem blade nomenclature [12].*

5. Governing Equations

The incompressible, two-dimensional, potential flow governed by the Laplace equation is solved numerically with a panel method, which provides the tangential external velocity. The pressure is then obtained using the Bernoulli equation, and C_l is derived by integrating the pressure over the airfoil.

5.1. Theoretical Equation of the Flow

For an irrotational flow, Figure 5, the velocity is the gradient of a quantity called the velocity potential ϕ.

$$V = (u, v) = \nabla \phi \qquad (3)$$

Substituting this into the continuity equation for an inviscid incompressible flow leads to:

$$\frac{\partial^2 \phi}{\partial x^2} + \frac{\partial^2 \phi}{\partial y^2} = 0 \ \text{ or } \ \Delta\phi = 0 \ \text{ (Laplace equation)} \qquad (4)$$

On the airfoil's surface "A", Figure 5, the external Neumann boundary condition must be satisfied:

$$\frac{\partial \phi}{\partial n} = V_n \qquad (5)$$

Where, V_n is set to zero, which represents the classical zero normal velocity condition.

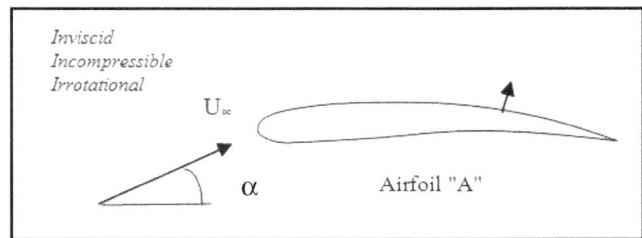

Figure 5. *Flow field of an airfoil.*

5.2. General Solution of the Incompressible Potential Flow

A general solution to the Laplace equation is obtained by adding a distribution of vortices γ on the airfoil's surface to the velocity potential of the free-stream, Figure 6. The solution at any field point P is thus given by [13]:

$$\phi_p = u_\infty . x + v_\infty . y + \int_A \gamma \phi_v ds \qquad (6)$$

Where ϕ_v is the potential of a unit strength vortex: (r, θ) are the polar coordinates of P relative to (ds).

$$\phi_v = -\frac{1}{2\pi}\theta \qquad (7)$$

This equation has to satisfy the boundary condition for every point on "A", which gives:

$$\nabla\phi_p.n = 0 \Rightarrow (u_\infty, v_\infty)n + \int_A \gamma\frac{\partial\phi_v}{\partial n}ds = 0 \qquad (8)$$

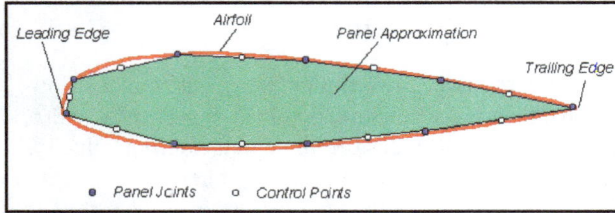

Figure 6. *Vortex-panel approximation to an airfoil* [13].

Moreover, the integration is performed on each panel and the boundary condition becomes:

$$(u_\infty, v_\infty)n + \sum_{i=1}^{n}\left(\int_{panel}\gamma\frac{\partial\phi_v}{\partial n}ds\right) = 0 \qquad (9)$$

5.3. Singularity Element

Now, the integral on each panel must be computed:

$$\int_{panel}\gamma\frac{\partial\phi_v}{\partial n}ds = \left(\int_{panel}\gamma\frac{\partial\phi_v}{\partial x}ds, \int_{panel}\gamma\frac{\partial\phi_v}{\partial y}ds\right).n \equiv (u,v).n \qquad (10)$$

We will compute u, v in the panel coordinate system, and then transform them back in the global coordinate system, Figure 7.

Consider the coordinates (x,y) in the panel system. They are obtained using the following transformation:

$$\begin{pmatrix} x \\ y \end{pmatrix}_p = \begin{bmatrix} \cos\alpha_i & -\sin\alpha_i \\ \sin\alpha_i & \cos\alpha_i \end{bmatrix}\begin{pmatrix} x-x_0 \\ y-y_0 \end{pmatrix}_G \qquad (11)$$

Where, (x_0, y_0) are the coordinates of the panel origin in the global coordinate system.

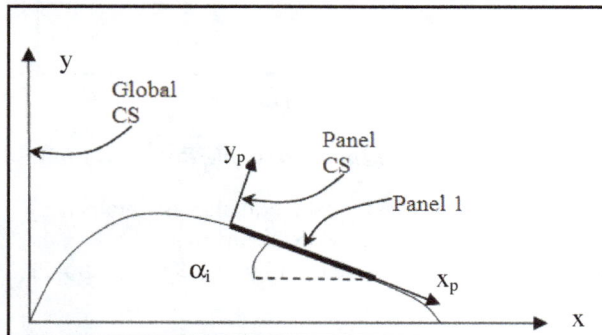

Figure 7. *Global and panel coordinate systems.*

On each panel, we choose a linear vortex distribution,

Figure 8, such as

$$\gamma(x) = \gamma_0 + \gamma_1 x \qquad (12)$$

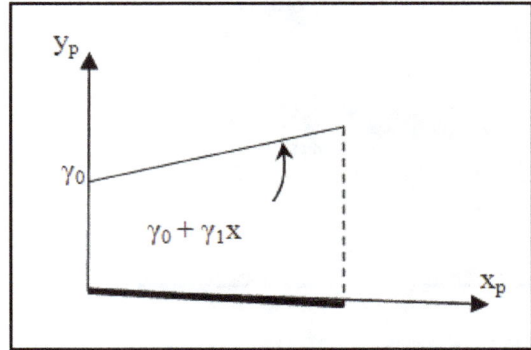

Figure 8. *Linear vortex distribution of a panel.*

This is simply the superposition of a constant strength element and a linearly varying element. For simplicity, we consider the two elements separately to compute the potential and the velocity induced by one panel.

$$\phi_{\gamma_0} = \frac{-\gamma_0}{2\pi}\int_0^L at\tan\frac{y}{x-x_0}dx_0 \qquad (13)$$

$$\text{so: } u_{\gamma_0} = \frac{\partial\phi_{\gamma_0}}{\partial x} = \frac{\gamma_0}{2\pi}\int_0^L\frac{y}{(x-x_0)^2+y^2}dx_0 \qquad (14)$$

$$\text{and: } u_{\gamma_0} = \frac{\partial\phi_{\gamma_0}}{\partial y} = \frac{\gamma_0}{2\pi}\int_0^L\frac{x-x_0}{(x-x_0)^2+y^2}dx_0 \qquad (15)$$

Integrating the above two equations we obtain:

$$u_{\gamma_0} = \frac{\gamma_0}{2\pi}(\theta_2-\theta_1) \text{ and } v_{\gamma_0} = \frac{\gamma_0}{2\pi}\ln\frac{r_2}{r_1} \qquad (16)$$

We consider the linear term now:

$$\phi_{\gamma 1x} = \frac{\partial\phi_{\gamma 1x}}{\partial x}\int_0^L x_0 a\tan\frac{y}{x-x_0}dx_0 \qquad (17)$$

$$\text{so: } u_{\gamma 1x} = \frac{\partial\phi_{\gamma 1x}}{\partial x} = \frac{\gamma_1}{2\pi}\int_0^L\frac{x_0 y}{(x-x_0)^2+y^2}dx_0 \qquad (18)$$

$$\text{and: } v_{\gamma 1x} = \frac{\partial\phi_{\gamma 1x}}{\partial y} = \frac{\gamma_1}{2\pi}\int_0^L\frac{x_0(x-x_0)}{(x-x_0)^2+y^2}dx_0 \qquad (19)$$

Solving the integrals gives:

$$u_{\gamma 1x} = -\frac{\gamma_1}{4\pi}\left[2z\ln\frac{r_1}{r_2} - 2x(\theta_2-\theta_1)\right] \qquad (20)$$

$$v_{\gamma 1x} = -\frac{\gamma_1}{2\pi}\left[x\ln\frac{r_1}{r_2} + L + z(\theta_2-\theta_1)\right] \qquad (21)$$

Now, what we want is a piecewise linear continuous vortex distribution on the whole airfoil surface. So, we have to set the strength γ at the beginning of each panel equal to the strength of the vortex at the end of the previous panel as shown in

Figure 9.

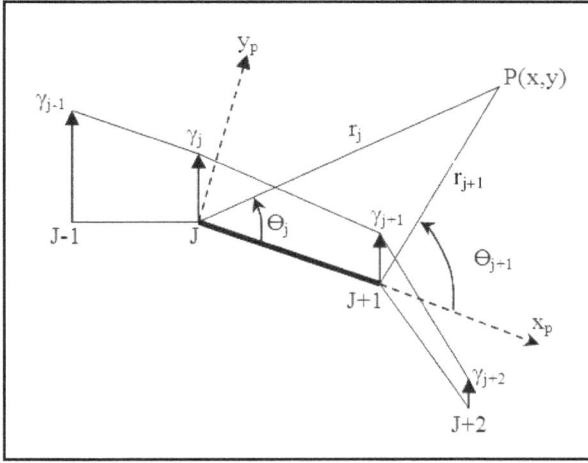

Figure 9. *Piecewise linear continuous vortex distribution.*

The relations between the vortex strengths of the elements shown in Figure 9 and the panel end values γ_0 and γ_1 are:

$$\gamma_j = \gamma_0 \quad \text{and} \quad \gamma_{j+1} = \gamma_0 + \gamma_1 L \quad (22)$$

Thus, rearranging the expressions for u and v in terms of γ_j and γ_{j+1} gives:

$$u = \frac{z}{2\pi} \left(\frac{\gamma_{j+1} - \gamma_j}{x_{j+1} - x_j} \right) \ln \frac{r_{j+1}}{r_j} +$$
$$\frac{\gamma_j (x_{j+1} - x_j) + (\gamma_{j+1} - \gamma_j)(x - x_j)}{2\pi (x_{j+1} - x_j)} (\theta_{j+1} - \theta_j) \quad (23)$$

$$v = \frac{\gamma_j (x_{j+1} - x_j) + (\gamma_{j+1} - \gamma_j)(x - x_j)}{2\pi (x_{j+1} - x_j)} \ln \frac{r_{j+1}}{r_j}$$
$$+ \frac{y}{2\pi} \left(\frac{\gamma_{j+1} - \gamma_j}{x_{j+1} - x_j} \right) \left[\left(\frac{x_{j+1} - x_j}{z} \right) + (\theta_{j+1} - \theta_j) \right] \quad (24)$$

These two equations can be divided into velocity induced by γ_{j+1} and γ_j such that:

$$(u,v) = (u^a, v^a) + (u^b, v^b) \quad (25)$$

Where, the superscripts "a" and "b" represent the contribution due to the leading and trailing singularity. By rearranging the equations, we obtain the "a" part of the velocity:

$$u^a = \frac{\gamma_j}{2\pi (x_{j+1} - x_j)} \left[-z \ln \frac{r_{j+1}}{r_j} + (x_{j+1} - x)(\theta_{j+1} - \theta_j) \right] \quad (26)$$

$$v^a = \frac{-\gamma_j}{2\pi (x_{j+1} - x_j)} \left[(x_{j+1} - x_j) \ln \frac{r_j}{r_{j+1}} + (x_{j+1} - x_j) + z(\theta_{j+1} - \theta_j) \right] \quad (27)$$

and the "b" part of the velocity:

$$u^b = \frac{\gamma_{j+1}}{2\pi (x_{j+1} - x_j)} \left[z \ln \frac{r_{j+1}}{r_j} + (x - x_j)(\theta_{j+1} - \theta_j) \right] \quad (28)$$

$$v^b = \frac{\gamma_{j+1}}{2\pi (x_{j+1} - x_j)} \left[-(x - x_j) \ln \frac{r_j}{r_{j+1}} + (x_{j+1} - x_j) + z(\theta_{j+1} - \theta_j) \right] \quad (29)$$

To transform these velocity components back to the global coordinate system, a rotation by the panel orientation angle α_i is performed as given by:

$$\begin{pmatrix} u \\ v \end{pmatrix}_G = \begin{bmatrix} \cos \alpha_i & \sin \alpha_i \\ -\sin \alpha_i & \cos \alpha_i \end{bmatrix} \begin{pmatrix} u \\ w \end{pmatrix}_P \quad (30)$$

The expressions above can be included in an induced velocity function F, which will compute the velocity (u, v) at an arbitrary point (x, y) in the global coordinate system due to the *j-th* panel.

$$\begin{pmatrix} u^a & v^a \\ u^b & v^b \end{pmatrix} = F(\gamma_j, \gamma_{j+1}, x, z, x_j, y_j, x_{j+1}, y_{j+1}) \quad (31)$$

5.4. Influence Coefficients

The zero normal flow boundary condition is implemented. For example the velocity induced by the *j-th* element with unit strength at the first collocation point is obtained by:

$$\begin{pmatrix} u^a & v^a \\ u^b & v^b \end{pmatrix}_{1j} = F(\gamma_j = 1, \gamma_{j+1} = 1, x_1, z_1, x_j, y_j, x_{j+1}, y_{j+1}) \quad (32)$$

This shows that the velocity at each collocation point is influenced by the two edges of the *j-th* panel, When adding the influence of the *j+1* panel on the local induced velocity will have the form:

$$(u,v)_1 = (u^a, v^a)_1 \gamma_1 + [(u^b, v^b)_1 + (u^a, v^a)_2] \gamma_2 +$$
$$\cdots + [(u^b, v^b)_{N-1} + (u^a, v^a)_N] \gamma_N + (u^b, v^b)_N \gamma_{N+1} \quad 33)$$

This equation can be reduced to the form:

$$(u,v)_1 = (u,v)_{11} \gamma_1 + (u,v)_{12} \gamma_2 + \cdots + (u,v)_{1N} \gamma_N + (u,v)_{1N+1} \gamma_{N+1} \quad (34)$$

Such that for the first and last terms:

$$(u,v)_{11} = (u^a, v^a)_1 \quad \text{and} \quad (u,v)_{1N+1} = (u^b, v^b)_N \quad (35)$$

and for all other terms:

$$(u,v)_{1j} = \left[(u^b, v^b)_{j-1} + (u^a, v^a)_{1j} \right] \quad (36)$$

The influence coefficient a_{ij} is defined as the velocity component normal to the surface. As see in Figure 10, the contribution of a unit strength singularity element j at collection point 1 is therefore:

$$a_{1j} = (u, v)_{1j} \cdot n_1 \quad (37)$$

$$\text{Where:} \quad n_i = (\sin \alpha_i, \cos \alpha_i) \quad (38)$$

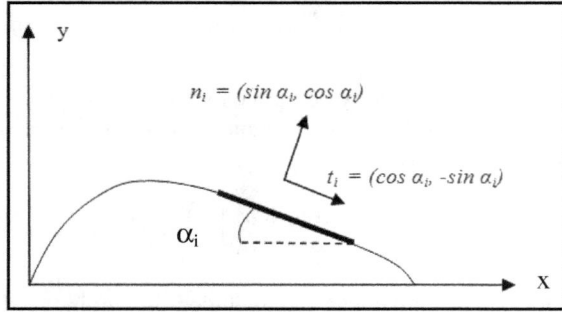

Figure 10. *Influence coefficient.*

5.5. Establishing Boundary Conditions

The free stream normal velocity component is found as:

$$RHS_i = -(u_\infty, v_\infty)(\sin\alpha_i, \cos\alpha_i) \qquad (39)$$

Specifying the boundary condition equation for each ($i = 1$ to N) of the collocation points results in N linear equations with the unknowns γ_j ($j = 1$ to $N+1$)

$$\begin{bmatrix} a_{11} & a_{12} & \cdots & a_{1N+1} \\ a_{21} & a_{22} & \cdots & a_{2N+1} \\ \cdots & \cdots & \cdots & \cdots \\ a_{N1} & a_{N2} & \cdots & a_{NN+1} \end{bmatrix}\begin{pmatrix} \gamma_1 \\ \gamma_2 \\ \cdots \\ \gamma_{N+1} \end{pmatrix} = \begin{pmatrix} RHS_1 \\ RHS_2 \\ \cdots \\ RHS_3 \end{pmatrix} \qquad (40)$$

An additional condition must be established in order to obtain a unique solution that is Kutta condition, which specifies that the circulation at the trailing edge must be zero: $\gamma_{TE} = 0$.

For our model, the circulation is given by: $\gamma_{TE} = \gamma_1 + \gamma_{N+1}$, and the Kutta condition is

$$\gamma_1 + \gamma_{N+1} = 0 \qquad (41)$$

This extra condition is added to the system of equations to give:

$$\begin{bmatrix} a_{11} & a_{12} & \cdots & a_{1N+1} \\ a_{21} & a_{22} & \cdots & a_{2N+1} \\ \cdots & \cdots & \cdots & \cdots \\ a_{N1} & a_{N2} & \cdots & a_{NN+1} \\ 1 & 0 & \cdots & 1 \end{bmatrix}\begin{pmatrix} \gamma_1 \\ \gamma_2 \\ \cdots \\ \gamma_{N+1} \end{pmatrix} = \begin{pmatrix} RHS_1 \\ RHS_2 \\ \cdots \\ RHS_n \\ 0 \end{pmatrix} \qquad (42)$$

The above set of equations can be solved for γ_i by using standard methods of linear algebra.

5.6. Calculation of the Velocity

The velocity is obtained by adding the tangential components of (u,v) of each panel to the tangential component of the external flow velocity.

So, we have to build the $N \times N+1$ matrix b of coefficients b_{ij} such that: $b_{ij} = (u,v)_{ij} \cdot t_i$

Where: $t_i = (\cos\alpha_i, -\sin\alpha_i)$ and the vector Ue_∞ of terms: $Ue_\infty = -(u_\infty, v_\infty).(\cos\alpha_i, -\sin\alpha_i)$

Then, we have

$$\begin{pmatrix} U_{e1} \\ U_{e2} \\ \cdots \\ U_{e4} \end{pmatrix} = \begin{bmatrix} b_{11} & b_{12} & \cdots & b_{1N+1} \\ b_{21} & b_{22} & \cdots & b_{2N+1} \\ \cdots & \cdots & \cdots & \cdots \\ b_{N1} & b_{N2} & \cdots & b_{NN+1} \end{bmatrix}\begin{pmatrix} \gamma_1 \\ \gamma_2 \\ \cdots \\ \gamma_{N+1} \end{pmatrix} + \begin{pmatrix} U_{e\infty 1} \\ U_{e\infty 2} \\ \cdots \\ U_{e\infty 4} \end{pmatrix} \qquad (43)$$

which gives the tangential velocity at each airfoil collocation point.

5.7. Computation of the Pressure

The Bernoulli equation applied to a streamline between the upstream infinity and a point on the airfoil's surface gives:

$$p + \frac{1}{2}\rho u_e^2 = p_\infty + \frac{1}{2}\rho_\infty u_\infty^2 \qquad (44)$$

so: $$p = p_\infty + \frac{1}{2}\rho_\infty u_\infty^2 - \frac{1}{2}\rho u_e^2 \qquad (45)$$

then: $$C_P = \frac{P - p_\infty}{0.5\rho_\infty U_\infty^2} = 1 - \frac{U_e^2}{U_\infty^2} \qquad (46)$$

We can thus compute the pressure coefficient at each airfoil collocation point.

5.8. Computation of the Aerodynamic Coefficients

As shown in Figure 11, the elementary forces f_{xj} and f_{yj} acting on panel j is obtained as:

$$f_{xj} = cp_j(y_{j+1} - y_j) \text{ and } f_{yj} = cp_j(x_{j+1} - x_j)$$

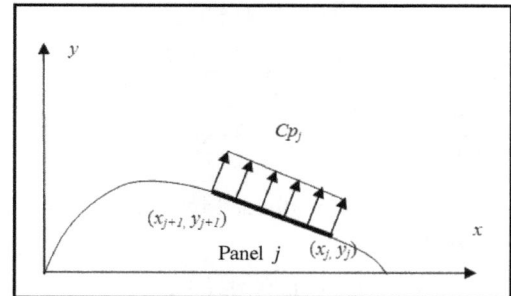

Figure 11. *Elementary forces on panel j.*

By doing this for each panel and by adding the elementary forces the total load applied to the airfoil is obtained (F_x, F_y).

The lift coefficient is then the component of F normal to the flow direction:

$$C_L = -\sin\alpha.F_x + \cos\alpha.F_y \qquad (47)$$

6. Computational Code and Test Cases

The flow along the tandem blade will be studied using different simulations for various relative blade positions. In all the tandem models, the front blade is fixed and the rear blade varies its position as a function of the axial and tangential displacements.

Two different approaches are chosen to identify the influence of location of the rear blade in the tandem arrangement:

- The blade is positioned in such a way that there are no gap nozzle effects between the blades.
- The blade is positioned in the vicinity of the trailing edge of the front blade in order to create a gap nozzle area.

A "Visual Basic" code was developed using linear strength vortex-panel method to predict the aerodynamic performance of tandem blades for various relative blade positions, Figure 12.

6.1. Initial Configuration

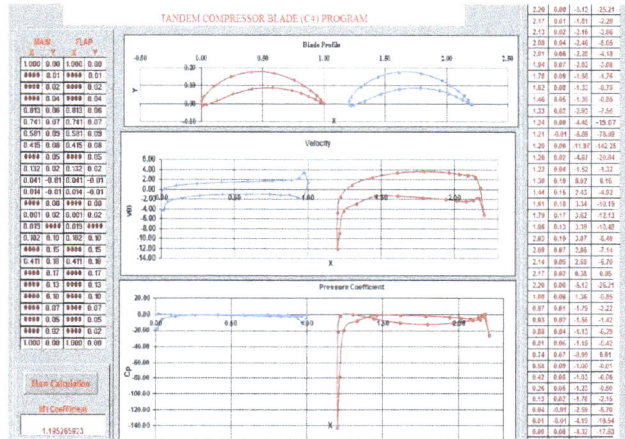

Figure 12. Tandem blade program profile.

The first step in the investigation is to characterize the flow when the second blade of the tandem blades is positioned in the "no gap nozzle effect", Figure 13. The interaction between the two blades is only evident in terms of the wake of the front blade that affects the flow behavior along the arrangement. To characterize the influence of the second blade position, different values are given for the axial and tangential displacements of the second blade. Figure 14 shows a graphical representation of the front blade's trailing edge with the possible second blade locations in terms of the axial and tangential displacements. There are a range of three axial displacements that varies between 0 and 0.2 and a range of five tangential displacements that varies between 0.15 and -0.15. These two ranges represent relative position of the rear

blade with respect to the front blade. All the displacements (axial/tangential) are measured from the trailing edge of front blade.

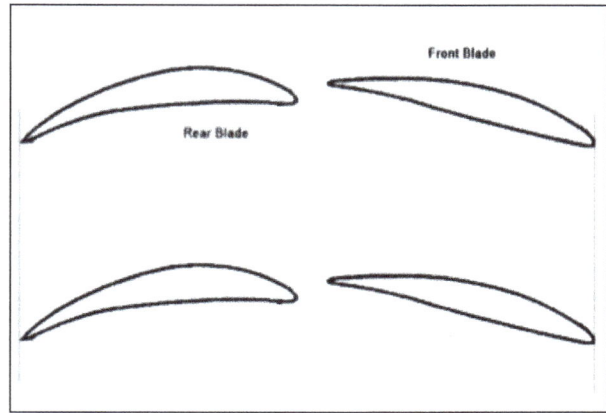

Figure 13. Tandem blade arrangement with "no gap nozzle area" between the profiles [7].

Figures 15-19 show the influence of the rear blade position on the pressure coefficient and velocity distributions. The axial displacement for all figures is set to 20% of the chord of the front blade. Table 3 illustrates the relative positions of the tandem blades for the case of "no gap nozzle effect".

6.2 Inlet Conditions

The incoming flow has a free-stream velocity of 1 m/s, an angle of attack of 8° and an inlet flow angel of 35°.

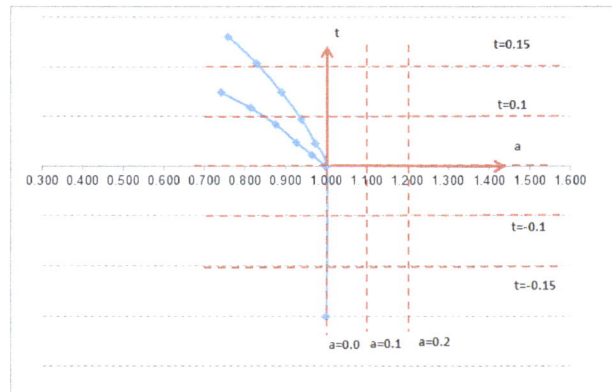

Figure 14. Terminology of relative positions of tandem blades.

Table 3. Relative positions of the tandem blades

Axial displacement [a]	Tangential displacement [t]	Comments
0.2	0.0	The second blade is positioned in the viscous-free region of the first blade
0.2	0.1	"No gap nozzle effect" between the two blades
0.2	0.15	"No gap nozzle effect" between the two blades
0.2	-0.1	"No gap nozzle effect" between the two blades
0.2	-0.15	"No gap nozzle effect" between the two blades

7. Results and Discussions

7.1. Original Blade Configuration

Figure 15a shows the relative positions of tandem blades for

a = 0.2, t = 0. Figure 15b demonstrates the velocity distribution of the tandem blades when the second blade is positioned in the viscous-free region of the front blade. The velocity distribution for the forward blade shows a steady constant acceleration on the suction surface over most of the chord length. On the pressure surface, the velocity drops to a low value (-4) and then increases up till $x/C = 0.3$. Then, for $0.3 < x/C < 0.97$, velocity is almost constant. A small sudden drop of velocity is found at $x/C = 0.97$, followed by a rise in velocity toward trailing edge.

For the rear blade, on the suction surface, the velocity drops to a low value (-12) followed by a steady rise up to $x/C = 0.97$. Then, a sudden drop occurs toward the trailing edge. On the pressure surface, the velocity is almost constant over most of the chord. Then, velocity drops suddenly towards the trailing edge. This is a consequence of the interaction between the wake of the front blade and the velocity field in the vicinity of the suction surface of the rear blade.

For the front blade, Figure 15c shows almost constant pressure on the suction surface due to the steady constant acceleration mentioned previously in Figure 15b. Similar behavior is noticed on the pressure surface except at the leading and trailing edges.

For the rear blade, on the suction surface, the pressure diffuses steadily till $x/C = 0.1$. Then, pressure drops in the range $0.1 < x/C < 0.6$ followed by a steady rise in the range $0.6 < x/C < 0.9$. Finally, the flow separates at the trailing edge due to the low flow-momentum produced by the low velocity in that region as shown in Figure 15b. On the pressure surface, the pressure is almost constant for most of the chord.

Figures 16-19 illustrate the velocity and pressure distributions of the tandem blades when the rear blade is positioned at $(a = 0.2, t = 0.1)$, $(a = 0.2, t = 0.15)$, $(a = 0.2, t = -0.1)$ and $(a = 0.2, t = -0.15)$. Generally, there is almost no difference in pressure and velocity distributions with the first case, when the rear blade is placed at the wake of the front blade $(a = 0.2, t = 0.0)$.

In Figures 15-19, it is clear that the flow behavior along the suction surface of the front blade is not affected by the relative position of the rear blade. This may be explained as a consequence of the no-interaction between the blades. However, there is a noticeable variation of the maximum negative pressure at the leading edge of the rear blade. This may be attributed to a higher inlet velocity to the rear blade due to its relative axial position.

The negative value of C_p for $t = -0.15$ is smaller than that of $t = -0.1$. Similar behavior is noticed for the two cases of $t = 0.1$ and $t = 0.15$.

It was not possible to obtain results for the tandem blades with $a > 0.2$ and $t > 0.15$ because the model did not converge due to massive flow separation on the suction surface of the rear blade.

Figure 20 illustrates the lift coefficient of the rear blade at fixed axial displacement $(a = 0.2)$ and different tangential displacements (t).

The lift coefficient increases as the rear blade gets away from the front blade. This may be attributed to the weak interaction between the two blades. Maximum value of lift coefficient is recorded at $t = -0.1$.

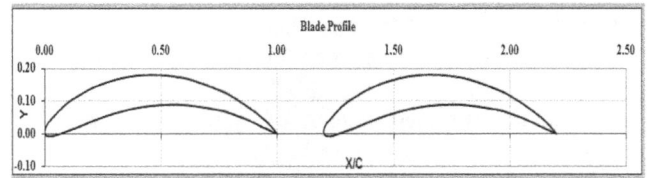

Figure 15a. Relative positions of tandem blades, a = 0.2, t = 0.

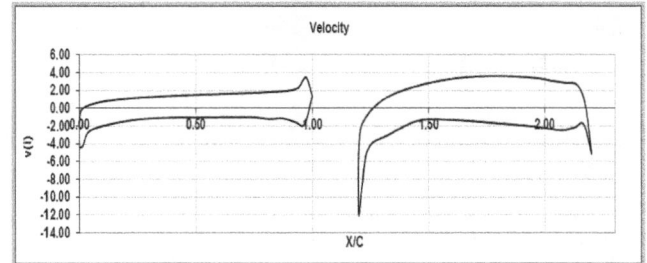

Figure 15b. Velocity distributions, a = 0.2, t = 0.

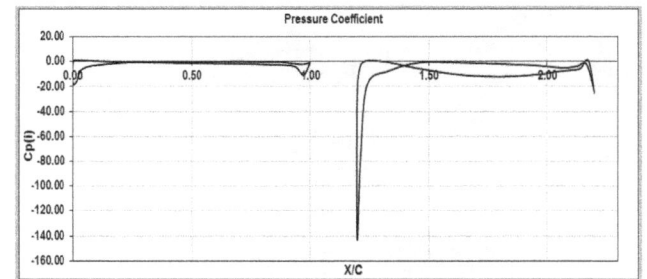

Figure 15c. Distributions of pressure coefficient, a = 0.2, t = 0.

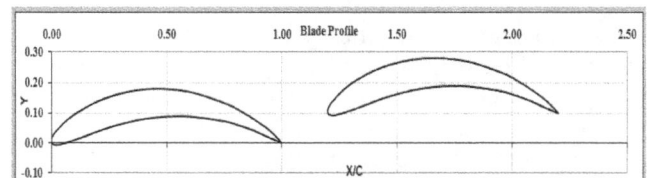

Figure 16a. Relative positions of tandem blades, a = 0.2, t = 0.1.

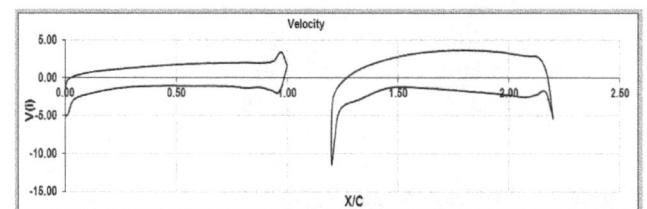

Figure 16b. Velocity distributions, a = 0.2, t = 0.1.

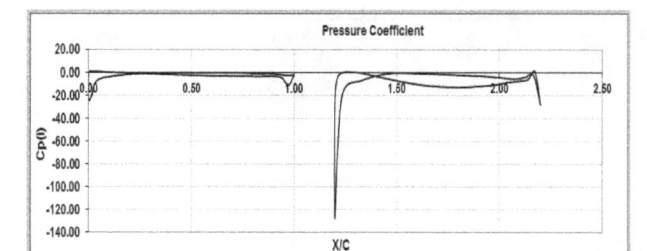

Figure 16c. Distributions of pressure coefficient, a = 0.2, t = 0.1.

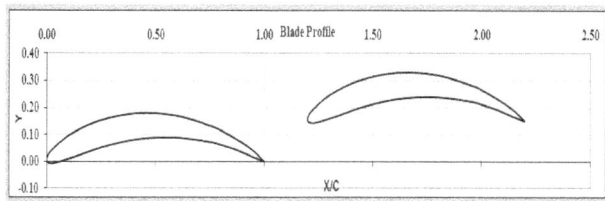

Figure 17a. Relative positions of tandem blades, a=0.2, t=0.15.

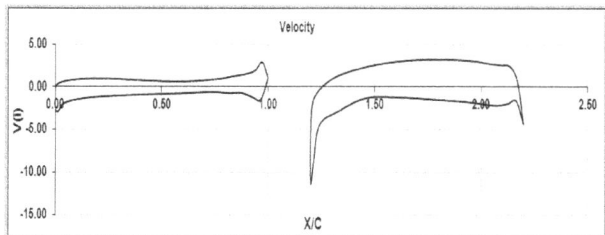

Figure 17b. Velocity distributions, a =0.2, t = 0.15.

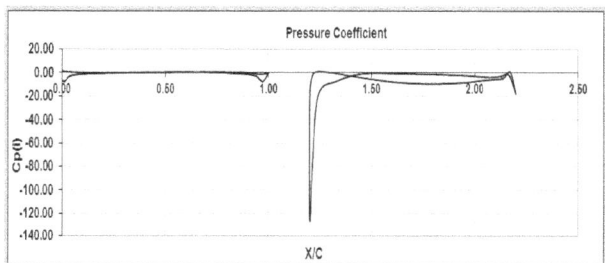

Figure 17c. Distributions of pressure coefficient, a=0.2, t= -0.1.

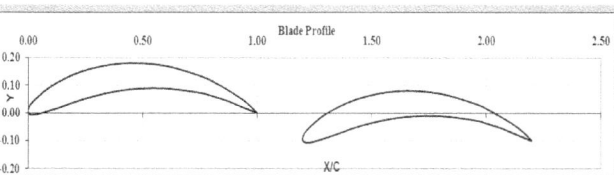

Figure 18a. Relative positions of tandem blades, a=0.2, t= -0.1.

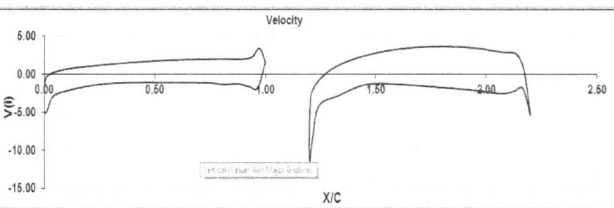

Figure 18b. Velocity distributions, a = 0.2, t = -0.1.

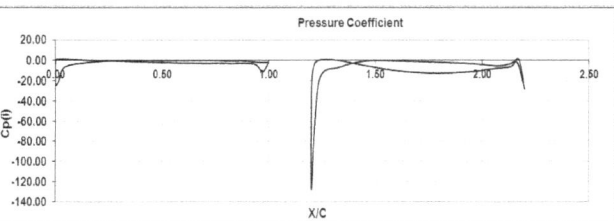

Figure 18c. Distributions of pressure coefficient, a=0.2, t=-0.1.

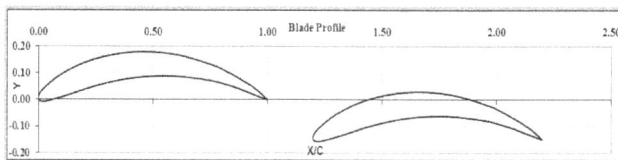

Figure 19a. Relative positions of tandem blades, a=0.2, t=-0.15.

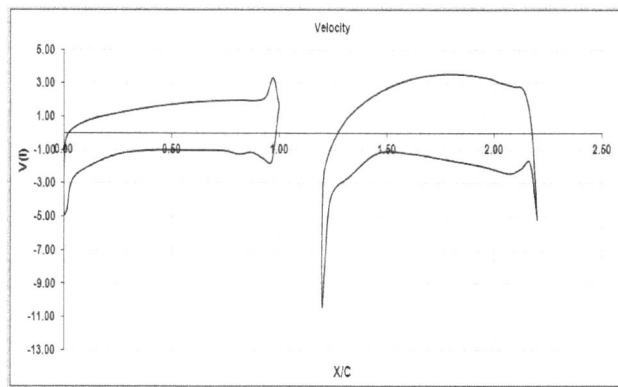

Figure 19b. Velocity distributions, a = 0.2, t = -0.15.

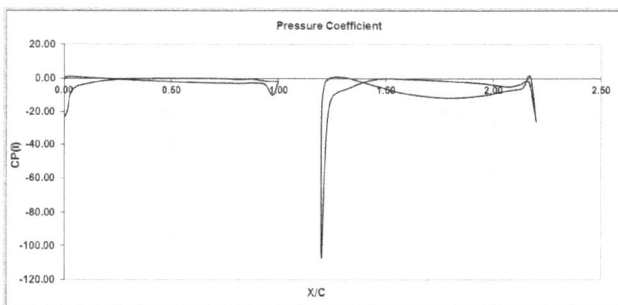

Figure 19c. Distributions of pressure coefficient, a=0.2, t=-0.15.

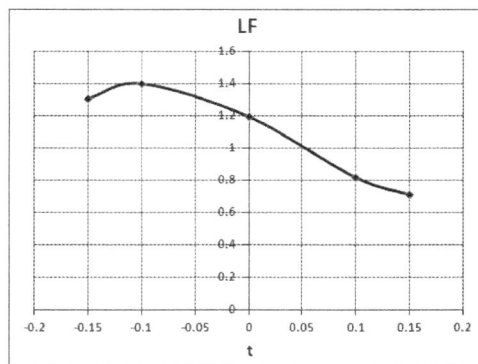

Figure 20. Lift coefficient of rear blade at fixed axial displacement a = 0.2 and different tangential displacements (t).

As a collection of the results that were shown in Figures 15-19, Figures 21a and 21b illustrates the velocity distributions on the front and rear blades, respectively. Generally, it is clear from Figure 21a that the flow along the pressure surface of the front profile is slightly decelerated in comparison to the case of tangential displacement ($t = 0$). However, the velocity on the suction surface varies considerably depending on the relative location of the rear blade. Figure 21b demonstrates that that there is a very limited effect of the relative location of the rear blade on the velocity distributions on both the suction and pressure surfaces of the rear blade.

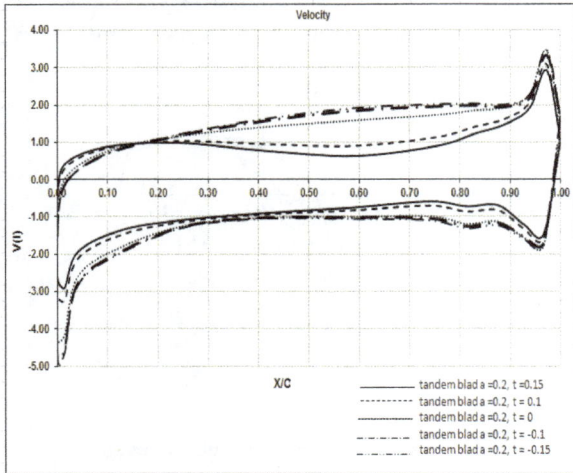

Figure 21a. Velocity distribution on front blade, "No gap nozzle effect".

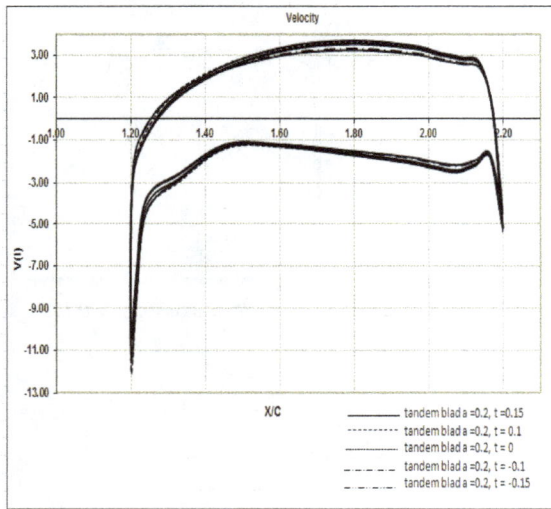

Figure 21b. Velocity distribution on rear blade, "No gap nozzle effect".

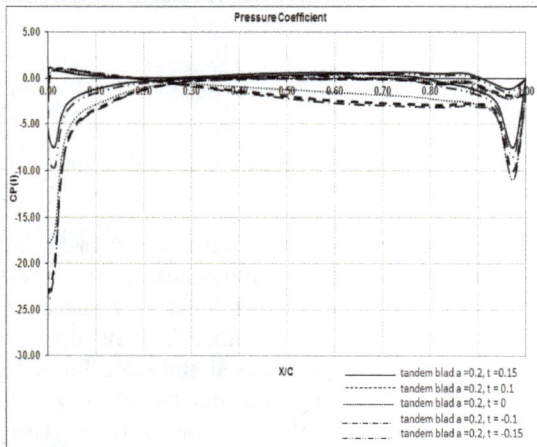

Figure 22a. Pressure distribution on front blade, "No gap nozzle effect".

Figure 22a shows the pressure distribution on the front blade. On the pressure surface, there is a big increase of the negative value of the pressure near the leading edge due to the changes of velocity in this region (Figure 21a). On the suction surface, the pressure values increase or decrease depending on the changes of velocity (Figure 21a) according to the relative position of the rear blade.

Again, Figure 22b shows minor effect of the relative position of the rear blade on the pressure distributions on both the suction and pressure surfaces (similar to Figure 21b).

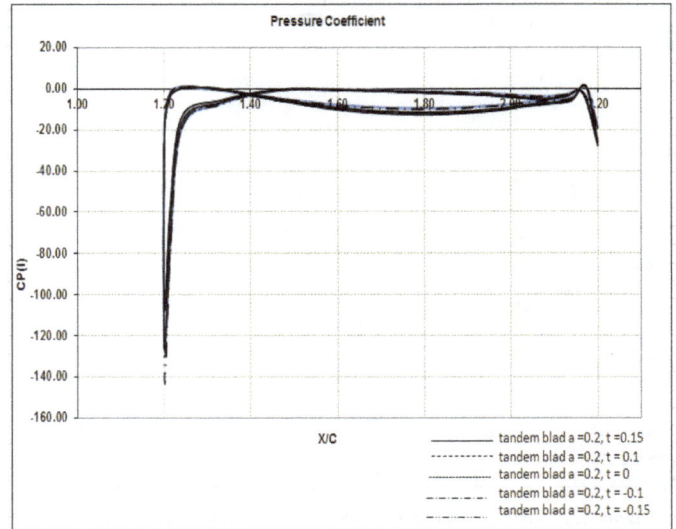

Figure 22b. Pressure distribution on rear blade, "No gap nozzle effect".

7.2. Variable Camber Blade

The concept of changing the camber of the rear blade is usually used to vary only the blade stagger of the stator rows in case of "no gap nozzle effect", Figure 23. The versatility of the tandem (rear) blade lies on its variable loading capability. This may be achieved by changing the stagger of the rear blade only; and thereby increasing the overall camber of the blades.

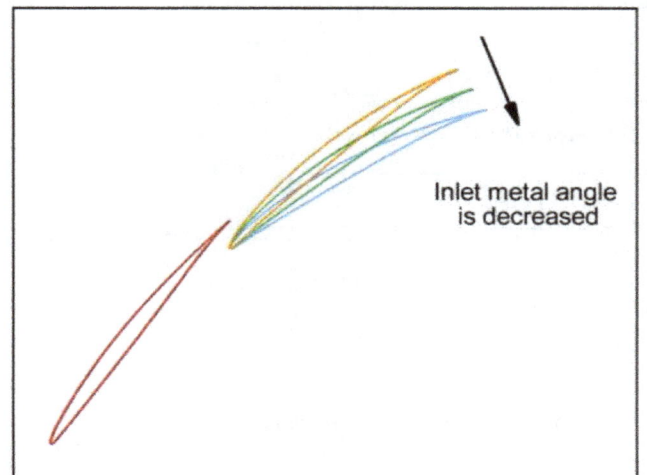

Figure 23. Increasing overall camber of the two tandem blades.

Several test cases were considered to demonstrate the increase in overall camber of the tandem blades by changing the stagger of the rear blade in the range from $10°$ to $60°$ relative to the front blade.

Figure 24a demonstrates the case of the stagger angle of the rear blade equals to $10°$ with relative position $a = 0.2, t = -0.1$ in respect to the front blade. The velocity distributions, Figure 24b, show an increase in velocity levels on both the front and rear blades as compared to the corresponding case with no change in the stagger angle of the rear blade (Figure 18b). The increase in the velocity level of the front blade is probably due to the increase of the loading of the front blade due to the change in the stagger angle of the rear blade. While the increase in the velocity level of the rear blade is due to better inlet flow angle to the rear blade, i.e., good flow path over the suction and pressure surfaces of the rear blade. As for the pressure distributions of the front and rear blades, Figure 24c ,there is a noticeable increase in the pressure values especially near the leading edge of the rear blade in comparison to the corresponding case with no change in the stagger angle of the rear blade (Figure 18c). This may be attributed to velocity changes (Figure 24b).

Figures 25-30 illustrate the change of the velocity and pressure distributions for different stagger angles ($15°$, $20°$, $30°$, $40°$, $50°$, $60°$) of the rear blade relative to the front blade.

Figure 31 shows the variation of the lift coefficient of the rear blade with the stagger angle. It is clear that the lift coefficient increases with the increase of stagger angle due to increase in flow attachment to the surface of the blade till $50°$, where it reaches a maximum value of 2.312. Then, the lift coefficient starts to decrease due to flow separation. Hence, the increase of camber becomes less beneficial. This value of 2.312 is a considerable increase of the lift coefficient when compared to the original configuration, Figure 20, where the maximum value was 1.4.

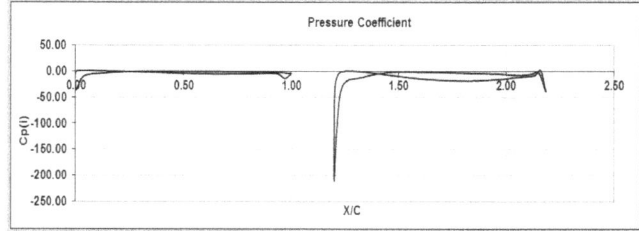

Figure 24a. *Relative positions of tandem blades, a=0.2, t= -0.1, rear blade stagger angle $\lambda = 10°$.*

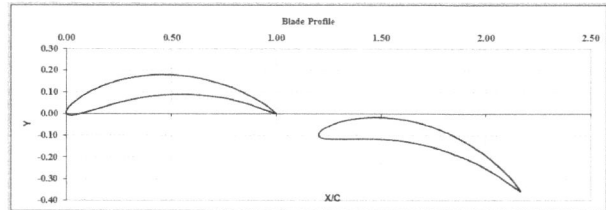

Figure 24b. *Velocity distributions, a = 0.2, t = -0.1, rear blade stagger angle $\lambda = 10°$.*

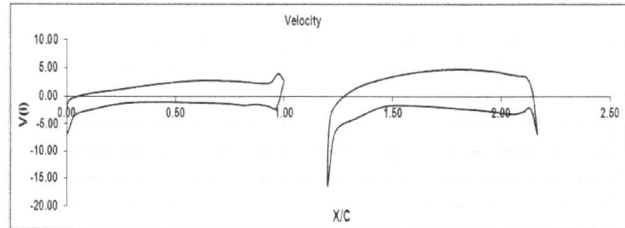

Figure 24c. *Distributions of pressure coefficient, a=0.2, t= -0.1, rear blade stagger angle $\lambda = 10°$.*

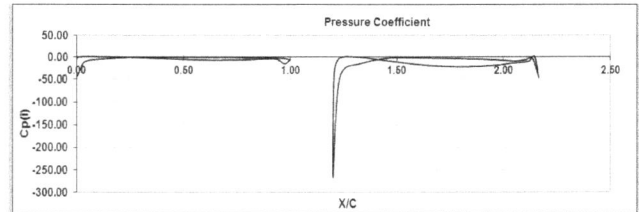

Figure 25a. *Relative positions of tandem blades, a=0.2, t= -0.1, rear blade stagger angle $\lambda = 15°$.*

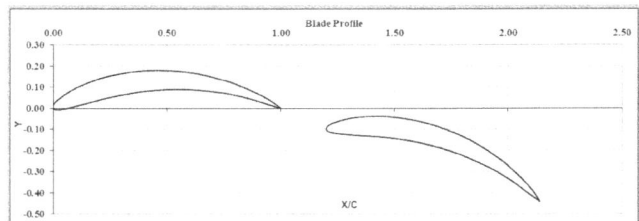

Figure 25b. *Velocity distributions, a = 0.2, t = -0.1, rear blade stagger angle $\lambda = 15°$.*

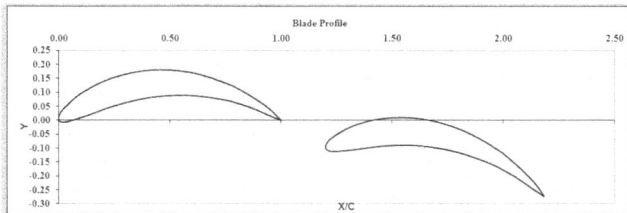

Figure 25c. *Distributions of pressure coefficient, a=0.2, t= -0.1, rear blade stagger angle $\lambda = 15°$.*

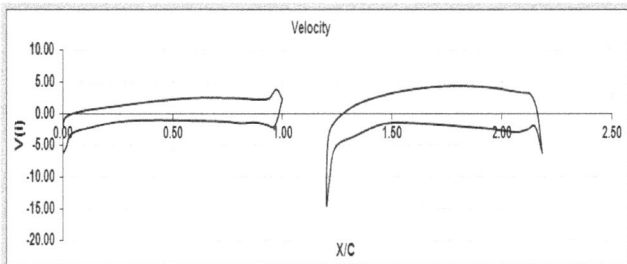

Figure 26a. *Relative positions of tandem blades, a=0.2, t= -0.1, rear blade stagger angle $\lambda = 20°$.*

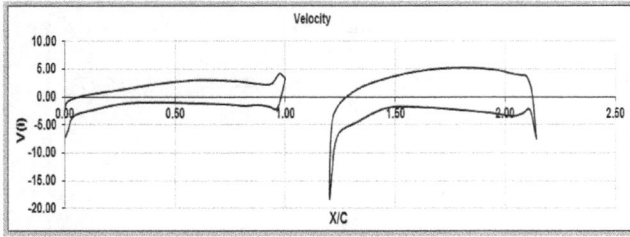

Figure 26b. *Velocity distributions, a = 0.2, t = -0.1, rear blade stagger angle λ = 20°.*

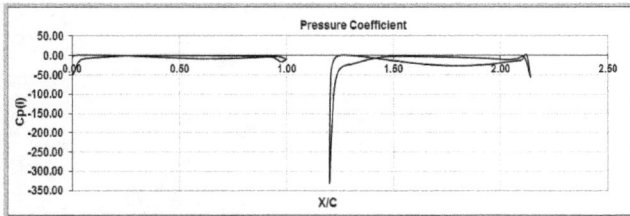

Figure 26c. *Distributions of pressure coefficient, a=0.2, t= -0.1, rear blade stagger angle λ = 20°.*

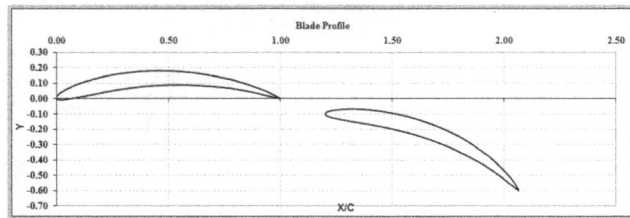

Figure 27a. *Relative positions of tandem blades, a=0.2, t= -0.1, rear blade stagger angle λ = 30°.*

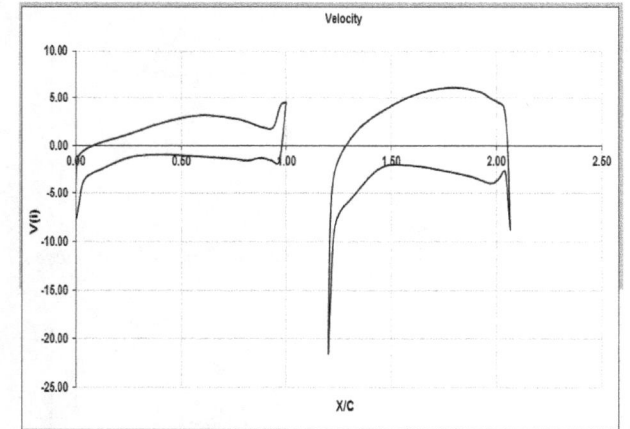

Figure 27b. *Velocity distributions, a = 0.2, t = -0.1, rear blade stagger angle λ= 30°.*

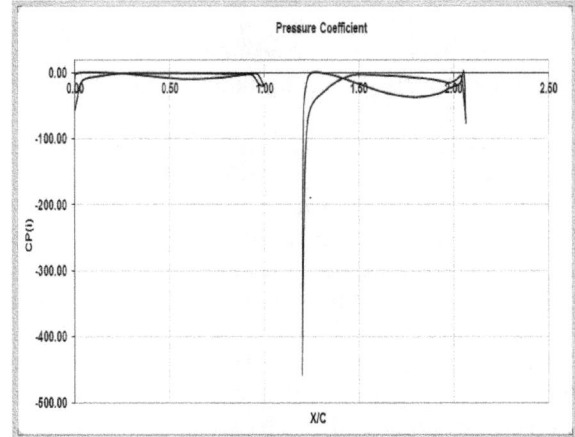

Figure 27c. *Distributions of pressure coefficient, a=0.2, t= -0.1, rear blade stagger angle λ = 30°.*

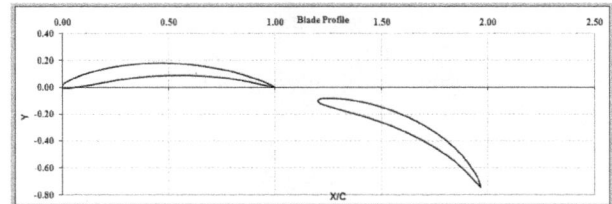

Figure 28a. *Relative positions of tandem blades, a=0.2, t= -0.1, rear blade stagger angle λ = 40°.*

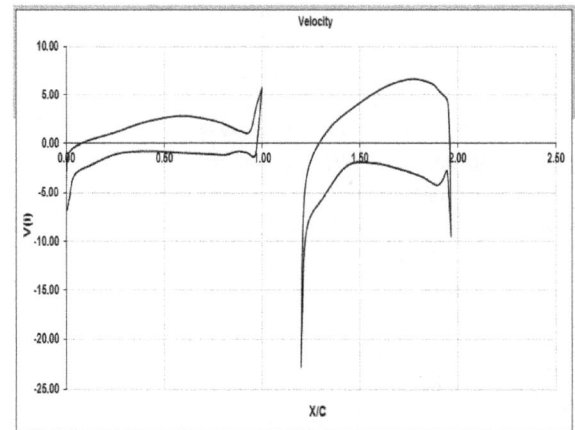

Figure 28b. *Velocity distributions, a = 0.2, t = -0.1, rear blade stagger angle λ = 40°.*

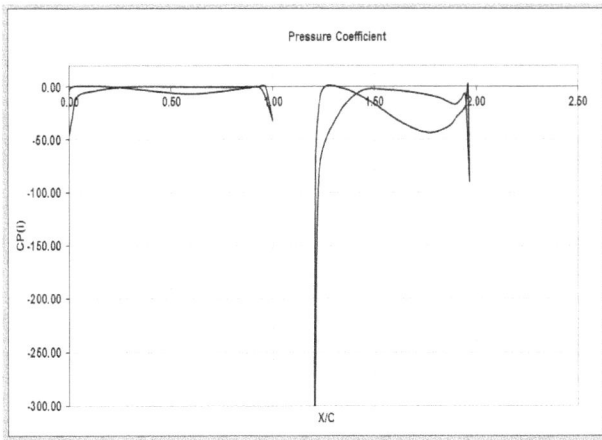

Figure 28c. Distributions of pressure coefficient, a=0.2, t= -0.1, rear blade stagger angle λ = 40°.

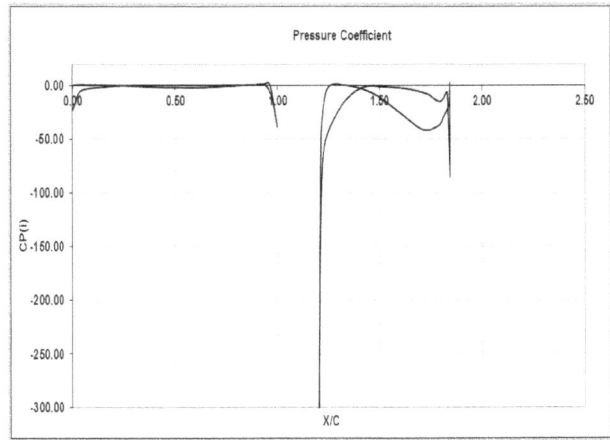

Figure 29c. Distributions of pressure coefficient, a=0.2, t= -0.1, rear blade stagger angle λ = 50°.

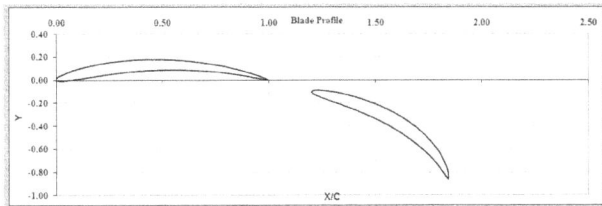

Figure 29a. Relative positions of tandem blades, a=0.2, t= -0.1, rear blade stagger angle λ = 50°.

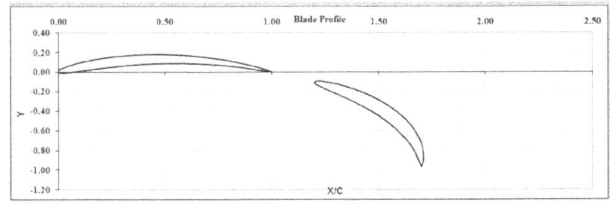

Figure 30a. Relative positions of tandem blades, a=0.2, t= -0.1, rear blade stagger angle λ = 60°.

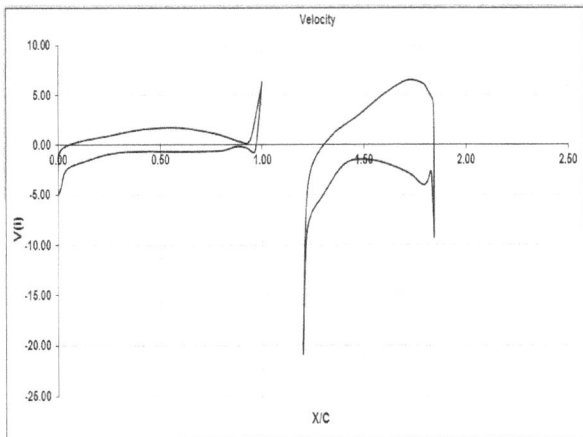

Figure 29b. Velocity distributions, a = 0.2, t = -0.1, rear blade stagger angle λ =50°.

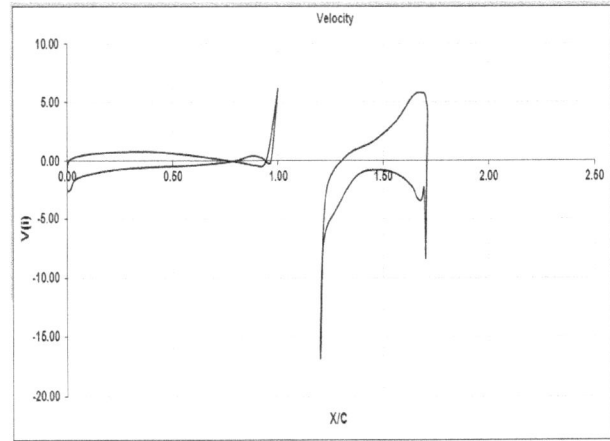

Figure 30b. Velocity distributions, a = 0.2, t = -0.1, rear blade stagger angle λ = 60°.

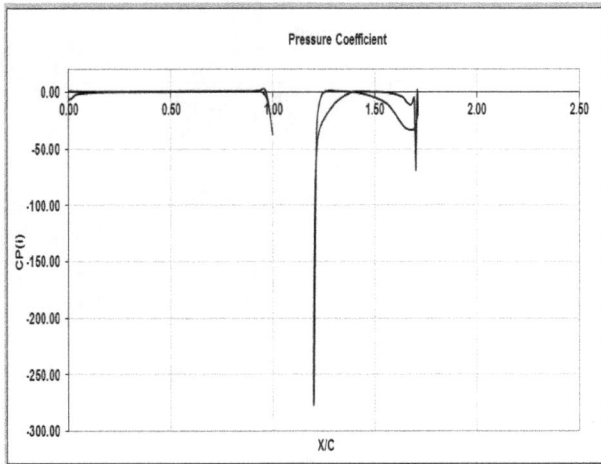

Figure 30c. *Distributions of pressure coefficient, a=0.2, t= -0.1, rear blade stagger angle λ = 60°.*

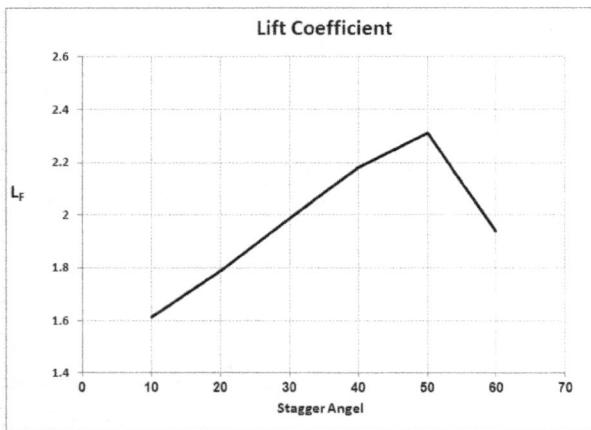

Figure 31. *Lift coefficient variation of the rear Blade with its stagger angle.*

Figure 32a illustrates velocity distribution on front blade for different staggered angles of the rear blade, "No gap nozzle effect". It is evident that the flow is highly influenced by the change in stagger angle of the rear blade. This influence is presented as a blockage effect that causes a lower flow velocity in the proximity of the trailing edge on the pressure surface with increasing the stagger angle. It is also shown that the lowest velocity on the pressure surface occurs when the rear blade is set at a stagger angle of 50°. This is due to the increase of the blockage effect. Generally, on the suction surface, the velocity increases with the increase in stagger angle of the rear blade.

Figure 32b shows velocity distribution along the rear blade. On the pressure surface the velocity decreases with the increase in stagger angle of the rear blade. With the increase in the stagger angle, the inlet flow angle to the rear blade increases. On the suction surface, the contrary to the pressure surface occurs. The velocity increases with the increase of the stagger angle of the rear blade and reaches a maximum value at a stagger angle of 50°.

Figure 33a shows the pressure distribution on front blade for different stagger angles of the rear blade, "No gap nozzle effect". On the pressure surface, it is shown that pressure

increases with increasing the stagger angle of the rear blade due to the decrease in velocity on that surface as explained previously in Figure 32a. While on the suction surface, the pressure increases by small amounts with increasing in stagger angle.

Figure 33b demonstrates the pressure distribution on rear blade for different stagger angles of the rear blade, "No gap nozzle effect". The pressure distributions on both the pressure and suction surfaces have the same pattern of the front blade except for the increase in pressure limits of the rear blade as compared to the front blade.

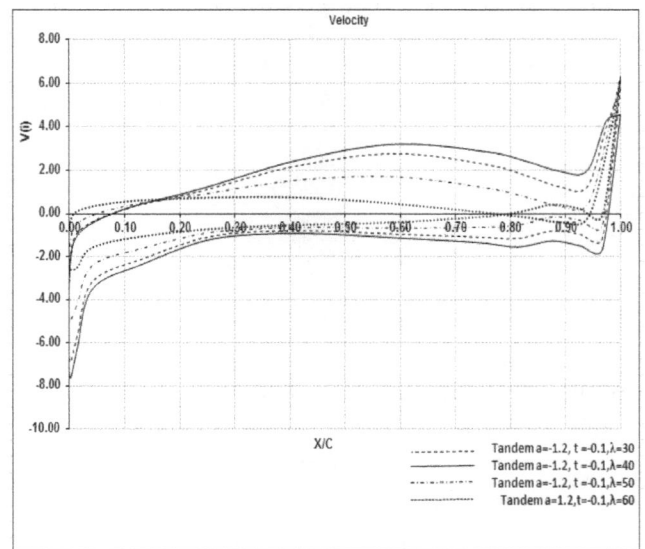

Figure 32a. *Velocity distribution on front blade for different stagger angles of the rear blade, "No gap nozzle effect".*

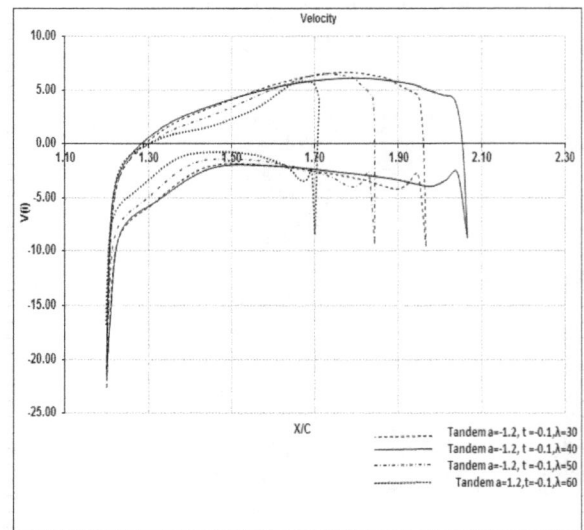

Figure 32b. *Velocity distribution on rear blade for different stagger angles of the rear blade, "No gap nozzle effect".*

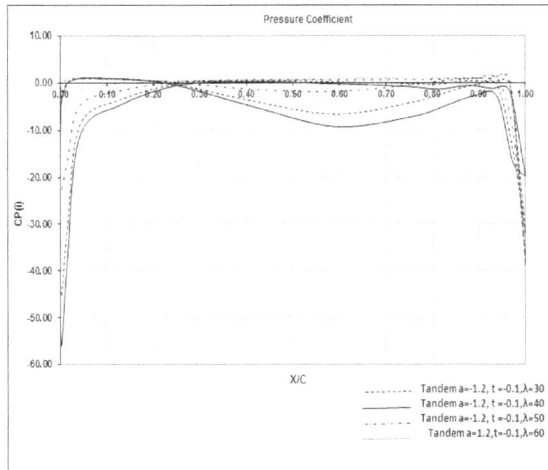

Figure 33a. Pressure distribution on front blade for different stagger angles of the rear blade, "No gap nozzle effect".

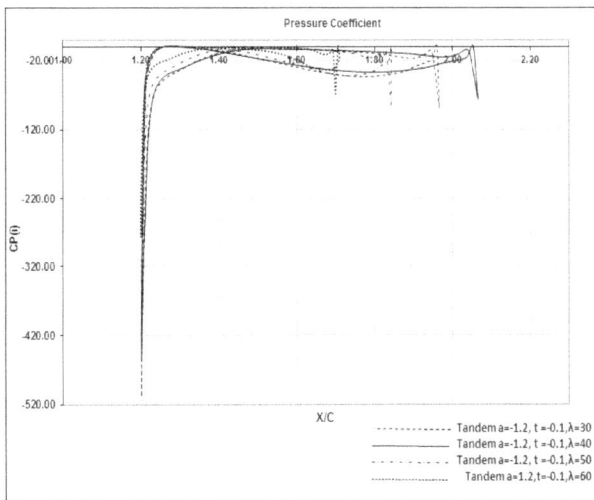

Figure 33b. Pressure distribution on rear blade for different stagger angles of the rear blade, "No gap nozzle effect".

7.3. Gap Nozzle Effect

7.3.1. Geometry Configuration

The gap nozzle effects appear in the tandem blades when the rear blade is positioned in the vicinity of the trailing edge of the front blade. The zone between the blades can be represented as a convergent gap with an inlet and outlet area characterized by the distances $F1$ and $F2$, respectively, Figure 34.

Therefore, the relation $F1/F2$ gives a measure of the flow acceleration by the presence of the nozzle configuration. In this section the ratio $F1/F2$ is used as a characteristic parameter for each test case of tandem blades.

The tangential displacements that satisfy the condition fixed by the gap nozzle inlet and outlet ratio were calculated graphically using Excel. An example of the procedure is depicted in Figure 35 for tandem blades with $F1/F2$ equal to 2.2 and axial distance of -0.87. In the Figure 35, the value of $F1=0.11$ and $F2=0.05$, thus the ratio is $F1/F2 = 2.2$.

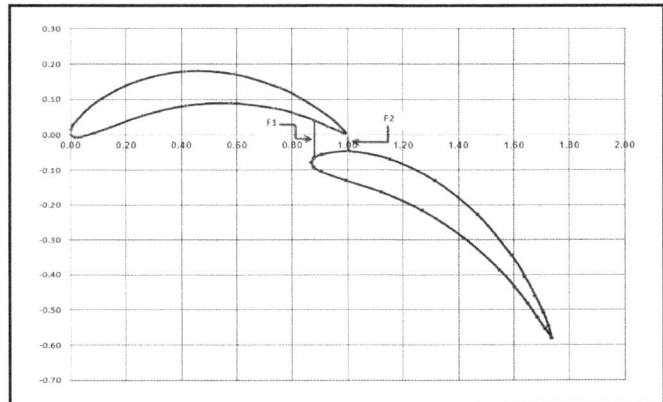

Figure 34. Geometry configuration, "Gap nozzle effects".

This part of the investigation was conducted to assess the advantage of a tandem cascade with the presence of gap nozzle geometry between the blades as opposed to the two single blades acting independently. The study is made on the bases of the static pressure distribution along the blades and the lift coefficient. The final gap-nozzle parameters that describe the area between the profiles are summarized in Table 4 for the various tested cases.

Table 4 Gap nozzle effect, tandem blades

Axial displacement [a]	Tangential displacement [t]	F1/F2
-0.87	-0.08	2.2
	-0.1	1.86
	-0.06	3.33

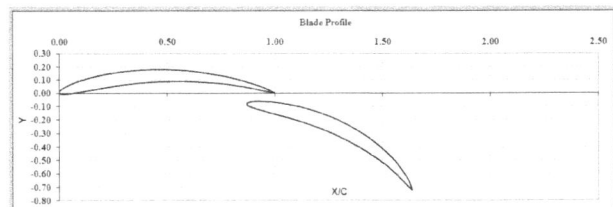

Figure 35a. Relative positions of tandem blades, a = -0.87, t= -0.08, rear blade angle relative to front blade=40, F1/F2=2.2.

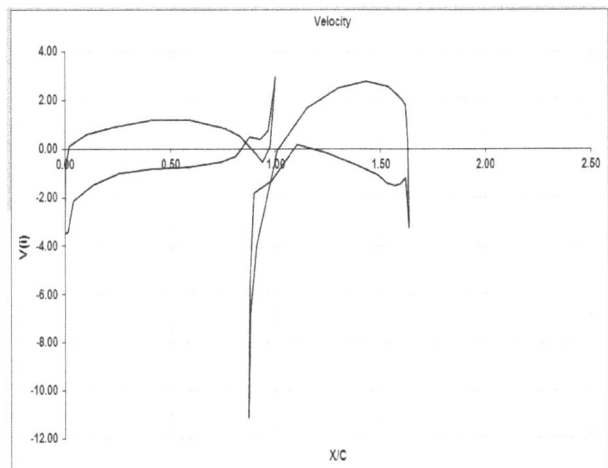

Figure 35b. Velocity distributions of tandem blades, a = -0.87, t = -0.08, rear blade stagger angle = 40, F1/F2 = 2.2.

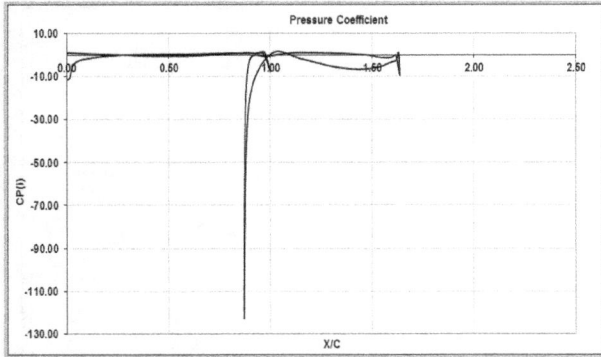

Figure 35c. *Pressure distributions of tandem blades, a = -0.87, t = -0.08, rear blade stagger angle = 40, F1/F2 = 2.2.*

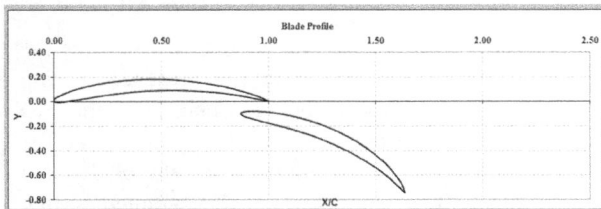

Figure 36a. *Relative positions of tandem blades, a = -0.87, t= -0.1, rear blade angle relative to front blade=40, F1/F2=1.86.*

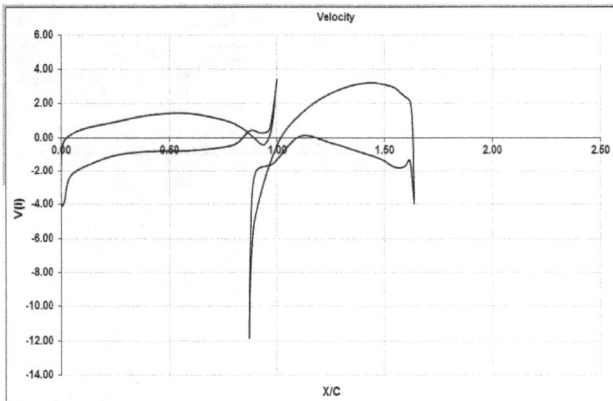

Figure 36b. *Velocity distributions of tandem blades, a = -0.87, t = -0.1, rear blade stagger angle = 40, F1/F2 = 1.86.*

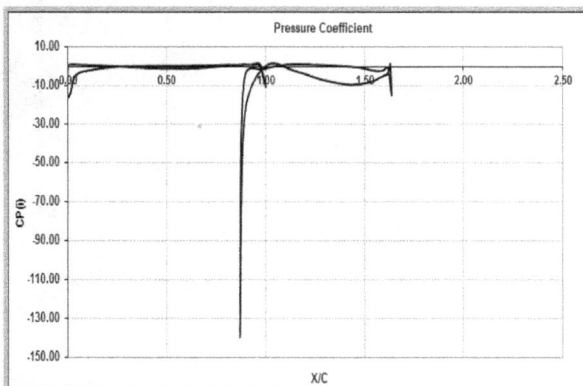

Figure 36c. *Pressure distributions of tandem blades, a = -0.87, t = -0.1, rear blade stagger angle = 40, F1/F2 = 1.86.*

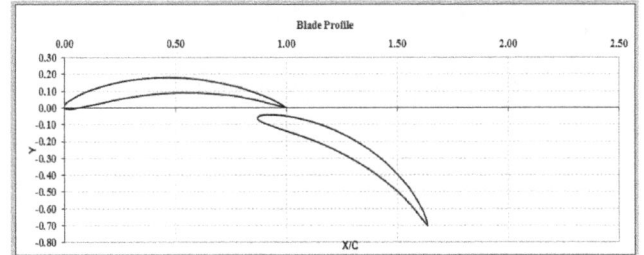

Figure 37a. *Relative positions of tandem blades, a = -0.87, t= -0.06, rear blade angle relative to front blade=40, F1/F2=3.33.*

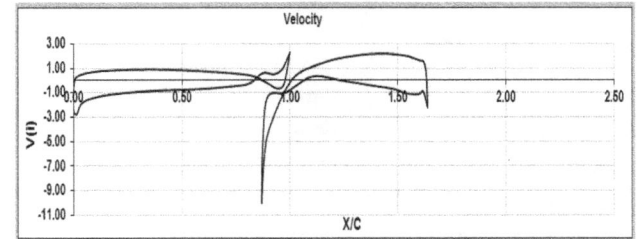

Figure 37b. *Velocity distributions of tandem blades, a = -0.87, t = -0.06, rear blade stagger angle = 40, F1/F2 = 3.33.*

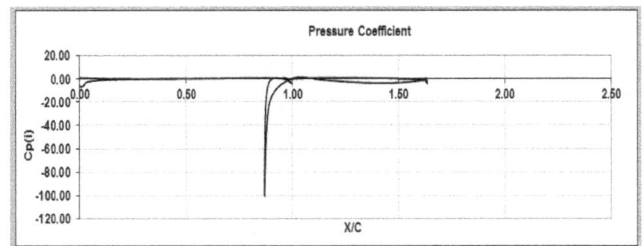

Figure 37c. *Pressure distributions of tandem blades, a = -0.87, t = -0.06, rear blade stagger angle = 40, F1/F2 = 3.33.*

7.3.2. Test Cases

From the results of Figures 35-37, it is noticed that the relative position of the rear blade does not disturb the flow along the suction surface of the front blade. This behavior is expected because the interaction between the two blades is inexistent in this surface. Thus, the flow behavior is quite similar to the flow along the blade acting alone.

So, it is noticed that there is an initial acceleration up to 60% of the chord from the leading edge; then, the flow starts decelerating in the range from $0.6 < x/C < 0.92$, followed by steep diffusion towards the trailing edge.

Nevertheless, it is appreciable that if the tangential displacement of the rear blade is quite large, which means low values of *F1/F2* ratio, the flow is slightly more decelerated than the corresponding flow on the suction surface of a rear blade with no gap nozzle. This fact is attributed to a mass flow balance between the gap nozzle area, where the flow has low momentum, and the flow channel between the two consecutive tandem blades.

On the pressure surface of the front blade, it is evident that the flow is highly influenced by the gap nozzle area. This influence is presented as a blockage effect that causes an increase in flow velocity in the proximity of the front blade trailing edge.

Figure 38a illustrates a comparison between the various values of $F1/F2$ for the front blade with a = -0.2. It is clear that the lowest velocity on the pressure surface occurs when the ratio $F1/F2$ is the highest. It is also appreciable that when the ratio $F1/F2$ increases, there is an increment in the blockage effect. The flow behavior, on this surface, shows a constant deceleration up to the gap channel limit. Thereafter, the flow enters into the gap nozzle area and the velocity increases towards the trailing edge of the blade. This behavior is due to the nozzle effect in the gap zone.

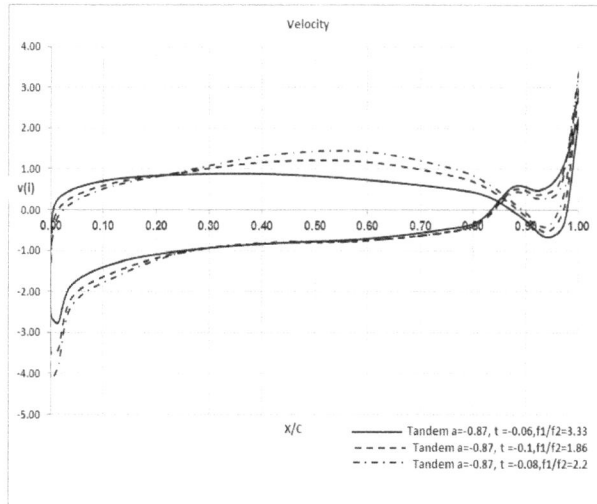

Figure 38a. Velocity distribution on front blade for different arrangements, "Gap nozzle effect".

Figure 38b shows that, on the pressure surface of the rear blade and for high values of $F1/F2$, the flow is forced to go along the pressure surface of the rear blade due to the blockage phenomenon in the gap nozzle area, which is presented as a decrement of the flow peak velocity.

Thus, the momentum transfer on the suction surface of the rear blade is not sufficient to ensure the no presence of flow disturbances. The low-momentum flow at the inlet of the gap zone is responsible for the flow separation at the rear part of the rear blade. Thus, an increase of the total losses is expected because the blade operates in stall conditions.

Concerning the pressure distribution on the front blade, Figure 39a, it is demonstrated that, as seen before in the velocity distribution, the relative position of the rear blade does not disturb the flow along the suction surface of the front blade. Thus, the flow behavior is quite similar to the flow along the blade acting alone. On the suction surface, the flow is slightly more accelerated with the larger values of the tangential displacement of the rear blade, *i.e.*, lower values of $F1/F2$. Also, the flow is more accelerated for all cases with "gap nozzle effect" in comparison to the corresponding flow on the suction surface of tandem blades with "no gap nozzle effect". This is due to the blockage effect; *i.e.*, as the ratio $F1/F2$ increases the blockage effect increases.

Figure 39b shows the pressure distribution on the rear blade. On the pressure surface, for high values of $F1/F2$, the flow is forced to go along the pressure surface due to the blockage

phenomenon, which is presented as a decrement of the flow peak velocity in the gap nozzle area.

For all cases, the gap-nozzle geometry prevents high flow acceleration at the suction surface of the rear profile. Therefore, it is said that the peak velocity of the rear blade is influenced by the relative position of the blades in a tandem cascade. In all cases, the flow is characterized by an initial acceleration, which is limited by the nozzle geometry, up to the gap-channel influence limit. Thereafter, a constant deceleration towards the trailing edge is evident.

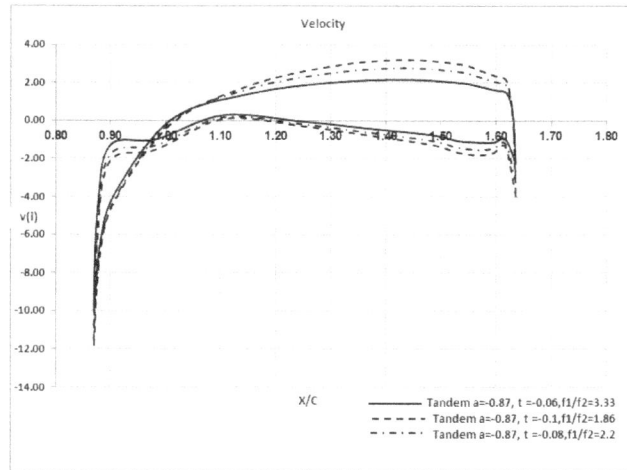

Figure 38b. Velocity distribution on rear blade for different arrangements, "Gap nozzle effect".

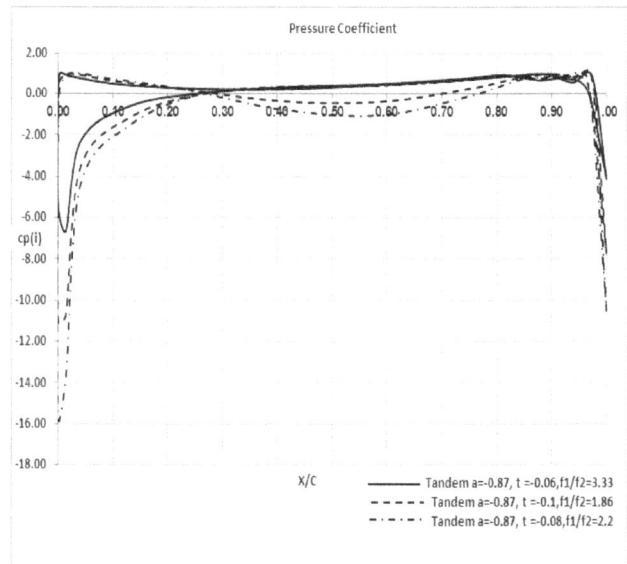

Figure 39a. Pressure distribution on front blade for different arrangements, "Gap nozzle effect".

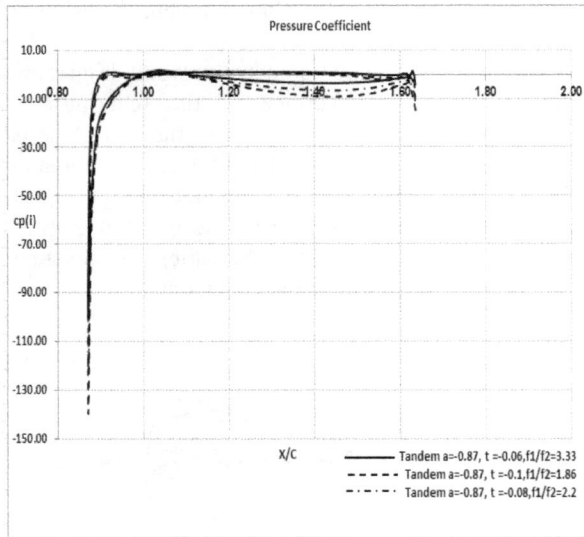

Figure 39b. Pressure distribution on rear blade for different arrangements, "Gap nozzle effect".

Figure 40 illustrates the lift coefficient of rear blade at different gap nozzles between the front and rear blade. In all cases, it is noticed that the gap-nozzle geometry highly influences the peak velocity on the suction surface of the rear blade. It is also concluded that if the rear blade is located in such a way that the gap-nozzle geometry promotes sufficient flow guidance for efficient momentum transfer on the rear blade suction surface, the rear blade shows higher lift.

This is very clear for the cases where the gap-nozzle area is 1.86 and 2.22. In these two cases, the gap-nozzle area promotes sufficient guidance on the rear blade that results in a higher pressure difference across the blade, Figure 39. However, the opposite occurs for gap-nozzle area of 3.33, which does not give enough flow guidance on the rear blade. Thus, lower pressure difference distribution is found and consequently, lower lift is obtained in comparison to the two cases that were mentioned previously.

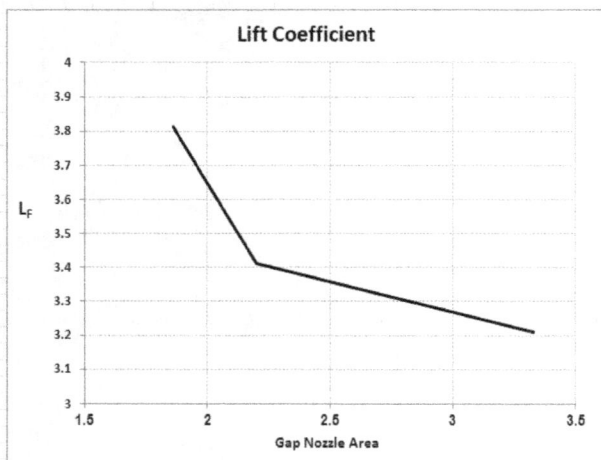

Figure 40. Lift coefficient of rear blade at different gap nozzles between the front and rear blade.

8. Conclusions

In this study, a numerical investigation of the steady two-dimensional flow for compressor tandem blades was carried out. The interaction mechanism between the two blades was inspected by varying the relative position of the two blades. Based on the above results and discussions, the following conclusions are obtained:

1. Although, it is not shown in the paper, validation of the present model and methodology was carried out with the study of [14]. Based on comparisons between the present results and those of [14], it was clear that the present model and methodology are suitable for the present study.

2. The deflection capability of tandem blades is higher than that of a single blade. This increase of deflection angle is attributed to the no-presence of flow separation with the increase of loading.

3. If the rear blade is positioned in the wake of the front blade (no gap nozzle effect), there is a decrease in lift coefficient as a consequence of the velocity deficit that causes flow disturbances on the suction surface of the rear blade.

4. When the rear blade is located in the viscous free region and there is "no gap nozzle effect", the tandem blades show an increase in the lift coefficient. This is a consequence of the interaction between the wake of the first blade and the velocity field in the vicinity of the suction surface of the rear blade.

5. In all cases, it was noticed that the gap nozzle geometry highly influences the peak velocity on the suction surface of the rear blade due to the accelerated flow in the gap nozzle channel.

6. If the second blade is located in such a way that the gap nozzle geometry promotes sufficient flow guidance for efficient momentum transfer on the suction surface of the rear blade, the rear blade shows higher lift than the case of two blades with "no gap nozzle effect".

7. The presence of the rear blade in the proximity of the pressure surface of the front blade causes a decrement in the flow velocity. This decrement is characterized as a blockage effect that increases with the increase of the ratio $F1/F2$ and vice versa.

8. For higher values of $F1/F2$, the gap nozzle geometry does not promote sufficient flow guidance for efficient momentum transfer on the suction surface of the rear blade. Therefore, the lift decreases for higher values of $F1/F2$. On the other hand, when the rear blade is positioned in such a way that the gap nozzle energizes the wake of the front profile and promotes sustained flow attachments on the suction surface of the rear blade (the wake having low momentum is filled up by the high momentum flow near the suction surface of the rear blade), the lift increases.

Nomenclature

a	Axial distance
C	Chord
C_1	Chord length of front blade
C_2	Chord length of rear blade
C_l	Lift coefficient
C_p	Pressure coefficient
C_t	Overall chord length
F_1	Inlet gap distance
F_2	Outlet gap distance
L	Lift
$\dot{m}$	Mass flow rate
P_{shaft}	Shaft power
S	Blade spacing
T	Tangential displacement
U	Velocity in x-direction
U_1	Blade speed at radius r_1
U_2	Blade speed at radius r_2
V	Velocity in y-direction
$V_{\Theta 1}$	tangential velocity at radius r_1
$V_{\Theta 2}$	tangential velocity at radius r_2
W_1	Inlet velocity
W_{12}	Inlet velocity at rear blade
W_2	Outlet velocity

Greek

α	Attack angle
β	Rear blade relative angle to front blade
β_1	Relative inflow angle
β_2	Relative outflow angle
δ	Deviation angle
γ	Vortex strength
Γ	Circulation
λ	Stagger angle
ρ	Density
θ	Camber angle

Abbreviations

FB	Front blade
RB	Rear blade

References

[1] A. J. Wennerstrom, "Low aspect ratio axial flow compressors: why and what it means," ASME Journal of Turbomachinery, Vol. 111, pp. 357-365, 1989.

[2] A. K. Saha, and B. Roy, "Experimental analysis of controlled diffusion compressor cascades with single and tandem airfoils," ASME paper number 95-CTP-41, 1995.

[3] A. K. Saha, and B. Roy, "Experimental investigations on tandem compressor cascade performance at low speeds," Experimental Thermal and Fluid Science, Vol. 14, pp. 263-276, 1997.

[4] E. Sheets, "Multiple row blades for blowers," ASME paper number 88-GT-124, 1988.

[5] A. Weber, and W. Steinert, "Design, Optimization, and Analysis of a High-Turning Transonic Tandem Compressor Cascade," ASME paper number 97-GT-412, 1997.

[6] General Electric Co. Manuals, Al-shabab Power Plant (1000 MW-simple gas turbine plant), Ismailia, Egypt.

[7] K. Bammert, and H. Beelte, "Investigations of an axial flow compressor with tandem cascades," ASME Journal of Engineering Power, pp. 971-977, October 1980.

[8] K. Bammert, and R. Staude, "Optimization for rotor blades of tandem design for axial flow compressors," ASME Journal of Engineering for Power, pp. 369-375, April 1980.

[9] J. McGlumphy, 2-D computational studies of subsonic axial rotors incorporating dual airfoils, UTSR Project Report for summer 2005.

[10] J. McGlumphy, W.-F. Ng, S. R. Wellborn, and S. Kempf, "Numerical investigation of tandem airfoils for subsonic axial-flow compressor blades," Journal of Turbomachinery, Vol. 131, pp. 1-8, April 2009.

[11] J. McGlumphy, W.-F. Ng, S. R. Wellborn, and S. Kempf, "3D numerical investigation of tandem airfoils for a core compressor rotor," Journal of Turbomachinery, Vol. 132, pp. 1-9, July 2010.

[12] G. A. Canon, Numerical investigation of the flow in tandem compressor cascade, Diploma Thesis, Vienna University of Technology, Austria, 2004.

[13] J. Katz, and A. Plotkin, Low-Speed Aerodynamics, 2nd ed., Cambridge University Press, 2001.

[14] C. A. Cox, Two Element Linear Strength Vortex Panel Method, Senior Project, Faculty of the Aerospace Engineering, California Polytechnic State University, San Luis, USA, March 2011.

18

Numerical solution of solar energy absorbed in porous medium with a new approach for vapor pressure calculation and consideration of solute crystallization

Sherif A. Mohamed[1], Ibrahim S. Taha[1], Mahmoud G. Morsy[1], Hany A. Mohamed[2], Mahmoud S. Ahmed[3]

[1]Department of Mechanical Engineering, Faculty of Eng., Assuit University, Assuit, Egypt
[2]Mech. Eng. Department, Faculty of Eng., Al Taif University, Al Taif, Saudi Arabia
[3]Faculty of Industrial Education, Sohag University, Sohag, Egypt

Email address:

sherifadham2000@yahoo.com (S. A. Mohamed)

Abstract: The goal of the study is to enhance the productivity of solar stills using an unsaturated porous medium initially saturated by salty water and using concentrating reflector. This paper concentrates only on the mathematical model for the porous medium and its solution using a finite-volume approach. The previous studies dealt with wick medium with high water content and liquid saturation in the wick medium was not determined. A physical model for the initially saturated porous medium was developed. The model takes into consideration the salt concentration in the solution, surface and internal water diffusions to humid air with vapor pressure determined from vapor mass balance. The system of transient one-dimensional differential equations was developed together with the boundaries and initial conditions. A finite-volume method was used for discretisation of the differential equations. A fully-implicit scheme was used for unsteady term discretisation while the convective terms (liquid solution, vapor and dry air) in the energy equation are handled by an upwind scheme method. The nonlinear equations are solved simultaneously by updating the coefficients matrix at one time step until the five variables converge to prescribed tolerance. Matlab was used as a programming tool. Solution of the model is obtained and discussed.

Keywords: Porous Medium, Solute Concentration, Vapor Pressure, Absorbed Solar Radiation

1. Introduction

The study of porous medium represents the importance in industry, drying, soil contaminants and solar still. The wick material can be used in solar still and it can be considered as a porous medium and the nature of study for porous medium is different from that of wick material. The consideration of porous medium helps to make a mathematical model of differential equations, the mathematical model gives the mass and heat transfer in porous medium also it gives the prediction for solute crystallization. Then the novel usage of porous medium is solar still applications. Then the mathematical model can be formed to know both heat and mass transfer knowledge

Ni et al. [1] had developed a multiphase porous media model to predict moisture transport during intensive microwave heating of biomaterials. They formulated the governing equations and boundary conditions for multiphase moisture transport. They solved the coupled equations numerically, compared the numerical results with convective "non-microwave" heating studies which were reported in their literature. Also they obtained experimental measurements of moisture loss in their study. They assessed the contribution of the convection terms to the energy equation.

Akbar and Haghi [2] studied the heat and mass transfer phenomena occurring within a carpet during combined microwave and convective drying. They analyzed the moisture, temperature, and pressure distributions generated throughout the process.

Costa et al. [3] studied porous media with high water content which can be successfully used as thermal barriers to operate under high exposure temperatures and/ or high heat fluxes. Modeling and simulation of thermal barriers consisting of highly humid porous media was a challenging task. Physical model needed took into account the heat transfer mechanisms, including radiation heat transfer and phase change. Liquid water and water vapor transfers were considered, including the capillary effects for the liquid phase, as well as the air transfer inside the porous medium. They concluded that the mass transfer rate from the thermal barrier to the environment is controlled by the vaporization process and vapor effusion, and not by the convection mechanism.

Neale et al. [4] studied the hygrothermal analysis of buildings which is becoming increasingly utilized for evaluating heat and moisture related problems within the building envelope. They concluded that the convective surface coefficients were particularly important for calculations involving boundary layer heat and mass transfer. The convective moisture coefficients were often calculated through analogy equations, i.e. Lewis and Chilton Colburn ones. However, these equations were not always valid. Therefore, a different approach is needed to accurately determine heat and mass convection coefficients. Murugesan et al. [5] studied numerically the evaporative drying of a two-dimensional rectangular brick. Average heat and mass transfer coefficients appropriate to the conjugate problem were defined based on constant temperature and moisture differentials between the solid and the ambient. Free convection effects on drying were also studied for some initial period for low Reynolds number. It was demonstrated that heat and mass transfer coefficients based on constant temperature and moisture potentials may be more representative of the conjugate heat and mass transfer process during drying. Hence, the conjugation of two coefficients is less representative based on instantaneous temperature difference and moisture difference.

Lee et al. [6] developed a transient two-dimensional mathematical model to simulate the through-air drying process for tufted textile materials. The heat and mass transfer in cylindrical porous medium and air flowing around it were analyzed separately. The resulting system of the three non-linear differential equations was numerically solved by an implicit finite-difference method. Duc Le et al. [7] solved the salt concentration equations in conjunction with the liquid saturation evolution in space and time using numerical and analytical methods for selected cases. The main objective of their study was to combine the advances in the understanding of flow-through evaporation from porous media with solute transport theory and develop a better understanding of evolution of crystallized salt saturation in rocks. The mass of solid salt crystallized was calculated by applying solubility limits on the liquid concentrations for the case of fast crystallization kinetics and capillary-dominated conditions. Koniorczyk and Gawin [8] had taken the salt phase change kinetics into account during the modeling of coupled moisture, salt, and heat transport. Their mathematical model was describing moisture and salt transport in non-isothermal conditions. The process of salt phase change in the pore solution was modeled using the non-equilibrium approach,

where the rate law describing mass change of the crystallized salt is known.

The previous works of wicked solar still did not deal with internal structure of porous media, solar irradiance penetration, utilization of drying period of porous media, crystallization inside pores and effect of partial saturation of the porous medium. However, the studying of heat and mass transfer is considered a surface phenomenon. A porous medium, being a heterogeneous system made of a solid matrix with its void filled with fluids, can be treated as a continuum by properly accounting for the role of each phase in transport through this system of phases [9]. In the previous work of porous medium, the vapor pressure is not taken as a variable in the differential Equations. However, it was not calculated as a variable in the differential equations and the source and sink terms were canceled each other by summing the liquid and vapor mass equations. In the present study, a new approach is presented by calculating the vapor partial pressure using suggested an evaporation model to calculate the vapor source term in the mass vapor equation. Also, the solar radiation attenuation is modeled in the present work to calculate an absorbed solar radiation at the porous medium different layers. Five one-dimensional differential equations expressing mass, momentum and energy equations are presented in five unknowns (S, C, P, P_v). The nonlinear equations are solved simultaneously by updating the coefficients matrix at one time step until the five variables converge to prescribed tolerance. The numerical code is written in Matlab. Solution of the mathematical model is obtained and discussed.

2. Physical Model

Figure 1 shows the system of the current work. The porous material is supposed rigid and unsaturated one with capillary behavior. Mass and heat transfers are supposed to be one dimensional. The closed side (z=0) of material is adiabatic and the open side (z=L) is subjected to natural convective flow and long wave radiation. The solar radiation $(\lambda = 0.3 - 3.0 \, \mu m)$ penetrates the porous material and it is absorbed by fluid and solid phases. The model considered in this paper is only that in the porous medium. Figure 2 shows the schematic sketch of porous medium phases.

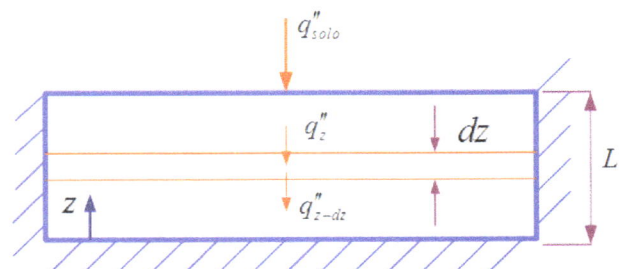

Fig 1. System of porous medium with absorbed solar energy

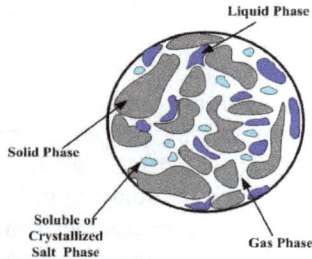

Fig 2. Physical model of non-hygroscopic porous medium

2.1. Assumptions

1. The skeleton of a porous body is a rigid solid which does not swell, absorb water, or dissolve in water and does not enter into chemical interactions with water solution (water/ NaCl solution)
2. The moisture adsorbed due to hygroscopicity of salt is neglected.
3. Transport in the material is simplified by making local thermal equilibrium.
4. The salt movement occurs by convection and Fickian's diffusion.
5. The effects of the salt crystallization on the moisture and ionic transport are taken into account by correcting the material transport properties as a function of the porosity in each time step.
6. The specific heat of the brine is assumed to be dependent on brine concentration.
7. The materials of the porous medium are treated as being continuous and having properties defined everywhere in space.
8. The porous material is homogeneous and isotropic.
9. The gas-phase is ideal in the thermodynamic sense.
10. The vapor mass is transferred by diffusion from porous medium to ambient.
11. The scattering from reflection in a porous medium bed is neglected.

2.2. Attenuation of Solar Radiation inside Porous Medium

The porous medium is assumed to be a semitransparent medium for solar energy which in turn is diffuse radiation. Internal reflection and scattering are neglected and only absorption is taken into consideration. At any location z, the change in radiation flux for the element, as a result of absorption, is a function of the flux reaching the element [10]. The change in intensity has been found experimentally to depend on the magnitude of the local intensity. If a coefficient of proportionality γ which depends on the local properties of the medium is introduced, then the decrease in solar flux in an element volume is given by:

$$\left. \begin{array}{l} dq''(z,t) = q''_z - q''_{z-dz} \\ dq''(z,t) = +\gamma_{eff}(z)q''(z,t)dz \end{array} \right\} \qquad (1)$$

The quantity γ_{eff} is called the effective absorption

coefficient of the material in the element layer which takes into consideration the absorption in the different components of the porous medium of element thickness dz. The absorption coefficient is a physical property of the material and has the units of reciprocal length as it is a volumetric property. The value of γ_{eff} is dependent on t and z as do the different component of the porous medium. The radiation flux along the path is attenuated exponentially while passing through an absorbing porous medium. The boundary conditions are:

$$at\ z = L, q''_z = q''_{solo} \Big\} \qquad (2)$$

Where:

q''_{solo} is the solar flux impinging on the top surface.
The solution is hence given by:

$$q''(z,t) = q''_{solo}\ \exp^{-\gamma_{eff}(L-z)} \qquad (3)$$

The energy absorbed by an elemental volume per unit volume is thus equal to $\dfrac{\partial q''_z}{\partial z}$ or

$$q'''_z = q''_{solo}\gamma_{eff}\ \exp^{-\gamma_{eff}(L-z)} \qquad (4)$$

The latter term is considered as energy generation term in the energy equation of the porous medium. The radiation flux reaching the bottom of the porous medium is assumed to be absorbed at the bottom surface (z=0) of the container holding the porous medium. The latter term is equal to:

$$q''_{sol}(0,t) = q''_{solo}(t)\exp^{-\gamma_{eff}L} \qquad (5)$$

2.2.1. Calculation of the Absorption Coefficient for Porous Medium Phases

The absorption coefficients for porous medium phases (solid, solution, gas) are functions of wavelength distribution. Thus, the total absorption coefficient for each phase of the porous medium is calculated from:

$$\gamma_{phase} = \frac{\displaystyle\int_{0.3}^{3.0} E_\lambda\gamma_\lambda d\lambda}{\displaystyle\int_{0.3}^{3.0} E_\lambda d\lambda} \qquad (6)$$

γ_{phase} is the medium phase total (average) absorption coefficient m^{-1}, γ_λ the medium phase spectral absorption coefficient, m^{-1}, E_λ spectral intensity $Wm^{-2}/\mu m$, λ is the solar irradiance wavelength, μm.

The effective absorption coefficient is the volume average of absorption coefficient of all components of porous medium, or

$$\gamma_{eff} = \left(\phi S_l\gamma_l + \phi\ w_v S_v\gamma_v + \phi S_p\gamma_{sal} + (1-\phi)\gamma_s\right) \qquad (7)$$

Where, γ_l is the water average absorption coefficient, γ_v is the vapor average absorption coefficient, γ_s is the sand average absorption coefficient, γ_{sal} is the salt average absorption coefficient.

The liquid water average absorption coefficient is calculated from the data of spectral absorption coefficient which is reported in [11]. Thus, the calculated average value is 350 m^{-1}. The water vapor absorption coefficient is calculated from the spectral absorption coefficient and the calculated average value is approximately 10 m^{-1} [12]. There is a little amount of carbon dioxide which is included to air components. Thus, the larger components of air are nitrogen and oxygen which have no absorption for solar energy. There is no data for absorption coefficient for sand and because 90% of glass is made of sand, the sand absorption coefficient was assumed to be equal the double of that of glass. Glass absorption coefficient is approximately 30 m^{-1} [10] and thus that of glass is taken 60 m^{-1}. The absorption coefficient of NaCl is 0.0007 m^{-1} [13] at wavelength of 1.06 μm. The total average absorption coefficient of NaCl is assumed 0.0007 m^{-1} through the solar wavelength band since no data is available at other wavelength values.

2.3. Water Vapor Diffusion inside Porous Medium

The vapor pressure is estimated from vapor mass balance and it is solved simultaneously with other four variables. The vapor mass transfer inside the porous medium consists of capillary mass rate, diffusion mass rate through gas (humid air) and evaporation mass rate by diffusion from water solution to porous medium humid air.

2.3.1. Diffusion of Vapor from Water to Solution to Gas

To calculate the mass rate of diffusion from water solution to vapor inside the porous medium, it is assumed that the salt solution in an element volume that has the profile of N cylindrical columns each of diameter (d) surrounded by a volume of gas making cylindrical annulus of outer diameter (d+y). The value of y represents the air space around water cylinder and was considered to be 0.0075 m. Thus, diffusion of water solution to humid air occurs from cylindrical surface areas to the surrounding air. The equations for mentioned evaporation model are as follows:

$$A_{cr}\phi S_g = N\left(\frac{\pi}{4}\left((d+y)^2 - d^2\right)\right) \tag{8}$$

$$A_{cr}\phi S_l = N\frac{\pi}{4}d^2 \tag{9}$$

From equations 8, 9, the following two equations are derived

$$N = \frac{4A_{cr}\phi S_l}{\pi d^2} \tag{10}$$

$$d^2 - \left(\frac{2yS_l}{S_g}\right)d - \left(\frac{S_l}{S_g}y^2\right) = 0 \tag{11}$$

The values of N and d can be obtained from equations (10) and (11). Since the temperature distribution is one dimensional, evaporation takes place by diffusion from saturated air at the surface of water to the unsaturated surrounding air at the same temperature. The rate of vapor mass diffused from water columns to the surrounding air per unit volume of the porous medium is calculated from:

$$\dot{m}'''_{evap} = A_{sr}h_{mp}\left(\rho_{vs} - \rho_v\right)/dz \tag{12}$$

Where:

A_{sr} is the ratio of water solution surface area A_{ls} to cross-section area of porous medium A_{cr} and h_{mp} is the mass transfer coefficient between solution surface and humid air. The liquid solution surface area is given by:

$$A_{ls} = N\pi d(dz) = A_{sr}A_{cr} \tag{13}$$

and from eq. (9)

$$\left.\begin{array}{l}\dfrac{N\pi d}{A_{cr}} = S_l\phi\dfrac{4}{d} \\[2mm] \therefore A_{sr} = \dfrac{N\pi d(dz)}{A_{cr}} = S_l\phi\dfrac{4}{d}(dz)\end{array}\right\} \tag{14}$$

Hence,

$$\therefore \dot{m}'''_{evap} = S_l\phi\frac{4}{d}h_{mp}\left(\rho_{vs} - \rho_v\right) \tag{15}$$

The mass transfer coefficient h_{mp} is calculated from [14]:

$$\left.\begin{array}{l}h_{mp} = Sh * D_{efg}/dz \\[1mm] Sh = 0.59(Gr*Sc)^{0.25}\ (10^5 < Gr*Sc < 10^9) \\[1mm] Sh = 0.1*(Gr*Sc)^{1/3}\ (10^9 < Gr*Sc < 10^{13}) \\[1mm] where, \\[1mm] Gr = g\left(\rho_{ha,s} - \rho_{ha}\right)L_c^3 \Big/ \dfrac{\left(\rho_{ha,s} + \rho_{ha}\right)}{2}v_a \\[1mm] Sc = v_a/D_{efg}\end{array}\right\} \tag{16}$$

The characteristic length L_c in eq. (16) is taken $(y/2)$.

2.3.2. Vapor Mass Transfer from Inside Porous Medium by Capillary and Diffusion

The total flux of vapor is composed of convective (Darcy flow) and diffusion flows respectively as these are given by the following equation [1]:

$$\dot{m}''_v = \left(-\rho_v\frac{KK_{r,g}}{\mu_g}\frac{\partial P}{\partial z} - \frac{C_g^2}{\rho_g}M_aM_vD_{eff}\frac{\partial(P_v/P)}{\partial z}\right) \tag{17}$$

2.4. Soluble and Crystallized Salt Model

There are two approaches for salt crystallization; namely, equilibrium approach and kinetic approach. In the first approach the equilibrium between the dissolved and precipitated salt is assumed. The amount of precipitated salt in the pores is defined by its saturation degree S_P, which is described by the salt binding isotherms. In this approach, it is assumed that salt solution is not saturated. Due to the physical adsorption, some of ions are captured by the solid skeleton surface [15]. In kinetic approach, it is assumed that there is no salt in the solid phase until the solution solubility limit C_{sat} (i.e. the salt mass concentration of the saturated solution at its temperature) is reached. After exceeding the maximum salt concentration, the solution is supersaturated. The supersaturation ratio (i.e. current concentration/concentration at saturation), the supersaturation is the driving force of the salt crystallization, which starts when the solution supersaturation ratio is greater than one. The first crystals are formed on the crystallization nuclei which are usually built by dust or other contaminations. An increase of solid salt mass is calculated on the basis of the supersaturation ratio, according to the following equation [15]:

$$\frac{dS_p}{dt} = \begin{cases} S_l K \left(C_s - A' C_{sat}\right)^p, & C_s \geq A' C_{sat} \\ -S_l K \left|C_s - A' C_{sat}\right|^p, & C_s < A' C_{sat} \end{cases} \quad (18)$$

$$\dot{m}'''_{prece} = \left(-\rho_p K'\right)\left(S_l \left(C_s - A' C_{s,sat}\right)\right) \quad (19)$$

In equation (18), it is assumed that there is no salt in the solid phase until reaches the solution solubility limit C_{sat} (i.e. the salt mass concentration of the saturated solution at a current temperature) is reached. After exceeding the maximum salt concentration, the solution is supersaturated. Equation (19) represents the precipitated salt mass when the solution is supersaturated. An increase of solid salt mass is calculated on the basis of the supersaturation ratio, according to equation (18). The process order, p, depends on the properties of porous body and the kind of salt [8]. There is a need to determine mass of precipitated solute in the solution (dissolved salt) or a mass crystallized solute. An increase of the mass of dissolved solute has a negative sign which physically means dissolution of salt crystals. It was found from the experiments that the form of rate law and the order of the process p = 1.9 was taken from previous work, where the salt crystallization in bricks was analyzed. The other constants of the rate law were assumed arbitrary due to the lack of the experimental data concerning cement mortar and the rate constant of crystallization of sodium chloride is given at [15].

2.5. Ambient Temperature and Humidity Calculation Method

The ambient temperature was assumed to be correlated as a cosine wave with time with a maximum at 3 p.m. and minimum at 3 a.m. The maximum and minimum temperatures are usually obtained from the meteorological data for different locations (Assuit city was considered for the current study). The relative humidity is known at 3 p.m. for Assuit. The humidity ratio is assumed to be constant during the day. These assumptions help in estimating the temperature and relative humidity of air at any time during the day. The data of Assuit, Egypt are used in this study.

3. Equilibrium State Laws and Properties

The input data are shown at table (1)

The total porous medium volume is:

$$\Delta V = \Delta V_s + \Delta V_w + \Delta V_g + \Delta V_p \quad (20)$$

The porosity is:

$$\phi = \frac{\Delta V_w + \Delta V_g + \Delta V_p}{\Delta V} \quad (21)$$

The water saturation, gas saturation and salt crystallized saturation are defined as:

$$S_l = \frac{\Delta V_w}{\phi \Delta V}, \ S_g = \frac{\Delta V_g}{\phi \Delta V}, \ S_p = \frac{\Delta V_p}{\phi \Delta V} \quad (22)$$

$$S_l + S_g + S_p = 1 \quad (23)$$

The mass density of water vapor and dry air and their mixture are:

$$\rho_v = \frac{P_v}{R_v T}, \ \rho_a = \frac{P_a}{R_a T}$$
$$\rho = \rho_v + \rho_a, \ P = P_v + P_a \quad (24)$$

The intrinsic permeability is given by [16]:

$$K(\phi) = K_0 \left(\frac{\phi(1 - S_p)}{\phi}\right)^3 \left(\frac{1 - \phi}{1 - \phi(1 - S_p)}\right)^2 \quad (25)$$

The effective gas diffusion coefficient is given by [1]:

$$D_i(\phi) = D_i^0 \left(\phi(1 - S_l - S_p)\right)^{\frac{4}{3}} \quad (26)$$

The effective salt diffusion coefficient is given by [17]:

$$D_{sal,ef} = S_l^{2/3}\left(0.779 + 0.027T + 3 \times 10^{-4} T^2\right) \times 10^{-9} \quad (27)$$

The capillary pressure is given by [9]:

$$\langle P_c \rangle = \frac{\sigma}{(K/\phi)^{\frac{1}{2}}} \left(0.364\left(1 - e^{-40(1 - S_l)}\right) + 0.221(1 - S_l) + \frac{0.005}{S_l - 0.08}\right) \quad (28)$$

The capillary diffusivity is given by [1]:

$$D_w = \left(\frac{K_{rl} K}{\mu_l} \frac{\partial P_c}{\partial S_l}\right) \quad (29)$$

The liquid viscosity, thermal conductivity, and surface tension are given by [18]:

$$\mu_l = 2.1x10^{-6} \exp\left(\frac{1808.5}{T}\right)\left(1+1.85C_s - 4.1C_s^2 + 44.5C_s^3\right) \quad (30)$$

$$K_l = 0.04C + K_w \quad (31)$$

$$\sigma = \left(0.03059 \exp\left(\frac{252.93}{273.15+T}\right) + 0.04055 \, C_s\right) \quad (32)$$

The liquid density is given by [19]:

$$\left.\begin{array}{c} \rho_l = \rho_o\left(1 + \varepsilon C\right) \\ \varepsilon = 0.7558 \end{array}\right\} \quad (33)$$

Relative permeability [9]:

$$\left.\begin{array}{c} S_{eff} = \dfrac{S_l - S_{lr}}{1 - S_{lr}} \qquad S_{eff} \leq 1 \\ K_{rl} = S_{eff}^3 \\ K_{rg} = \left(1 - S_{eff}\right)^3 \end{array}\right\} \quad (34)$$

The effective heat capacity and thermal conductivity:

$$\left(\rho c_p\right)_{eff} = \left(\phi S_l \rho_l C_{p,l} + \phi\left(1-S_l-S_p\right)\rho_v C_{p,v} + \phi\left(1-S_l-S_p\right)\rho_a C_{p,a} + \phi S_p C_{p,cr} + \left(1-\phi\right)\rho_s C_{p,s}\right) \quad (35)$$

$$K_{eff} = \left(\phi\left(1-w_v\right)\left(1-S_l-S_p\right)K_a + \phi w_v\left(1-S_l-S_p\right)K_v + \phi S_l K_l + \phi S_p K_{cr} + \left(1-\phi\right)K_s\right) \quad (36)$$

Where:

$C_{pl} = 4180 - 4.396\left(C/100\right)\rho_l + 0.0048\left(C/100\right)^2\rho_l^2$ [20], $\left(\rho C_p\right)_{eff}$ is the porous medium effective heat capacity, K_{eff} is the effective thermal conductivity of porous medium components, C_{pl} is liquid solution specific heat, K_l is the liquid solution thermal conductivity.

4. Mass and Heat Transfer

Transport in the material is simplified by making the assumption of local thermal equilibrium. This means that at any cross section, the solid, liquid, gas and precipitated salt phases are at the same temperature. This allows the thermal transport to be characterized by a single equation for conservation of energy. Heat transfer model includes conduction, radiation, enthalpy convection, sensible heating and phase change. The partially saturated media raises liquid water solution by capillary action. There is a phase change for liquid water and soluble salt. For the liquid water is being evaporated by diffusion for some periods of time, the salt concentration increases to reach the supersaturation condition. When supersaturation occurs, the salt will be crystallized at dry region of porous media. A realistic model is considered at the exposed boundary in what concerns mass transfer and the outflow mass transfer of vapor is dictated by convective diffusion.

4.1. Governing Equations of Heat and Mass Transfer

The mass and energy conservations are given by the following five equations:

$$\phi\frac{\partial}{\partial t}\left(\rho_l S_l\right) + \dot{m}_{evp}''' = -\left(\frac{\partial}{\partial z}\left(-\rho_l \frac{KK_{r,l}}{\mu_l}\left(\frac{\partial P_l}{\partial z} - \rho_l \vec{g}\right)\right)\right) \quad (37)$$

$$\phi\frac{\partial}{\partial t}\left(\rho_v S_g\right) - \dot{m}_{evp}''' = -\frac{\partial}{\partial z}\left(\begin{array}{c} -\rho_v \dfrac{KK_{r,g}}{\mu_g}\dfrac{\partial P}{\partial z} \\ -\dfrac{C_g^2}{\rho_g}M_a M_v D_{eff}\dfrac{\partial\left(P_v/P\right)}{\partial z} \end{array}\right) \quad (38)$$

$$\phi\frac{\partial}{\partial t}\left(\rho_l C_s S_l + \rho_p S_p\right) = -\frac{\partial}{\partial z}\left(\begin{array}{c} -\rho_l C_s \dfrac{KK_{r,l}}{\mu_l}\left(\dfrac{\partial P_l}{\partial z} - \rho_l \vec{g}\right) \\ -\phi\rho_l S_l D_{sal}\dfrac{\partial C_s}{\partial z} \end{array}\right) \quad (39)$$

$$\phi\frac{\partial}{\partial t}\left(S_g \rho_a\right) = -\frac{\partial}{\partial z}\left(\begin{array}{c} -\rho_a \dfrac{KK_{r,g}}{\mu_g}\dfrac{\partial P}{\partial z} \\ -\dfrac{C_g^2}{\rho_g}M_v M_a D_{eff}\dfrac{\partial\left(P_a/P\right)}{\partial z} \end{array}\right) \quad (40)$$

$$\left(\rho C_p\right)_{eff}\frac{\partial T}{\partial t} + \frac{\partial}{\partial z}\left(\begin{array}{c} -K_{eff}\dfrac{\partial T}{\partial z} \\ + q_{conv}'' \end{array}\right) = \left(\begin{array}{c} -\dot{m}_{evp}''' h_{fg} - \dot{m}_{crs}''' h_{cr} \\ + q_{solo}'' \gamma_{eff}\exp\left(-\gamma_{eff}\left(L-z\right)\right) \end{array}\right) \quad (41)$$

The mass balance equation (37) describes the liquid solution mass transfer. The salt solution leaves the volume across its boundary due to flow moving at Darcy's velocity and takes into consideration gravity force. The mass balance represented by equation (38) describes the vapor mass transfer. The mass balance equation (39) represents the solute mass transfer. The first (LHS) term represents the stored solute and the second term represents the precipitated salt which results from dissolution or crystallization. The mass balance equation (40) describes the dry air mass transfer. Fick's law governs mass fluxes due to concentrations gradients (diffusion part) which are often the most significant driving force as shown in equations (38), (39) and (40). The modified Navier–Stokes equation for porous media reduces to Darcy's equation inside the porous media for low-permeability systems [21]. The Darcy's velocity is applied at mass balance equations (37), (38), (39) and (40). The energy balance, equation (41), describes heat transfer inside porous medium. The first (LHS) term is the storage term; the second is the net conduction and convection flux of fluid phases (solution+ vapor+ air). The (RHS) term is the local absorbed solar energy per unit volume. The convection term in the energy equation is defined by the following equation,

$$-\frac{\partial q_{conv}''}{\partial z} = \left(C_{pv}\vec{m}_v'' + C_{pa}\vec{m}_a'' + C_{pl}\vec{m}_l''\right)\frac{\partial T}{\partial z} \quad (42)$$

4.2. Initial and Boundary Conditions

4.2.1. Initial Condition

$$
\left.
\begin{aligned}
S(z,0) &= S_i \\
C(z,0) &= C_i \\
P(z,0) &= P_i \\
T(z,0) &= T_i \\
P_v(z,0) &= P_{vi}
\end{aligned}
\right\}
\tag{43}
$$

Where:

S_i , C_i , P_i , T_i , and P_{vi} are the saturation, solute concentration, pressure, temperature, and concentration respectively at time zero.

4.2.2. Boundary Conditions

The bottom (closed) and top (open) boundaries are given as follows:

4.2.2.1. Closed Boundary Conditions

At z=0,

$$
\left.
\begin{aligned}
\dot{m}_l'' &= 0 \\
\dot{m}_v'' &= 0 \\
\dot{m}_{salt}'' &= 0 \\
\dot{m}_a'' &= 0
\end{aligned}
\right\}
\tag{44}
$$

The energy balance at closed boundary is given by:

$$
\left(-K_{eff} \frac{\partial T}{\partial z} \right) = q_{solo}'' \exp\left(-\gamma_{eff} L\right)
\tag{45}
$$

4.2.2.2. Open boundary Condition

At z=L

It is assumed that $P = P_{amb}$, P_{amb} is the ambient pressure. The mass balance for liquid, and vapor are given by:

$$
\left.
\begin{aligned}
\dot{m}_l'' &= \dot{m}_{evap1}'' \\
\dot{m}_v'' &= \dot{m}_{evap2}'' \\
\dot{m}_{salt}'' &= 0
\end{aligned}
\right\}
\tag{46}
$$

Where:

$\dot{m}_{evap1}''$ is mass flux diffused from solution surface to ambient air, and $\dot{m}_{evap2}''$ is mass flux diffused from inner volume humid air to ambient air at surface. The two mentioned fluxes are calculated from:

$$
\left.
\begin{aligned}
\dot{m}_{evap1}'' &= \left(\phi S_l h_{mlv} \left(\rho_{v,sat} - \rho_{amb} \right) \right) \\
\dot{m}_{evap2}'' &= \dot{m}_{vD}''
\end{aligned}
\right\}
\tag{47}
$$

$$
\dot{m}_{v,D}'' = \frac{2}{\Delta z}
\left(
\begin{aligned}
&\left(\frac{P M_v M_a D_{eff}}{RT(M_v P_v + M_a (P - P_v))} \right)_N (P_{v,N} - P_{v,ha}) \\
&+ \left(-\frac{M_v M_a D_{eff} P_v}{RT(M_v P_v + M_a (P - P_v))} \right)_N (P_N - P_{ha})
\end{aligned}
\right)
$$

h_{mlv} is the mass transfer coefficient for vapor diffusion from porous medium to ambient, $\dot{m}_{vD}''$ is mass vapor diffusion from humid air in porous medium's surface to humid air region, h_{mlv} is calculated from [14],

$$
h_{mlv} = \frac{hc}{\rho_a} \left(\frac{D_{efg}}{\alpha} \right)^{\frac{2}{3}}
\tag{48}
$$

It is assumed that the vapor mass is transferred from porous medium and vapor mass is diffused from porous medium surface to ambient. Wherever the liquid mass is reached to porous medium surface by capillary action, the diffused surface liquid is compensated by it. The water vapor mass transfer coefficient from porous medium to ambient is unknown and there is no experimental data for its value. However, the boundary node for vapor mass transfer is taken as shown in Fig. 3 [22].

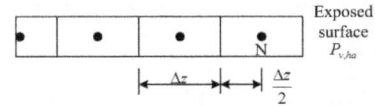

Fig 3. The boundary element for vapor mass transfer

The energy balance at open boundary is given by:

Energy flux in = Energy flux out

Figure 4 shows the energy fluxes at open surface boundary.

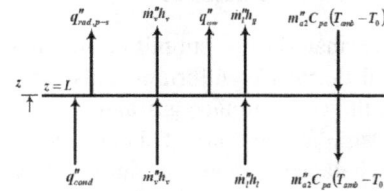

Fig. 4. Open surface energy boundary condition

The heat fluxes enter the top boundary are:

q_{cond}'' the conduction heat flux

$\dot{m}_l'' h_l$ the enthalpy flux of liquid

$\dot{m}_{evp2}'' h_v$ the enthalpy flux of vapor diffused from gas to ambient

The energy fluxes out from the top boundary are:

q_{conv}'' the convection flux to ambient

$q_{rad,sur-s}''$ the radiation flux to sky

$\dot{m}_{evp1}'' h_g$ the enthalpy flux of vapor diffused from liquid to ambient air

The energy balance at open boundary can be derived from:

$$\left(-K_{eff.sur}\frac{\partial T}{\partial z} = \dot{m}''_{evp1}h_{fg} + q''_{rad} + q''_{conv} \right) \quad (49)$$

Where:

$$K_{eff.sur} = \left(\phi(1-w_v)(1-S_l-S_p)K_a + \phi v_v(1-S_l-S_p)K_v + \phi S_l K_l + \phi S_p K_{cr} + (1-\phi)K_s \right) \quad (50)$$

The convection and radiation fluxes are given by:

$$\left. \begin{array}{l} q''_{conv} = h_c \left(T_{sur} - T_a \right) \\ q''_r = f\sigma \left(T_{sur}^4 - T_s^4 \right) \end{array} \right\} \quad (51)$$

The heat transfer coefficient from porous medium surface to ambient is given by [14]

$$hc = Nu\left(K_a(1 - w_{sur}) + K_v w_{sur} \right)/L_c \quad (52)$$

$$\left. \begin{array}{l} Nu = 0.54(R_{aL})^{1/4} \left(10^4 \le R_{aL} \le 10^7 \right) \\ Nu = 0.15(R_{aL})^{1/3} \left(10^7 \le R_{aL} \le 10^{11} \right) \\ R_{aL} = \dfrac{g(\rho_{ha,sur} - \rho_{ha,amb})L_c^3}{\left(v_a\left(\dfrac{K}{C_p} \right)_{ha} \right)} \end{array} \right\} \quad (53)$$

Both air and vapor properties are calculated at film temperature. Wherever, the concentration difference is between air at the porous medium surface and ambient air, the mass can be transferred from porous medium surface to ambient. Thus, the Grashof number which depends upon the density differences is applicable here [14]. Prantdl number and thermal diffusivity α are calculated at humid air conditions.

5. Numerical Solution

The finite-volume method is applied to a conservation statement for a control volume.

$$\int_v Div \, \vec{V} \, dv = \oint_S \vec{V} \cdot \hat{n} \, dS \quad (54)$$

Where V the volume of interest, S is is the boundary surface of the volume, $\vec{V}$ is a vector function with continuous first spatial derivatives within the volume, and n is the unit vector normal to the surface oriented to the positive pointing outward from the volume. In Eq. 54, the integral on the left is evaluated over the control volume, which may be a one-, two-, or three-dimensional region, whereas the integral on the right is a contour integral to be evaluated over the boundaries of the control volume [23]. The finite-volume method is applied to a conservation

statement for a control volume. A finite-volume method was used for discretisation of the differential equations. A fully-implicit scheme was used for unsteady term discretisation while the convective terms (liquid solution, vapor and dry air) in the energy equation are handled by an upwind-scheme method. The system of equations is solved by direct method (inverse matrix). The coefficients matrix (K-coefficients) is a function of five dependent variables (liquid saturation, salt solution concentration, gas pressure, temperature, vapor pressure). The nonlinear equations are solved simultaneously by updating (iteration using new solution of variables) the coefficients matrix at one time step until the five variables converge to prescribed tolerance. The numerical code is written in Matlab. The K-coefficients are calculated by an arithmetic average at the interfaces [24]. The convergence is ensured by five convergence criteria for the variables. Figure 5 shows a flow chart for the program designed for numerical solution.

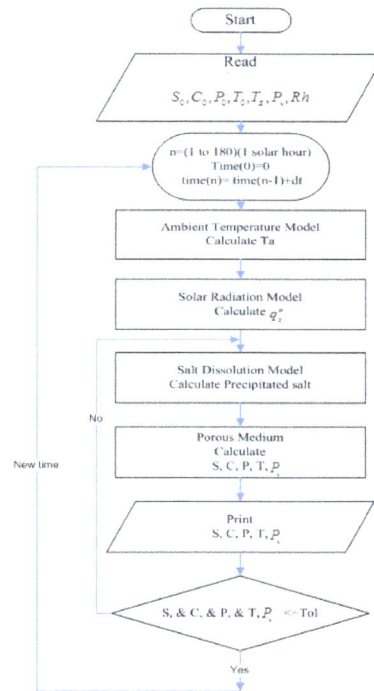

Fig 5. *Programming flow chart.*

5.1. Time Step and Number of Nodes

The convergence is occurred at time step of 20 seconds. Another time step is used when the vapor pressure equation is unstable for high evaporation rate and the second time step is 10 seconds. The number of nodes was found to be 31 which, gives the best results with selected time steps. The mass of different phases that were calculated for the porous medium, were checked at the end of each interval; also, the energy balance for the porous medium was checked for each time interval to release the convergence criteria by under and over relaxation. The liquid solution mass reduction and masses of water vapor and solute are given respectively by the following equations:

$$mass_l = \left(\sum_1^N \rho_l \phi S_l \Delta z A_{cr}\right)\Bigg|_{t_2} - \left(\sum_1^N \rho_l \phi S_l \Delta z A_{cr}\right)\Bigg|_{t_1} \tag{55}$$

$$mass_v = \int_{t_1}^{t_2}\left(m_{evp1} + m_{evp2}\right)A_{cr}\,dt \tag{56}$$

$$\left.\begin{array}{l} m_{solute}\big|_{t=0} = \sum_1^N C\rho_l S\phi A_{cr}\,z \\[2mm] m_{solute}\big|_t = \sum_1^N C\rho_l S\phi A_{cr}\,\Delta z \end{array}\right\} \tag{57}$$

The energy balance is given by:

$$\left.\begin{array}{l} E_{in} = \int_{t_1}^{t_2} q_c''\big|_{z=0} + \int_{t_1}^{t_2} q_{ca}''\big|_{z=N} + \int_{z=0}^{z=L}\int_{t_1}^{t_2} q_{sol}''\,dt\,dz \\[2mm] q_{ca}''\big|_{z=N} = \dot{m}_{evpl}''/M_v * M_a C_{pa}\left(T_a - T_0\right) \\[2mm] I.E = \sum_{phases}\sum_{z=0}^{z=L}\left(\left(\rho C_p\right)_{eff}\left(T - T_o\right)\right)\bigg|_{t=t_1} \\[2mm] E_{out} = \int_{t_1}^{t_2}\left(q_{cov}'' + q_{rad}'' + q_{evp}''\right)_{z=L}\,dt \\[2mm] I.E = \sum_{phases}\sum_{z=0}^{z=L}\left(\left(\rho C_p\right)_{eff}\left(T - T_o\right)\right)\bigg|_{t=t_2} \end{array}\right\} \tag{58}$$

6. Results

The porous medium used in this analysis is sand of an average diameter of 0.63 mm. The sand porosity is 0.3. The sand permeability is 2.1×10^{-11} m^2 which is considered to be a high permeability [22]. The results of the model were obtained for Assuit location (longitude $31^o 10' 58'' E$ and latitude $27^o 10' 58'' N$) in a summer day (May 22- 2013) from 7:00 a.m. to 2:00 p.m. (solar time). The local solar noon is at 12:03:21. The maximum and minimum ambient temperatures are 311.15 K and 299.15 K and relative humidity is 0.4 at 3:00 p.m. These measurements were obtained from the meteorological data at website for Assuit location in Egypt [25]. The initial condition for liquid saturation, salt concentration and gas pressure are considered to be 0.99, $0.035\,kg_{solute}/kg_{solution}$ and 101.325 kPa, respectively. The ambient temperature and vapor pressure at 7:00 a.m. estimated to be 302.2 K and 1.58 kPa, respectively, considered to be the initial temperature for the porous medium's components and the initial vapor pressure in the medium's gas.

Figures 6 to 10 show the results of the distributions of the five variables (saturation, concentration, pressure, temperature, vapor pressure) at different times. Figure 6 shows very slight decrease in liquid saturation in the porous medium at the first two hours. However, the saturation values decrease more as time passes almost linearly with z for the same time. The liquid saturation is relatively high in the distance from medium's center to the medium's bottom. The liquid saturation decreases at the front region mainly due to the vapor mass transfer to the surrounding. Wherever, the liquid saturation is decreased, the solar radiation penetrates

more through medium thickness. This causes solar energy absorption to spread deeper to the interior region. The latter explains the increase in the evaporation rate by diffusion in the interior region. This is confirmed by the relatively large vapor pressure in the interior region as shown in Fig. 9 (water vapor pressure distribution curve).

Figure 7 shows the solution concentration profiles. It is clear that the maximum concentration values are at the upper region of the medium. The salt concentration is initially the same for the medium and by water evaporation the salt concentration increases. Firstly, the concentration is approximately taken the initial concentration value because the vapor pressure is approximately equal to the vapor saturation pressure. This leads to small diffusion of vapor from solution to gas for the interior region. At the top Region the concentration increases with increase in z. At the last three hours, there is a relatively large difference in solution concentration between top and interior regions with the maximum concentration occurring at the surface. The soluble solute starts to crystallization at the hour 14 for z=L since the supersaturation ratio has reached to a value of 2.6 and this value is larger than crystallization supersaturation ratio for sodium chloride (2.31) [26].

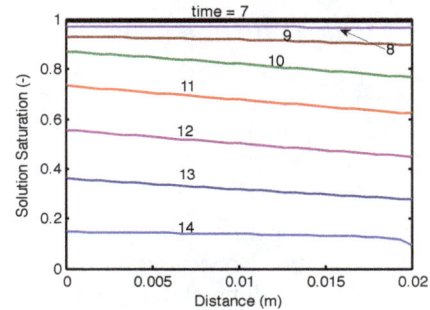

Fig 6. *Saturation distribution –date 22/05/13;Time: 7:00- 14:00*

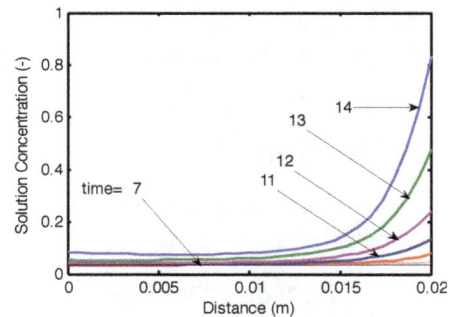

Fig 7. *Concentration distribution –date 22/05/13;Time: 7:00- 14:00*

In Fig. 8, the temperature profiles show that the maximum temperature is not at the surface. In the first two hours, the maximum temperature is at the top half region of the porous medium. However, for the following hours, the maximum temperature is at the bottom of the medium. The cooling effect at the surface causes its temperature to be always lower than that of the rest of the medium. The absorption coefficient for each of liquid and solid (sand) is larger than gas. Hence, as the liquid saturation decreases, the absorption coefficient decreases and the solar flux penetrates deeper in porous medium. This causes the maximum temperature to move deeper in the medium. The maximum temperatures occur at the hour 12 coping with time of almost maximum solar flux.

Figure 9 shows the partial water vapor pressure. The vapor pressure is changed at the end of each time interval according to the value of the stored vapor mass in the interval. In Fig. 9, it is clear that the water vapor pressure at surface node has lowest value due to diffusion of vapor to the atmospheric air. However, the surface pressure is changed from 1.58 kPa to 5.0038 kPa and the surface vapor pressure returns to reduce to 4.64 02 kPa at 13 p. m. It is generally clear from Fig. 9 that the water vapor pressure is mainly function of temperature which is responsible for increasing vapor mass by diffusion. The maximum value for vapor pressure is obtained at about the hour 12 (noon) at the bottom region of the medium. This is due to the maximum solar flux beside its deeper penetration in the medium. At this time the lowest vapor pressure is at the surface as expected and discussed before. It is also clear that the vapor gradient in the lower region of medium is relatively small at the first three hours and after that it increases with time till around solar noon. On the other hand, for top region there is always a pressure gradient that increases near the top and increases with time till solar noon. The results obtained for total vapor transferred to atmospheric air was about 8.3 kg for a time period of 7 hours.

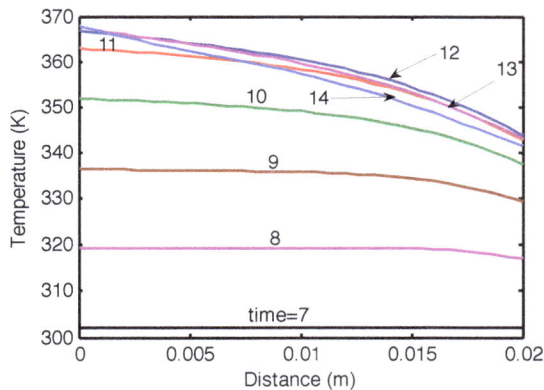

Fig 8. *Temperature distribution– date 22/05/13; Time: 7:00- 14:00*

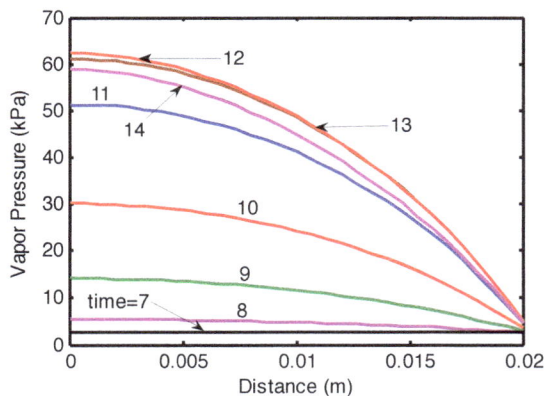

Fig 9. *Water Vapor Pressure distribution– date 22/05/13; Time: 7:00- 14:00*

Figure 10 shows the total pressure profiles. It shows an increase in pressure in the first three hours especially in the interior region. The value of increase in total pressure is not exceeding 4 Pa. At first hour, the gas relative permeability K_{rg} is very small since S_l is relatively large (Eq. 34). The gas capillary mass flux term is decreased as $KK_{rg}\dfrac{\partial P}{\partial z}$ is decreased leading to an increase in the stored mass of vapor and dry air (Eqs. 38 and 40, respectively). At the first three hours, especially for the interior region, the values of K_{rg} and K are small since the liquid saturation is high (see Eq. 34). Also Fig. 10 shows that the total pressure gradient is small. Thus, $KK_{rg}\dfrac{\partial P}{\partial z}$ is small at the first hours; giving high storage terms for vapor and dry air. Hence, the pressure is relatively high at the first hours.

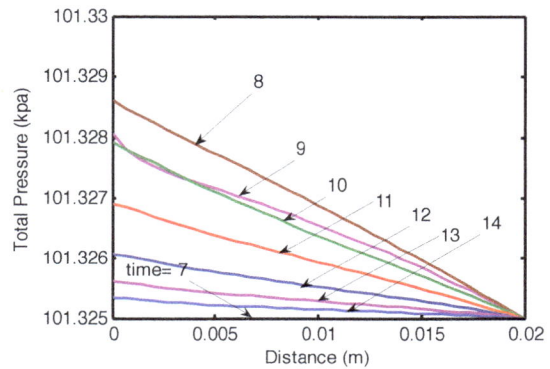

Fig 10. *Pressure distribution –date 22/05/13;Time: 7:00- 14:00*

7. Conclusions

The physical model takes into account the heat transfer mechanism, absorbed solar radiation; relative humidity and phase change inside the porous medium, the latter is the reason of building up water vapour pressure inside the medium. The solar radiation term is solved by considering it as energy generation in the elements's volumes. The mass transfer occurs in the liquid and vapor phases, each phase phenomenon is used alone to calculate the liquid saturation and water vapor pressure, respectively. The evaporative term is estimated and the water vapor pressure is calculated depending upon the physical phenomenon of water vapor mass transfer inside porous medium. The previous studies estimated the water vapor pressure using thermal equilibrium inside the porous medium thickness and the authors did not take into consideration the source term (evaporation term) of water vapor mass transfer inside the domain's elements. Whenever the surface water vapor pressure will be changed the inner water vapor pressures are affected. Then the water vapor mass in porous medium will be changed up or down according to the changing in surface vapor pressure. This scenario helps for controlling the phenomena of mass diffusion in porous medium in an opposite side of other studies which calculated the water vapor pressure from empirical equations. The change in ambient water vapor pressure affects the water vapor pressure inside the porous medium's elements. The previous studies did not take into consideration the salt concentration with water vapor pressure as variables. This porous medium system can be used with another system such as humid air region in solar still application and it has high dynamically response to the change in the water vapor pressure in the humid air region. The model's equation has taken highly nonlinearity and they strongly have linked, linearization of matrix coefficients is used. The coefficient matrix linearization is occurred by substituting the updated variables into it until the convergence criteria is occurred. Numerical solution gave converging results by applying mass and energy balances.

Table (1). Input constant data

Parameter	Value	Unit	Source	Parameter	Value	Unit	Source
Porosity (ϕ)	0.37-0.5	-	[9]	Air molecular weight (M_a)	28.97	$kg/kmol$	[4]
Permeability (K_o)	$2x10^{-11} - 1.8x10^{-10}$	m^2	[9]	Vapor molecular weight (M_v)	18.015	$kg/kmol$	[4]
Gas viscosity (μ_g)	$1.8x10^{-5}$	Pa-s	[1]	Salt diffusion coefficient (D_i^o)	$1.5x10^{-9}$	m^2/s	
Ambient pressure (P_o)	101325	Pa		Vapor diffusion coefficient (D_{av})	$2.6x10^{-5}$	m^2/s	
Air kinematics viscosity (υ_a)	1.6x10^-5	m^2/s	[8]	Irreducible saturation (S_{ir})	0.08	-	[9]
Air thermal conductivity (K_a)	0.028	$W/(m.K)$	[8]	Initial temperature	Ta	K	
Vapor thermal conductivity (K_v)	0.0212	$W/(m.K)$	[8]	Initial Saturation (S_i)	0.99	-	
Sand thermal conductivity (K_s)	0.27	$W/(m.K)$	[9]	Stefan-Boltzmann constant (σ)	$5.67x10^{-8}$	$W/(m^2K^4)$	[4]
Sand density (ρ_s)	1530	kg/m^3		Sand emissivity (ε_s)	0.95	-	
Vapor specific heat (C_{pv})	1410	$J/(kg.K)$	[8]	Gravity acceleration (g)	9.81	m/s^2	
Air specific heat (C_{pa})	1004	$J/(kg.K)$	[14]	Pure water density (ρ_w)	998.2	kg/m^3	[15]
Sand specific heat (C_{ps})	800	$J/(kg.K)$	[4]	Sand absorption coefficient γ_s	60	m^{-1}	
Salt specific heat $(C_{p,sal})$	850	$J/(kg.K)$	[27]	water absorption coefficient γ_l	350	m^{-1}	
Vapor gas constant (R_v)	0.4615	$kJ/(kg.K)$	[4]	vapor absorption coefficient γ_v	10	m^{-1}	
Air gas constant (R_a)	0.287	$kJ/(kg.K)$	[4]	Sand emissivity (ε_s)	0.95	-	

Nomenclature

A'	supersaturation parameter
C	salt mass concentration (kg of solute/kg of solution)
c	phase content (kg/m^3)
$(C_p)_{eff}$	effective specific heat capacity $(kJ/kg.K)$
$D_{eff},$	effective-diffusivity (m^2/s)
D_{sal}	salt diffusivity (m^2/s)
D_w	capillary diffusivity (m/s^2)
q''	radiation flux (W/m^2)
$\vec{g}$	gravity acceleration (m/s^2)
Gr	Grashof number;
h	enthalpy (kJ/kg)
h_c	convective heat transfer coefficient $(W/m^2 K)$
$h_{m,vv}$	mass transfer coefficient inside porous medium (m/s)
h_{mlv}	mass transfer coefficient from surface porous medium (m/s)
I.E	internal energy (kJ/m^3)
K	thermal conductivity $(kW/m.K)$ and intrinsic permeability (m^2)
$K_{r,g}, K_{r,l}$	gas relative permeability (-), liquid relative permeability (-)
K'	NaCl kinetic parameter
L, L_C	Porous medium thickness (m), characteristic length (m)
$\dot{m}''$	mass flux $(kg/m^2.s)$
$\dot{m}'''_{evp}$	volumetric evaporation rate $(kg/m^3.s)$
$\dot{m}'''_{prece}$	volumetric salt precipitation rate $(kg/m^3.s)$
M_a, M_v air, vapor	molar mass $(kg/kmol)$
$\left(Nu = \frac{h_c L_c}{K}\right)$	Nusselt number
P	gas pressure (kPa)
P	Precipitated salt
P_c, P_v	capillary pressure (kPa), vapor pressure (kPa)
$q''_{r,0}$	input solar flux (W/m^2)
R_j	gas constant for gas $(kJ/kg.K)$
Rh	relative humidity

$\left(Sc = \frac{\upsilon}{D}\right)$	Schmidt number.
$\left(Sh = \frac{h_m L_c}{D}\right)$	Sherwood number;
S_g , S_l	gas saturation (-), liquid saturation (-)
S_p	precipitated salt saturation (-)
T , T_s	temperature (K), sky temperature (K)
V	volume (m^3)

Greek Symbols

α	thermal diffusivity (m^2/s)
β	volumetric thermal expansion (1/K)
Δ	difference value
γ	absorption coefficient $(1/m)$
υ	kinematic viscosity (m^2/s)
μ	dynamic viscosity $(kg/m.s)$
w_a	humidity ratio of ambient air
w_{sur}	humidity ratio at medium surface.
ρ_l , ρ_v	liquid density (kg/m^3) , Water vapor density (kg/m^3)
ρ_p	salt crystal density (kg/m^3)
ρ_{vs}	vapor density of water vapor at saturation (kg/m^3)
ϕ	porosity (-)

Symbols

eq	equation
t	time

Subscripts

a	air
amb	ambient
l	liquid
v	vapor
sal	salt
sat	sat
$tras$	transparent
eff	effective value
cry	crystallization
s	sky
sur	surface

References

[1] H. Ni, A. K. Datta and K. E. Torrance, Moisture transport in intensive microwave heating of biomaterials: a multiphase porous media model. International Journal of Heat and Mass Transfer, 42 (1999) 1501- 1512.

[2] A. Khodaparast and Haghi, Relations for water vapor transport through fibers. Journal of Computational and Applied Mechanics, 5 (2) (2004) 263- 274.

[3] V.A.F. Costa, M.L. Mendonca, and A.R. Figueiredo, Modeling and simulation of wetted porous thermal barriers operating under high temperature or high heat flux. International Journal of Heat and Mass Transfer; 51(2008) 3342–3354.

[4] A. Neale, D. Derome, B. Blocken and J. Carmerliet, Coupled Simulation of Vapor Flow between Air and a Porous Material, ASHRAE, 2007.

[5] K. Murugesan, H. N. Suresh, K. N. Seetharamu, P. A. Aswatha Narayana and T. Sundararajan, A theoretical model of brick drying as a conjugate problem, International Journal of Heat and Mass Transfer, 44 (2001) 4075- 4086.

[6] H. Stephen Lee, Wallance W. Carr, Haskell W. Beckham, and Johannes Leisen, A model of through-air drying of tufted textile materials", International Journal of Heat and Mass Transfer, 45(2002) 357- 366.

[7] D. Le, H. Hoang, and J. Mahadevan, Impact of capillary driven liquid films on salt crystallization ". Transp. Porous Med, 80 (2009) 229–252.

[8] M. Koniorczyk, and D. Gawin, Numerical modeling of salt transport and precipitation in non-isothermal partially saturated porous media considering kinetics of salt phase changes. Transp. Porous Med, 87 (2011) 57–76.

[9] M. Kaviany, Principles of heat transfer in porous media, Handbook of Heat Transfer, Second edition, McGraw-Hill, New York, (1995) 2, 28, 479, 491.

[10] M. M. Elsayed, I. S. Taha, and J. A. Sabbagh, Design of Solar Thermal Systems. Scientific Publishing Centre; King Abdulaziz university, Saudi Arabia, (1994).

[11] D. Kraus, Two phase plow in homogenous porous media- The role of dynamic capillary pressure in modeling gravity driven fingering, Master's Thesis, 2011.

[12] Wikipedia, "Electromagnetic absorption by water" (http://en.wikipedia.org/wiki/Electromagnetic_absorption_by_water)

[13] H. H. Li, "Absorption Coefficients", Int. J. Therm., Vol1, No. I, (1980).

[14] Y. A. Cengel, Heat and mass transfer", Hand book of Heat and Mass Transfer, Third edition, A practical approach, McGraw Hill, New York, 2006.

[15] M. Koniorczyk, Modelling the phase change of salt dissolved in pore water– Equilibrium and non-equilibrium approach, Construction and Building Materials, 24 (2010) 1119–1128.

[16] T.Q. Nguyen, J. Petkovic, P. Dangla and V. Baroghel-Bouny, Modeling of coupled ion and moisture transport in porous building materials, Construction and Building Materials (2007) 1- 11.

[17] J. Bear and A. Gilman, Migration of salts in the unsaturated zone caused by heating, Letters in Mathematical Physics, 19 (1995) 139-156.

[18] S. O. Pstalle, Non-isothermal multiphase flow of brine and gas through saline media, Doctoral-Barcelona, Universitat Politecnica de Catalunya, 1995.

[19] M.C. Boufadel, M.T. Suidan, and A.D. Venosa, Numerical modeling of water flow below dry salt lakes: effect of capillarity and viscosity, Journal of Hydrology, 221 (1999), 55–74.

[20] A. Ramalingam and S. Arumugam, Experimental Study on Specific Heat of Hot Brine for Salt Gradient Solar Pond Application, International Journal of Chem. Tech Research, 4 (3) (2012), 956-961.

[21] A. Haldera and A. K. Dattab, Surface heat and mass transfer coefficients for multiphase porous media transport models with rapid evaporation, Food and Bio-products Processing , 90(2012) 475–490.

[22] V.A.F. Costa, M.L. Mendonca and A.R. Figueiredo, "Modeling and simulation of wetted porous thermal barriers operating under high temperature or high heat flux". International Journal of Heat and Mass Transfer; Vol. 51; 2008; 3342–3354.

[23] G. F. Pinder, W. G. Gray, Essentials of multiphase flow and transport in porous media, A John Wiley & Sons, Inc., Hoboken, New Jersey, 2008.

[24] S. V. Patankar, Numerical heat transfer and fluid flow, Book, Publishers, Talyor & Francis, 1980.

[25] Website of " Egyptian meteorological Authority " (http://ema.gov.eg/articles?menu=62&lang=eg)

[26] H. Na, S. Arnold, and A. S. Myerson, Cluster formation in highly supersaturated solution droplets, Journal of Crystal Growth, 139 (1994) 104-112.

[27] W. Scott Pegau, Deric Gray, and J. Ronald V. Zaneveld, Absorption and attenuation of visible and near-infrared light in water: dependence on temperature and salinity, Applied Optics, 36 (24) (1997) 6035- 646.

19

Towards the Capabilities of Rocket Engines with Solar Heating of Working Fluid

V. M. Kotov

Republican State Enterprise "National Nuclear Center of the Republic of Kazakhstan", Kurchatov, Republic of Kazakhstan

Email address:

kotovvm@nnc.kz

Abstract: The purpose of this research is to show the potentiality of the rocket with solar heating of working fluid. The study describes design of the heater with tungsten pipes, ensuring full-range application of focusing radiation (95 %) under the hydrogen heating to 2900 K; gives an examples of optimization of rocket engine thrust by the flight time with predetermined fuel weight and specific weight of engine with mirror and payload; illustrates design of the engine and mirrors providing the conjugation between the thrust vector and solar radiation vector; demonstrates design of power-supply source with high-efficient solar energy transformation into electric one and compares various types of the rockets.

Keywords: Nuclear Rocket Engine (NRE), Working Fluid, Solar-Heated Rocket Engine, Mass of Engine and Mirrors, Open Space, High-Temperature Reactor Components

1. Introduction

The rocketry is a pioneer industry of human activity. In up-to-date perception of its objectives and uses the initial studies are not even hundred years. During this time it became clear enough to understand the potential of rocket technology in the development of the nearest cosmic space. Nevertheless, the big financial assessments and relatively unreliable equipment interfere implementing above tasks although a lot of these are able to be solved both by the international community and separate countries. Search for new technical solutions providing reduced costs and increased reliability is of first-priority in this regard.

The modern rocket engines are the engines using chemical components in their work. This leads to a relatively low specific impulse followed by big gross launching mass, high start-up expenses and low operational reliability. Therefore, the development of the engines with a higher specific fluid impulse is a key direction to improve the technique of space exploration.

Substantial growth of specific impulse is achieved using nuclear rocket engines (NRE), which became possible after becoming the technology of nuclear reactors in the 40-50 years of the 20th century [1, 2]. Examples of this type modern device are described in [3] Specific impulse becomes equal to ~900s when hydrogen is used as the working fluid of

the engine and it is heated to 3000 K.

The same characteristics can be obtained by heating the hydrogen with solar radiation. The advantage of using the solar radiation engines is a lack of problems with storing energy source and transforming one type of the energy into another one. One of example of rocket engine with solar heating of working fluid is described in [4]. There are some problems in the technology of rocket engines with solar heating of working fluid. Let's consider possible solutions to some of them.

2. Material and Methods / Experimental Details / Methodology

2.1. Design of Solar Heater

Let's use some tungsten pipes irradiated with solar flux from one side in order to heat hydrogen. Let us assume inner pipe diameter of 4 mm, wall thickness of 1 mm and pipe length of 1 m.

Let's make performance prediction of such a system by hydrogen heating to the temperatures close to 3000 K according to the program [5]. The hydrogen pressure at the pipe inlet will be assumed as 0.5 MPa, and the power applied to each pipe will be 4.28 kW. Fig. 1 shows the temperature distribution of the pipes' and working fluid surfaces along the length of the pipes calculated from the condition of uniformity of solar flux along the pipe. It can be seen that the temperature difference between the pipe surface and hydrogen in cross section is reduced when approaching the gas towards the outlet.

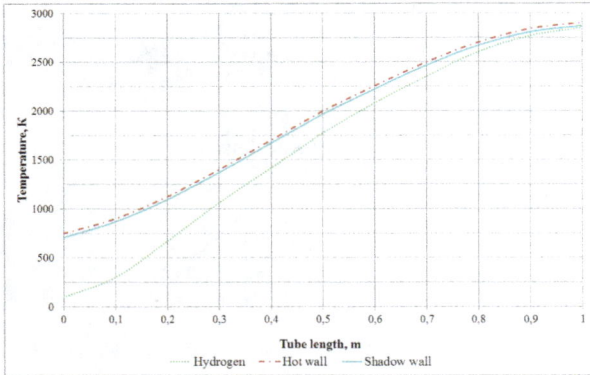

Figure 1. *Temperature distribution of the walls and hydrogen along the length of the pipe here.*

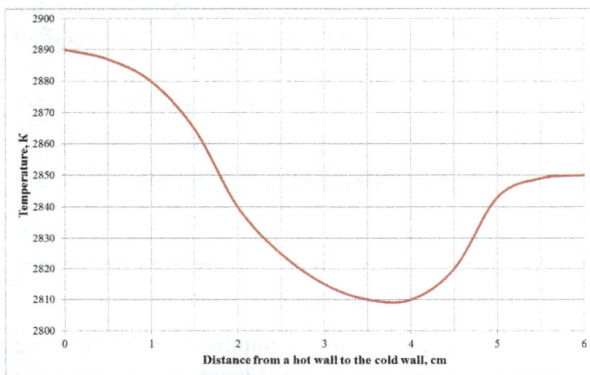

Figure 2. *Temperature distribution by diameter of outlet pipe section.*

Fig. 2 shows the temperature distribution in the pipe outlet section. It is seen that there is certain contribution into the heat transfer to the hydrogen from the shady pipe side.

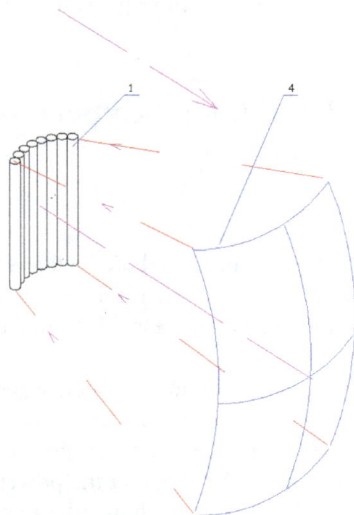

Figure 3. *Simplex heater.*

Fig. 3 shows a simplex hydrogen heater with tungsten pipes. It is equipped with the set of pipes (1) and focusing mirror (4). The focus of this mirror is behind the place of the pipes. The heater feature is the high losses of thermal energy due to energy radiation in accordance with the Stefan-Boltzmann law.

The heater showed in Fig. 4 provides solar radiation from the mirror 4 into the closed cavity 2 through an orifice of its surface 3. Heat insulator covers the walls of the cavity 2. The radiation scattered by pipes is absorbed within the cavity (and on the pipes themselves) reducing the proportion of irretrievably lost energy [6].

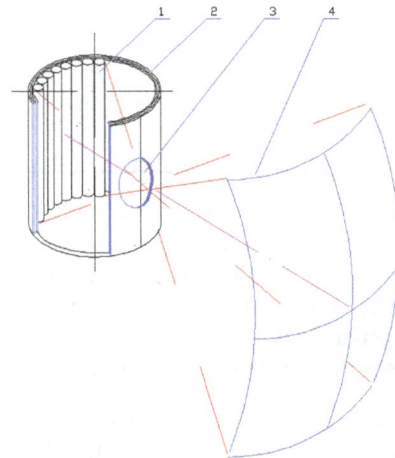

Figure 4. *Heater with minimized losses.*

Let us estimate energy loss in the heater according to Fig. 3. With given value of solar radiation flux the partial consumption decreased by radiation value according to the Stefan -Boltzmann law having determined by the pipe surface temperature will spend on the hydrogen heating.

Fig. 5 shows the dependences of the fraction of thermal energy transferred from the solar flux to hydrogen in a tubular receiver of this embodiment taken into account temperature increase in pipes' surface under two values of the heat flux density having removed by the hydrogen (200 and 500 kW/m^2).

It is seen that the efficiency of energy transfer to the working fluid decreases with increasing temperature and decreasing heat flux density of solar radiation having impacted on the surface of the pipes. The overall efficiency of energy transfer by heating up to 3000 K and a flow of 500 kW/m^2 is ~ 18%, and at a flow of 200 kW/m^2 is ~ 8%.

According to Fig. 4, the losses in the heater will be determined by the probability of scattering the Stefan-Boltzmann radiation from the surface of the pipes in the window 3 of a closed cavity 2. In the first approximation, this probability can be calculated as the ratio of the surface of the window to the irradiated surface of the pipes. By the Fig. 4, losses in real heaters will not exceed 5%.

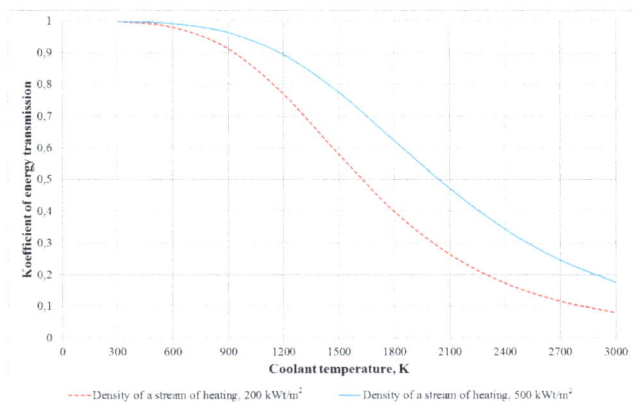

Figure 5. *Dependences of the efficient solar energy transfer to the hydrogen in simplex heater under densities of extraction of energy 200 and 500 kW/m².*

2.2. Optimization of Solar Heated Rocket Engine Thrust

By the parameters of tungsten pipe described above it is easy to estimate the minimum specific weight of the engine. If the weight of the pipe is 304 grams, transmitted to hydrogen energy is 4.28 kW, hydrogen heating is 2850 K, hydrogen flow rate is 0.085 g/sec, so the thrust will be 76 grams and specific impulse of the engine will be ~ 900 s.

When intensity of solar radiation at the Earth's orbit is 1.37 kW/m² and the specific weight of the focusing mirror is ~ 4 kg/m², so required for mirror mass to heat hydrogen in one pipe will be 12.5 kg. The total mass of the mirrors and the pipe in equivalent of the thrust of 1 g/s will be ~ 150 kg.

In case we take the rocket engines with solar heating thrust equal to the nuclear rocket engine thrust (6400 kg) for rocket [3], the mass of the solar engine will be close to 1000 tons. This option is completely irrational.

For open space it is rationally to reduce the engine thrust increasing its operation time. The trust value should be optimized according to the flight parameters. In this case, the determinant is not in the maximum achievable speed of the rocket but the flight time from the starting point to the goal.

Dependences of the in-flight time, initial rocket mass and its maximum speed on the engine hydrogen flow were calculated in different options of the initial hydrogen mass in the rocket under specified masses of fuel tanks - hydrogen storages, payload.. Fig. 6 shows the dependences of the initial mass, in-flight time and maximum speed during the flight in the range of hydrogen consumption close to optimal intended for the following initial data:

- Flight distance in the free space (L)   100 m. km;
- Initial and final rocket speed   0 km/h;
- Hydrogen mass at the beginning
  of the flight (Mfuel)   150 tons;
- Mass of hydrogen tanks (Mfuel tanks)   4.69 tons;
- Mass of the payload (Mpayload)   6.0 tons.

The optimal hydrogen consumption on the basis of minimum fly time under initial conditions is ~ 0.05 kg/s. The more flow rate, the more required heating power and additional weight of the engine and the mirror. An increase in the duration of the acceleration of the rocket becomes important with a decrease in the flow.

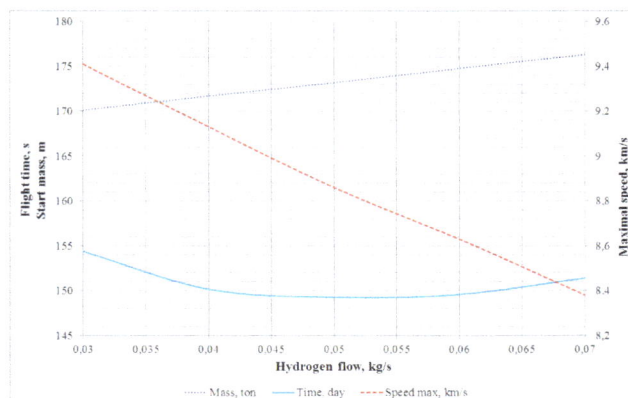

Figure 6. *Dependences of in-flight time, initial rocket mass and its maximum speed on hydrogen flow rate. L = 10⁸ km, Mfuel= 150 t, Mfuel tanks =4.69 m, Mpayload = 6.0 t, V0 = Vk = 0.*

2.3. Comparison of the Rocket with Solar Heating of Working Fluid (SRE) and Nuclear-Heated Propellant (NRE)

Fig. 7 provided minimum in-flight times of SRE and NRE, and its mass, under the same initial conditions but with different initial hydrogen masses [3].

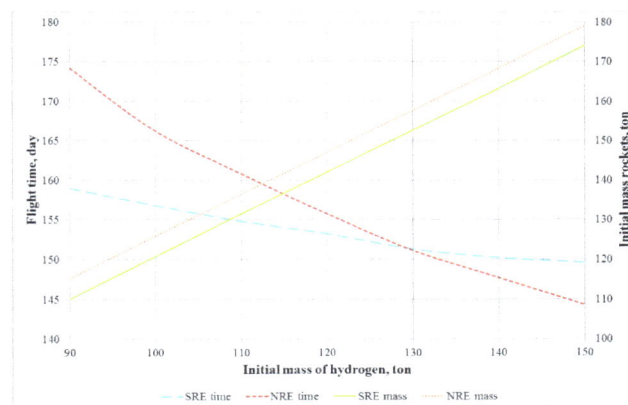

Figure 7. *Dependences of in-flight time and rocket mass with SRE and NRE on hydrogen mass. L = 10⁸ km, Mfuel tank =4.69 t, Mpayload = 6.0 t, V0 = Vk = 0*

It appears, that solar rocket engine benefits the in-flight time when hydrogen mass is less than 130 tons. In-flight time change in given range of hydrogen mass is 3.2 times less for SRE than for NRE.

It should be noted that increase in hydrogen mass is lead to both growing the total initial rocket mass, its value grain and loss in operation reliability. It follows, that the version of SRE rocket and initial hydrogen mass of 90 tons looks more preferable rather than NRE rocket with initial hydrogen mass of 150 tons.

2.4. Item Design of the Rocket with Solar Heating of Working Fluid

In comparison with other rockets, the rocket with solar

heating of working fluid has one feature as presence of a radiating scarce (the Sun) used as an orienting point for focusing mirror if it is necessary to orient the engine power in any direction toward the source. The task is to provide the rocket nozzle having oriented in other direction with solar radiation energy produced from one direction in one form or another.

This problem can be solved by various ways. For this, paper [4] uses transformation of solar radiation into electrical current, heat of high-temperature heat-exchanging unit - accumulator with the current. It isn't complicated to transfer the power from solar panel to the heat-exchanging unit rigidly bound with rocket effluent nozzle. However, this design has low conversion efficiency of solar radiation into electric power.

Fig. 8 shows a scheme in which the conjugation of the vectors is achieved by means of movable connections of the hydrogen supplying line to the heater and supply line from the heater to the nozzle. Since, there are two mating vectors, there is always will be a plane in which they are both located. The use of one group of such connections lead to the plane of the vectors will pass through the body of the missile. The decision of the problem is to separate focusing mirror and the engine for two symmetric objects [7].By the parameters of tungsten pipe described above it is easy to estimate the minimum specific weight of the engine. If the weight of the pipe is 304 grams, transmitted to hydrogen energy is 4.28 kW, hydrogen heating is 2850 K, hydrogen flow rate is 0.085 g/sec, so the thrust will be 76 grams and specific impulse of the engine will be ~ 900 s.

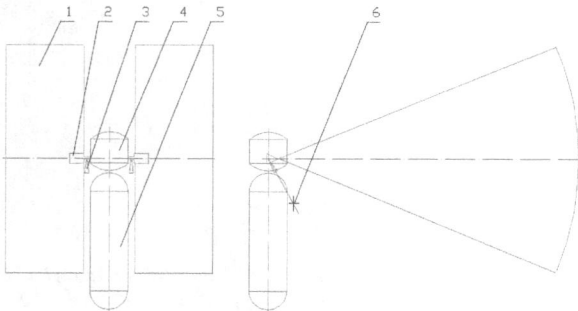

Figure 8. *Solar –heated rocket. 1 – Focusing mirror, 2 –heaters, 3 – engine nozzles, 4 – payload module, 5 – fuel tank, 6 – rocket centre of gravity.*

Heaters 2 in this scheme are made by drawing 4. Connection of hydrogen supply line from the missile body to the highway entrance to the solar heater and line of hydrogen release from the heater to the highway entrance to the rocket nozzle is shown in Fig. 9.

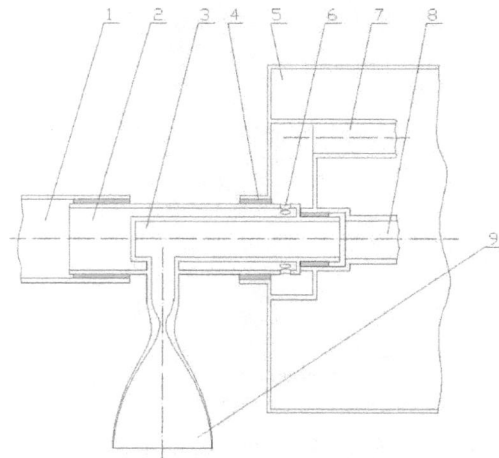

Figure 9. *Diagram of junction the rocket vessel elements with the heater and nozzle. 1 - stationary propellant feed pipe, 2 - moving feed pipe, 3 - pipe feeding the propellant to the nozzle, 4 – seal, 5 - heater body, 6 – holes, 7 - pipe feeding the propellant to the heater, 8 – hot propellant pipe, 9 - nozzle.*

2.5. Moon Mission Parameters

Let's consider options for flight to the Moon, which are different by composition of the payload and the conditions approaching to the Earth on the return flight:

1. Flight of the crew from the Earth's orbit to the Moon's orbit and return to the Earth at a speed of 11 km / s.

2. Flight of the crew from the Earth's orbit to the Moon's orbit and return to the Earth's orbit.

3. Shipping to the Moon's orbit landing module (lunar module) and return the rocket to the Earth's orbit.

4. Flight to the Moon's orbit landing module and a crew and the crew's return to the Earth's orbit.

Table 1. *Main characteristics of flight options to the Moon*

Parameter	Option 1	Option 2	Option 3	Option 4
Start mass on the Earth, ton	310	400	1100	1400
Mass on the Earth's orbit, kg	7965	10342	28050	35700
Lunar module mass, kg	0	0	14000	14000
Fuel mass on the Earth's orbit, kg	3165	5542	10650	14700
Mirrors, heater and engine mass, kg	1200	1200	3400	3400
Dwelling unit mass, kg	3600		0	3600
Mirrors area, m²	229,4	229,4	659,5	659,5
Engine thrust, N	80	80	230	230
T1, hours	81,25	103	98	124,5
T2, hours	17,8	18	6	9,3
T3, hours	0	55	13	25

Here: T1 – engines working time in flight to the Moon, T2 - engines working time in flight from the Moon's orbit, T3 – engines working time while deceleration during the return to Earth's orbit.

May be options 2 and 3 will be more preferred, they are carried out separately delivery of the crew and lunar modules with reusable solar rocket engines in flight to the Moon and return to the Earth's orbit.

The main thing is that it reduces the cost of the Project on flights to the Moon and makes possible to improve performance of the lunar rocket by decreasing its mass and increasing security at the expense of low- encroaching speed to the Earth with earth satellite arrival in the orbit.

2.6. Power-Supply Source

Long-running solar rocket engines during the flight to the Mars make it possible to develop generators with a very high efficiency of conversion of thermal energy into mechanical one [8]. Figure 10 provides a diagram of the generator. According to this diagram a refrigerator of a heat machine is used to heat a cryogenic working body so the thermal energy of all the processes fully utilizes.

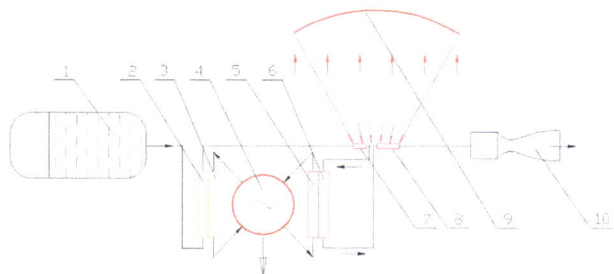

Figure 10. *Rocket with solar heating of working fluid power-supply source. 1 – vessel with a working fluid, 2 – working fluid heater, 3 – heat machine refrigerator, 4 – heat machine, 5, 6 – heat exchanger of heat machine, 7 – solar heater of a heat machine, 8 – engine heater, 9 – focusing mirror, 10 – engine nozzle. –heated rocket*

The rocket with mirror area of 229 m2 when coefficient of efficiency of 20 % and minimum cycle temperature of 300 K is able to achieve electric power of 7.8 kW at engine burn.

The 2.5 % of the solar flux (approximately 6 m^2 of the mirror) is tapped herewith off. Using solar panels with the same electrical power would require additional panel with an area of ~40 m^2.

3. Conclusion

Heater with tungsten pipes ensures hydrogen heat up to 2900 K with high density of focused solar energy flux and hydrogen pressure at inlet 0.5 MPa.

Heater with closed cavity design, which ensures high effectiveness of solar energy transfer to hydrogene (>95 %), is presented.

Necessity of the trust value optimization according to the flight parameters with determined fuel mass, specific weight of the engine with the focusing mirror, and mass of payload, is shown.

The engine with the focusing mirror design, which ensures solar flux and rocket propulsion vectors conjugation, is presented.

The power-supply source design with high-efficient solar energy transformation into electric one is demonstrated.

Comparison of various types of the rockets is shown.

References

[1] Bussard R., De Lauer R. Los Alamos Scientifie University of California. McGrow-Hill Book Company, Inc. New York-Toronto-London; 1958.

[2] Bussard R., De Lauer R. Fundamentals of Nuclear Flight. New York; 1966.

[3] Romadova E.L., Smetannikov V.P., Cherepnin Yu.S. etc. Nuclear power and power propulsion systems for space exploration. Nuclear Energy in the Republic of Kazakhstan. International Scientific and Technical Conference. Kurchatov; 2005.

[4] Koroteev A.S., Akimov V.N., Arkhangelsky V.I., Kuzmin E.P. Solar thermal rocket engine. Patent of Russia No. 2126493 of Feb. 20; 1999.

[5] Fluent version 6.3.26 User Reference. Fluent, Inc.; 2006.

[6] Kotov V.M. High-temperature solar heater. Innovative patent of the Republic of Kazakhstan. No. 22587 of Jun 15, 2010.

[7] Kotov V.M. Rocket with a solar fluid heating. Innovative patent of the Republic of Kazakhstan. No. 22933 of Sept 15, 2010.

[8] Kotov V.M. Energy source for a solar-heated rocket engine. Innovative patent of the Republic of Kazakhstan. No.22809 of Aug 15, 2010.

The Techniques to Control a Space Laboratory Orbital Motion During Conducting of Gravity-sensitive Processes on Its Board

A. V. Sedelnikov

Departament of Space Mechanical Engineering of Samara State Aerospace University, Samara, Russia

Email address:
axe_backdraft@inbox.ru

Abstract: The following article deals with three different techniques to control orbital motion of spacecraft with big flexible structures during conducting of gravity-sensitive processes on its board. These processes require low level of microaccelerations. There were given examples of application of the techniques, recommendations for application of these techniques in practice, discussed advantages and disadvantages of each technique.

Keywords: Control of Orbital Motion, Level of Microaccelerations, Gravity-Sensitive Processes

1. Introduction

Conducting of gravity-sensitive processes on a board of spacecraft requires low level of microaccelerations [1]. The technological processes which require level of microaccelerations about 10^{-7} g are worked out. It is very difficult to support such level. Most of these processes are energy-intensive. It is necessary to use big solar paddles for its conducting. The constant orientation of spacecraft on its orbit is necessary for effective use of solar paddles. Running of attitude engines has a negative effect on facilities for technological processes [2]. The firing of engines causes a sudden increase of microacceleration level. It leads to unsuccessful carrying out of the processes. That is why the problem of optimal control of spacecraft on its orbit is of great urgency.

2. Total Experience, Taking into Account the Last Projects

2.1. Unoriented Orbital Flights

The practice in conducting of unoriented orbital flights of technological spacecraft from line of "Foton" ("Foton-1", "Foton-M" No 3 (figure 1)) showed two significant disadvantages of this approach.

1. Short operational term of spacecraft on its orbit.

In case of the longest flight operational term was less than 18 days [1]. The term is not enough for conducting of long gravity-sensitive processes on a board of spacecraft.

Figure 1. Space laboratoty "Foton-M" No 3.

2. Unoriented flight leads to promotion of spacecraft.

A rotational speed of "Foton-M" No 2 (figure 1) rose in 10 times in the end of the operational term, speed of «Foton-M» No 3 rose in 3 times [3]. This fact significantly involves an estimation of microacceleration level and prediction of probability of successful carrying out of technological processes [4].

These disadvantages are inappropriate for the most of gravity-sensitive processes. To eliminate defects it is necessary to control spacecraft orbital motion.

For maintenance of required orientation it is necessary to use different facilities such as gyroscopes and gyrodines. These facilities accumulate an angular momentum from factors which disturb a motion of spacecraft. The main problem of this orientation scheme is to search the methods to reduce this momentum.

2.2. The Technique 1. the Application of Low Thrust Liquid Rocket Engines

These engines run in pulse mode. When an angular velocity of wheel of gyroscope or gyrodine becomes critical, the engine starts. It causes a reactive moment around center of mass of spacecraft (about 4-10 $N \cdot m$) and stimulate reduction of angular momentum of wheel of gyroscope or gyrodine. The technique was used on "Skylab" (figure 2), "Bion-M" No 1 (figure 3) and other spacecraft.

Figure 2. Space laboratoty "Skylab".

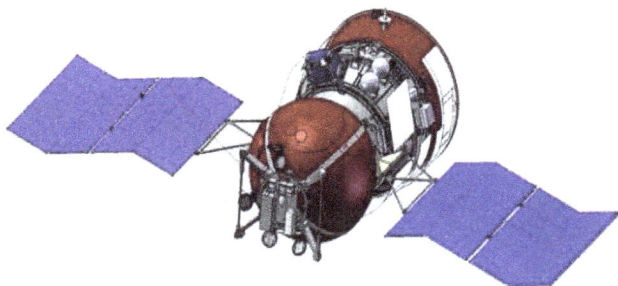

Figure 3. Space laboratoty "Bion-M" No 1.

The advantages of this technique are listed below:
- effectiveness of orbital motion control;
- small fuel consumption for orientation control.

However, there are some disadvantages:
- engine firing can cause an increase of microaccelerations to 10^{-3} g;
- engine firing causes oscillations of big flexible structures.

2.3. The Technique 2. the Application of Low Thrust Liquid Rocket Engines and Electric Jet Engines

This perspective technique to control orbital motion for successful conducting of gravity-sensitive processes is described in article [5]. The advantages of this technique are listed below:
- a constant running of electrojet engine (figure 4) allows to raise a time interval between firings of liquid rocket engines;
- it is possible to use solar wind as a fuel in the long term.

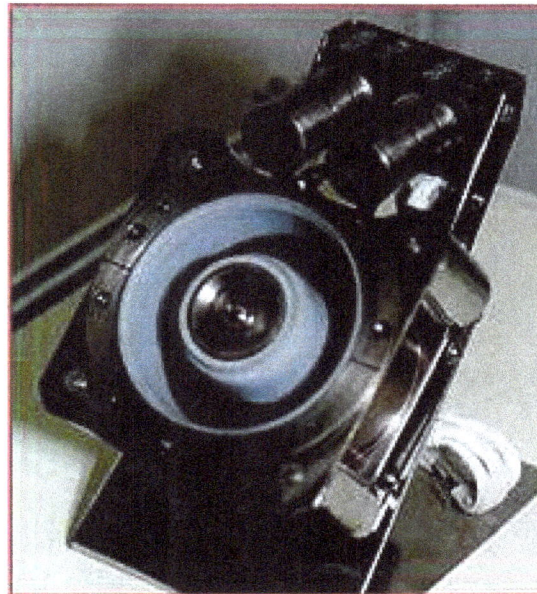

Figure 4. Electrojet engine SPD-100 [6].

Table 1. Main Characteristics of SPD-100

Caracteristic, dimension	Value
Draft, mN	83
Power consumption, kW	1,2
Work resource, h	7500
Mass, kg	3,5
Specific impulse, s	1100
Coefficient of efficiency	0,3
Level of microaccelerations, $\mu m/s^2$	0,01-0,03

2.4. The Technique 3. the Application of Low Thrust Liquid Rocket Engines and Flywheel Engines

Recently completed space project "Foton-M" No 4 was oriented by this new method. The low thrust liquid rocket engines stared only two times during the flight: after an orbital injection of spacecraft and before a launching of reentry vehicle. The spacecraft was oriented on the orbit by the system of attitude-control flywheel engines (figure 5) during all operational term. Indeed an advantage of this technique is that there is not any recent increase of microacceleration level which is usually caused by firings of attitude engines. However techniques 1 and 2 have a common disadvantage. During impulse firings a level of microaccelerations can decrease to 10^{-7} g. A constant running of electric jet engine or system of flywheel engines can lead to minimal microacceleration level about $(1...3) \cdot 10^{-6}$ g. Unfortunately, there were some problems during realization of "Foton-M" No 4. That is why we can't consider this project to be absolutely successful. "Foton-M" No 4 wasn't launched to the operational orbit because of loss of communication with spacecraft in the beginning of the flight. According to preliminary estimates microacceleration level at perigee could rise to 10^{-4} g instead of 10^{-5}-10^{-6} g because of great aerodynamic disturbance. As the result duration of the experiments was shorten. The operational term of the spacecraft was diminished from 60 to 42 days.

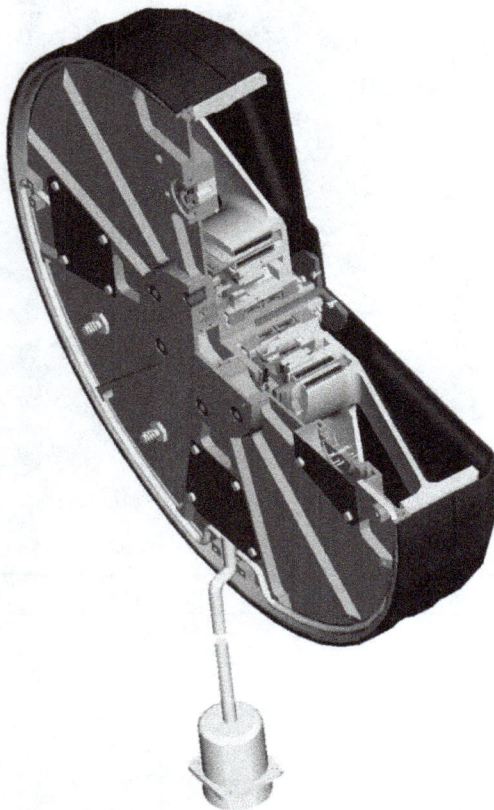

Figure 5. *Flywheel engines [7].*

2.5. Space Laboratory "Foton-M" No 4

In spite of the difficulties "Foton-M" No 4 (figure 6) nowadays is the most progressive realized project. The realization of this project allowed conducting of the longest biological experiments on the board of autonomous unmanned space laboratory. New control system of orbital motion allows maintaining low microacceleration level about 10^{-6} g during all operational term of space laboratory (to 60 days). Test of this control system will allow using of it also in more significant space projects [8,9]. For example, the project of reusable space laboratory "OKA-T" (figure 7) is being developed in Russia at the present time [1]. Laboratory

will be tended by international space station. It will allow raising its operational term to 5 years. At the same time "OKA-T" could be in an autonomous controlled flight to 90-100 days. Achievable level of microaccelerations on space laboratory "OKA-T" will be about 10^{-7} g.

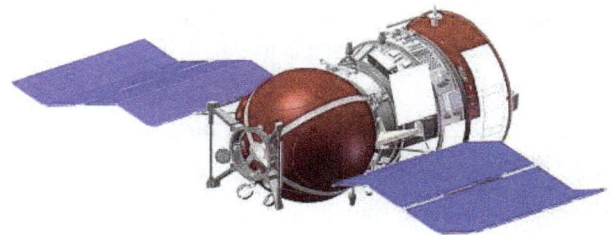

Figure 6. *Space laboratoty "Foton-M" No 4 [10].*

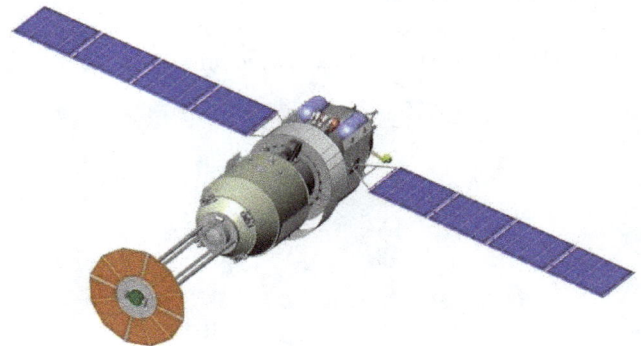

Figure 7. *Space laboratoty "OKA-T" [11].*

If the space laboratory is tended by international space station, time of return of experimental results to the Earth will rise. But some of the results should be delivered to the Earth quickly. That is why a project of small spacecraft "Vozvrat-MKA" is being developed now for quick launching of the experimental results.

References

[1]   A.V. Sedelnikov The Problem of Microaccelerations: from Comprehension up to Fractal Model, (Moscow, Russian Academy of Sciences: The Elected Works of the Russian School, 2012), p. 277.

[2]   A.I. Belousov and A.V. Sedelnikov "Problems in Formation and Control of a Required Microacceleration Level at Spacecraft Design, Tests, and Operation," Russian Aeronautics, vol. 57, No 2, 2014, pp.111–117.

[3]   T. Beuselinck, Van C. Bavinchove, V.I. Abrashkin, A.E. Kazakova and V.V. Sazonov "Determination of the Spacecraft Foton M-3 Attitude Motion on Measurements of the Earth Magnetic Field," Preprint, Keldysh Institute of Applied Mathematics, Russia Academy of Sciences, No 80, 2008.

[4]   A.I. Belousov and A.V. Sedelnikov "Probabilistic Estimation of Fulfilling Favorable Conditions to Realize the Gravity-sensitive Processes Aboard a Space Laboratory," Russian Aeronautics, vol. 56, No 3, 2013, pp.297–302.

[5] A.V. Sedelnikov and A.A. Kireeva "Alternative solutions to increase the duration of microgravity calm period on board the space laboratory," Acta Astronautica, vol. 69, 2011, pp. 480–484.

[6] A.V. Sedelnikov, E.Yu. Sygurova and A.A. Kireeva "The Use of Electric Engine to Reduce the Microaccelerations Level of Space lab," Messenger the Samara State Aerospace University name of the academician of S. P. Korolev (national research university), No 3 (34), part 2, 2012, pp. 16–20.

[7] A.G. Brovkin, B.G. Burdygov and S.V. Gordiyko Anboard control systems of spacecrafts, (Moskow, Publishing house of MAI-PRINT, 2010), p. 304.

[8] A.V. Sedelnikov "Fractal quality of microaccelerations," Microgravity Scienes and Technology, vol. 24, No 5, 2012, pp.345–350.

[9] A.V. Sedelnikov "The usage of fractal quality for microacceleration data recovery and for measuring equipment efficiency check," Microgravity Scienes and Technology, vol. 26, No 5, 2014, pp.327–334.

[10] V.I. Abrashkin "Preparation of the TUS space experiment for UHECR study," International Journal of Modern Physics A, vol. 20, No 29, 2005, pp. 6865–6868.

[11] Patent RU 2181094 C1 29.08.2000.

21

Basic Development Stages of the Algorithms Applied to Recover Lost Microacceleration Data and Check Efficiency of Measuring Equipment on the Board of Space Laboratory

K. I. Potienko

Faculty of Computer Information Technologies and Automatic Equipment, Student of Donetsk National Technical University, Donetsk, Ukraine

Email address:

potienko97@mail.ru

Abstract: The following article deals with the algorithm applied to recover lost microacceleration data. The algorithm allows recovering on the basis of available part of measuring data. It is based on the fractal quality of constructive part of microacceleration field. The algorithm can be applied for technological space laboratories only if a constructive part of microacceleration field is prevalent.

Keywords: Space Laboratory, Algorithm Applied to Recover Data, Microaccelerations

1. Introduction

Microacceleration level in the working area of technological equipment is one of the most important characteristics of modern space laboratory [1]. The problem to measure microaccelerations during orbital flight of space laboratory correctly is very difficult. The causes of it are mentioned below:

1. Highly sensitive equipment can fail because of great overloads which can achieve 10g during launching of unmanned space laboratory.

For example, during different sets of experiments of Microgravity Isolation Mount (MGIM) in the board of space station "MIR" amplitudes of oscillations differed in 20 times [1,2]. The data of microaccelerations which were measured by French microaccelerometers "BETA" on the board of uncontrolled and unmanned space laboratory "Foton-11" which are shown in figure 1 were similar [3]. Readings of accelerometers for three different axes of a bound coordinate system differed also in 20 times.

2. There are some difficulties to measure microaccelerations directly in contrast to temperature, pressure etc. That is why different gauges give different data, especially if these gauges are based on different methods to estimate microaccelerations and have different sensitivity. Readings of French equipment "Alice-2" and Perm Convection Sensor "DACON" on the

board can exemplify it [1,4].

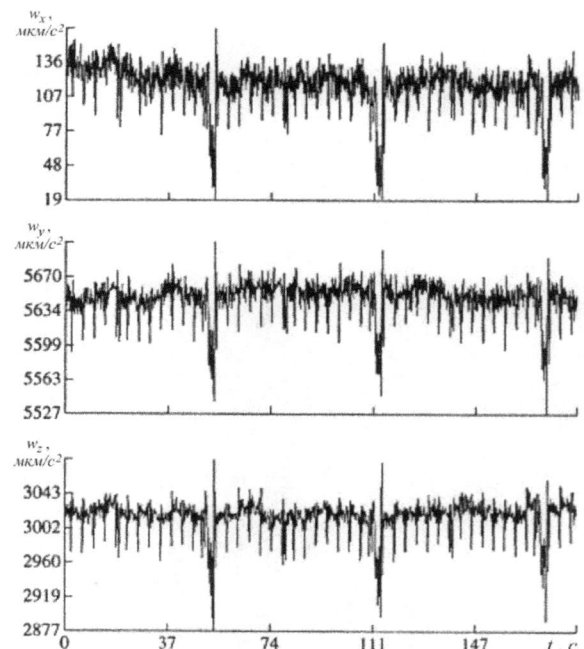

Figure 1. The estimation of microaccelerations on the board of space laboratory "Foton-11" by French microaccelerometers "BETA".

Valid data are necessary to explore how microacceleration

field effects on gravity-sensitive processes. Some articles [5-7] show ambiguity of this effect by the example of oriented crystallization. That is why the problem to increase data reliability is of great urgency for development of space technologies and space laboratories.

2. Microaccelerometers

2.1. Classification of Microaccelerometers

Microaccelerometers can be categorized on the basis of their operating principle. There are three groups of ones: electric, mechanical and electromechanical.

Operating principle of electric accelerometers (figure 2) is based on changing of electro-capacity between static and dynamic states.

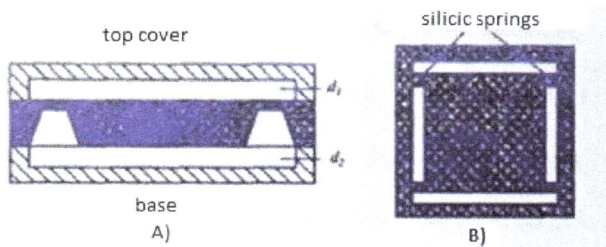

Figure 2. Capacitive accelerometer with differential capacitor: A) cross-section of accelerometer; B) top view..

Thereafter a capacitor must be included in construction of the detector. Measuring of microscopical amplitudes of vibration oscillations requires high-accuracy detector of such movements. Maximum movement which can be measured by electric accelerometers is about 20 μm. This condition significantly effect on range of application of these accelerometers. Electric microaccelerometers are used for measuring of low-frequency vibrations, movement and constant accelerations. The range of application of these gauges is very comprehensive. However, there is a significant disadvantage: electrostatic forces changes running of accelerometer. So these gauges hardly can be tested on the Earth [8].

Mechanical microaccelerometers (figure 3) significantly differ from electric ones. The basis constructional part of mechanical detector is a pendulum. The sensor records its displacement from point of equilibrium. Each pendulum accelerometer has elastic suspension, which compensate a torque caused by microaccelerations. Elastic suspension moves pendulum because they are connected. The construction of accelerometer also should involve some element, which recover pendulum, such as return induction coil. Thus, microaccelerations can be estimated on the basis of certain displacement of elastic suspension caused by small forces. However, presence of elastic suspension in construction of accelerometer puts on certain application conditions and application sphere [9].

Recently developments, which include pendulum accelerometers as well as capacitive, have appeared. Sensor of these accelerometers is made from monocrystalline silicon and glass.

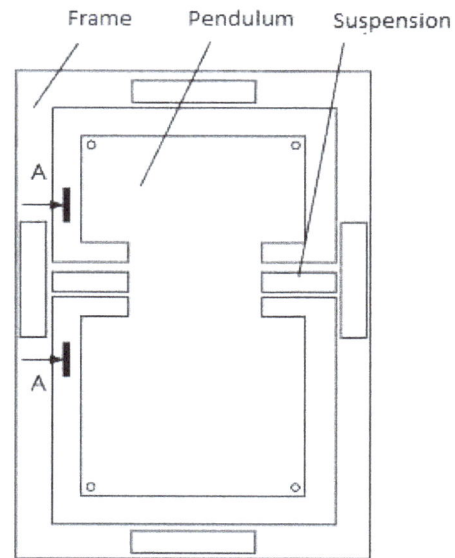

Figure 3. Space laboratoty "Foton-M" No 3.

It guarantees high reliability, accuracy, time and temperature stability of the gauges. If the sensor has measurement range about ±1g, detector can withstand loads about 50000g. At the same time it is sensitive to static acceleration and vibration. Moreover, important criteria of accelerometers is its measuremenrt, frequency, temperature range, maximum shock load, number of axes, access speed, size and mass. For example, great overloads have their action on space laboratory during its going to orbit. Usually, low-sensitive capacitive accelerometers can support such overloads. These gauges run on the board of space laboratory until it has gone to its orbit. High-accuracy pendulum accelerometers are running during orbital flight. So microaccelerometers based on microelectromechanical system are the most practically feasible.

2.2. Examples

Microgravity Acceleration Measurement System (MAMS) studies the small forces, or vibrations and accelerations, on the International Space Station (ISS) that result from the operation of hardware, crew activities, dockings and maneuvering. Results are used to generalize the types of vibrations affecting vibration-sensitive experiments. Investigators seek to better understand the vibration environment on the space station. Vibrations exist on the ISS from a variety of sources, such as equipment operation, life-support systems, crew activities, aerodynamic drag, gravity gradient, rotational effects and the vehicle structural resonance frequencies. The quasi-steady acceleration is caused by forces from aerodynamic drag, gravity gradient effects, centripetal (rotational) motion, spacecraft propulsion, and vehicle orientation control actions. MAMS consists of two sensors, the Orbital Acceleration Research Experiment (OARE) Sensor Subsystem (OSS) and the High Resolution Accelerometer Package (HiRAP), monitor these disturbances. The OARE OSS measures low range frequency (up to 1 Hz). The HiRAP characterizes the ISS vibratory environment from 0.01 Hz to 100 Hz. These quasi-steady state accelerations occur in the frequency range below 1 Hz. MAMS consists of a low-frequency triaxial accelerometer, the Miniature Electro-Static Accelerometer (MESA), a high-frequency accelerometer, the High-Resolution Accelerometer Package (HiRAP), and associated computer, power, and signal processing subsystems contained within a Double Middeck Locker enclosure.

The DIMAC (Direct-measurement Microaccelerometer) system developed for the "Foton-M" № 3 flight is a modular inertial monitoring system which measures vibrations, quasi-steady accelerations and the Earth magnetic field. It was built at RedShift Design and Engineering BVBA of Sint-Niklaas, Belgium.

The DIMAC system consists of DIMAC Control Unit, DIMAC Sensor Head and three DIMAC Magnetometer Modules (figure 4).

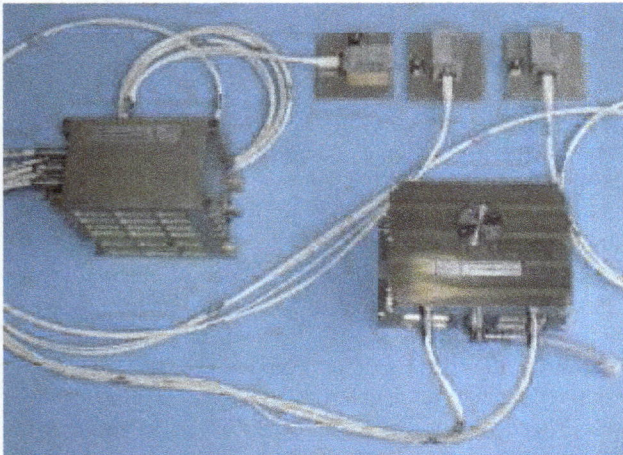

Figure 4. Overview of DIMAC.

The DIMAC Control Unit is an enhanced version of the TAS3 device featuring: autonomous acquisition of internal tri-axial vibration sensor (based on Honeywell QA3000 series sensors), acquisition of vibration data (resolution 1μg, bandwidth 0.01Hz to 200Hz, sample rate 1000Hz) and sensor temperature data. The DIMAC Sensor Head is a proto-flight version of the experimental quasi-steady acceleration sensor developed during the DIMAC study phase. For the "Foton-M" №3 mission the DIMAC Sensor Head was located at the bottom of the re-entry capsule, close to the center of gravity of the spacecraft. This sensor was implemented in order to correct for the minor magnetic field susceptibility of the acceleration measurements that was detected during the processing of TAS3 measurement data after the "Foton-M" №2 mission.

Tri-axial magnetometer modules are measuring the Earth magnetic field (and hence indirectly the spacecraft attitude). Three modules have been placed at different locations inside the "Foton-M" №3 re-entry capsule in order to compensate for any local magnetic field disturbances caused by payloads or spacecraft subsystems.

The stored data have been downloaded, post-processed and analysed after the mission. During the mission subsets of the DIMAC measurement data have been downloaded using the telemetry interface. These data have been processed immediately and preliminary results have been available to the scientists and operators of other "Foton-M" №3 payloads. The flexible implementation of the telecommand interface allowed telemetry data bandwidth optimisation and DIMAC system troubleshooting. However, the flight results show no major disturbances. This indicates that if an appropriate location is selected a single magnetometer module might be sufficient for future missions. The modularity of the DIMAC system allows easy implementation of additional sensor modules. The resulting system could provide full inertial and environmental monitoring of a spacecraft for future missions. The major parts of this system have been qualified on the "Foton-M" №3 mission.

The majority of microgravity experiments show their highest sensitivity to residual acceleration in the low frequency range, typically <0.01 Hz, where atmospheric drag, gravity gradient and centrifugal forces are pre-dominant. QSAM (Quasi-Steady Accelerometer Measurement System) is an instrument especially developed to detect this range where conventional methods are hampered by bias and noise problems. Bias is generally varying in time due to unknown dependence on temperature, ageing and other effects precluding pre-mission calibration. In QSAM, signal modulation is applied to cancel the instantaneous bias during the measurement cycle. The modulation is achieved by turning the sensor's sensitive axis. QSAM is capable to detect the frequency range between 0 and 0.02 Hz with a resolution better than $10^{-7}g_0$.

STAR (Space Three-axis Accelerometer for Research mission) which is shown in figure 5 is an accelerometer system provided by CNES and developed by ONERA in France.

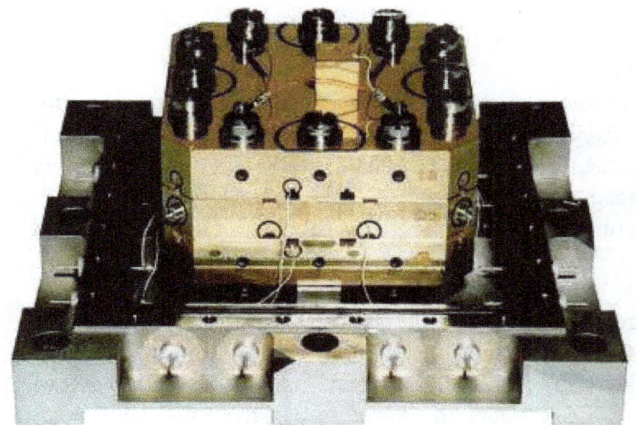

Figure 5. Overview of STAR.

The objective is to measure all non-gravitational accelerations of the satellite (drag, solar and Earth radiation pressure) in order to determine the Earth's gravity field from purely gravitational orbit perturbations (orbit from BlackJack).

The accelerometer measurement principle is based on electrostatic suspension of a proof-mass in a cage. Instantaneous position of the proof-mass is measured by three capacitive sensors which permit a determination of the acceleration vector. The instrument has a dynamic range of $\pm 10^{-4}$ ms^{-2}, a resolution of better than $\pm 3 \cdot 10^{-9}$ ms^{-2}, and a frequency range of 10^{-1} to 10^{-4} Hz. The STAR instrument is positioned at the center of gravity of CHAMP to minimize the influence of measurement disturbances due to rotational accelerations and gravity gradients. STAR is also connected to a star sensor (ASC of DTU, Denmark) with two heads to provide the accelerometer's axes orientation. The accelerometer proof-mass is positioned at the center of gravity of CHAMP.

Convection sensor DAKON-M is a cavity cylinder filled with carbon dioxide. Diameter and height of cylinder have the same value L=10 cm. The set temperature difference $\Delta T = 60°C$ is kept fixed on the opposite ends of the sensor. There are two differential thermocouples to measure temperature differences in the two pairs of fixed points inside the cavity. The sensor shows these differences. Convection sensor DAKON-M is an improved version of DAKON sensor, which was tested on the orbital station "MIR" [1]. According to [2-5] data from the sensor depend on microaccelerations experienced by it. This sensor is an example of a gravitationally sensitive system.

It is worth to say that on the orbital station "MIR" was tested French accelerometer Alice-2 along with DAKON (figure 6). However, the data of two different sensors were different.

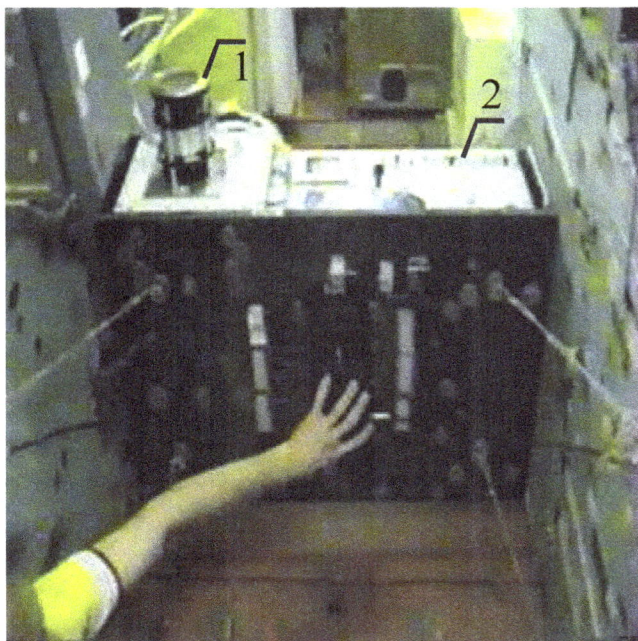

Figure 6. Convection sensor DAKON (1) and equipment Alice-2 (2).

During the last experiments on the board of "Foton-M" №4 was applied equipment to measure microaccelerations "GRAVITON" which is shown in figure 7. Its measurements are based on analyze of effect of microaccelerations on terrestrial magnetic field.

Figure 7. Overview of "GRAVITON".

Thus, accelerometers based on the different operational principle gives different data. It is one of the main problems how to measure accelerometers correctly.

3. Fractal Quality of Constructive Microaccelerations

Classification of microaccelerations according to ways of its control is shown in the article [1]. It divides all disturbing factors into three categories:

-metastable (outside disturbing factors and running of technological equipment);

-random (random oscillations of disturbing factors, micrometeorites, etc.);

-constructive (inside disturbing factors).

In some cases constructive microaccelerations mainly contribute to microacceleration field. Base model of such cases can be oriented orbital motion of space laboratory with big flexible elements [10]. Fractal (scaling) quality of microaccelerations was opened for similar cases [11]. It consists in invariability of microaccelerations dependence on time during scaling of sizes and inertia-mass characteristics of space laboratory. Method applied to recover lost microacceleration data was developed on the basis of this quality [12].

The purpose of the following article is to develop algorithms on the basis of the method [12].

4. Development of Algorithms for Different Problems

4.1. The Problem to Recover Lost Part of Data

Description: A part of data was lost during exploitation of space laboratory because of failure of gauges or problems with transmission of telemetric data.

Urgency: Solution of the problem will allow recovering of lost data on the basis of available part of data. This make possible to analyze facilities for realization of technological processes during all operational term. For instance, there was no communication with spacecraft from 19.07 to 26.07.2014 during realization of the project "Foton-M" №4. This circumstance caused some difficulties with telemetric data processing.

Algorithm:

1. Check availability of fractal quality by means of estimation of contribution of constructive part to microacceleration field.

2. Check possibility of recovering with prescribed accuracy by comparison of lost part with available part of data.

3. Estimate constructive part of microaccelerations from available part of data using fractal quality.

4. Check accuracy of estimation by comparison of model data with the available part.

5. Recover lost part of data using generated model.

This algorithm is shown in figure 8.

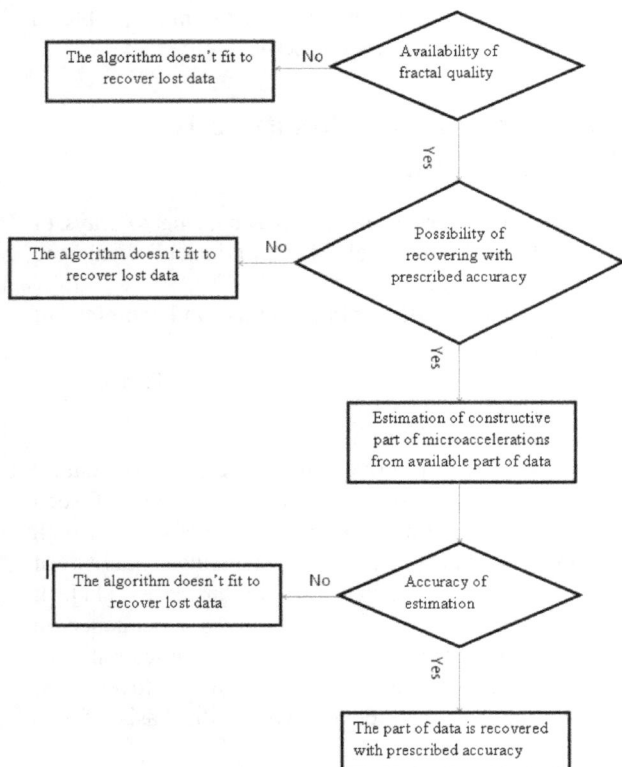

Figure 8. *Algorithm 4.1.*

4.2. Correction of Data from Defective Gauges

Description: Great launching overloads caused some gauges to fail during exploitation or software which is assigned to processing of information from gauges was running incorrectly.

Urgency: The practice of realization of space projects

shows that in some cases gauges runs incorrectly. In figure are showed data of French accelerometers "BETA" for space laboratory "Foton-11". There were no physical cases for such significant differences of data from gauges [13]. That is why false run of gauges can be considered to be one of the possible cases. Incorrect data on the state of microacceleration field in the indoor environment of space laboratory doesn't allow analyzing of effect of microaccelerations on experiment. Requirements for successful realization of this experiment also will be wrong. Thus, the problem solution will allow:

-increasing of data accuracy;

-correcting of data from damaged gauges;

-analyzing of effect of microaccelerations on experimental results.

Algorithm:

1. Uncover significant differences of data from gages by test of statistic hypothesis of homogeneity.

2. Test availability of fractal quality by means of estimation of contribution of constructive part to microacceleration field.

3. Gather source data from each gauge to control them.

4. Estimate constructive part of microaccelerations on the basis of data from each gauge using fractal quality.

5. Generate secondary data in the form of deflection of source data from estimation of microaccelerations to control each gauge.

6. Analyze compliance between estimation of microaccelerations and source data by means of test of statistic hypothesis by fitting criterion.

7. Uncover damaged gauges.

8. Replace data by fractal estimation for each damaged gauge.

This algorithm is shown in figure 9.

4.3. Efficiency Check of Gauges

Description: It is necessary to control technological state of gauges on the stage of test and exploitation [14]. The control can be done by the compliance test of source data with fractal quality of microaccelerations.

Urgency: Efficiency check of gauges raises reliability of data. It makes possible to do passive control of gauges on the stage of test and exploitation. Moreover disadvantages of gauges can be detected by the efficiency check for the purpose of its development.

Algorithm:

1. Check availability of fractal quality by means of estimation of contribution of constructive part to microacceleration field.

2. Gather source data from gauges to control them.

3. Estimate constructional part of micriaccelerations using fractal quality.

4. Generate secondary data in the form of deflection of source data from estimation of microaccelerations.

5. Analyze compliance between estimation of microaccelerations and source data by means of test of statistic hypothesis by fitting criterion.

6. Draw a conclusion on correctness of data and technological state of gauges.

This algorithm is shown in figure 10.

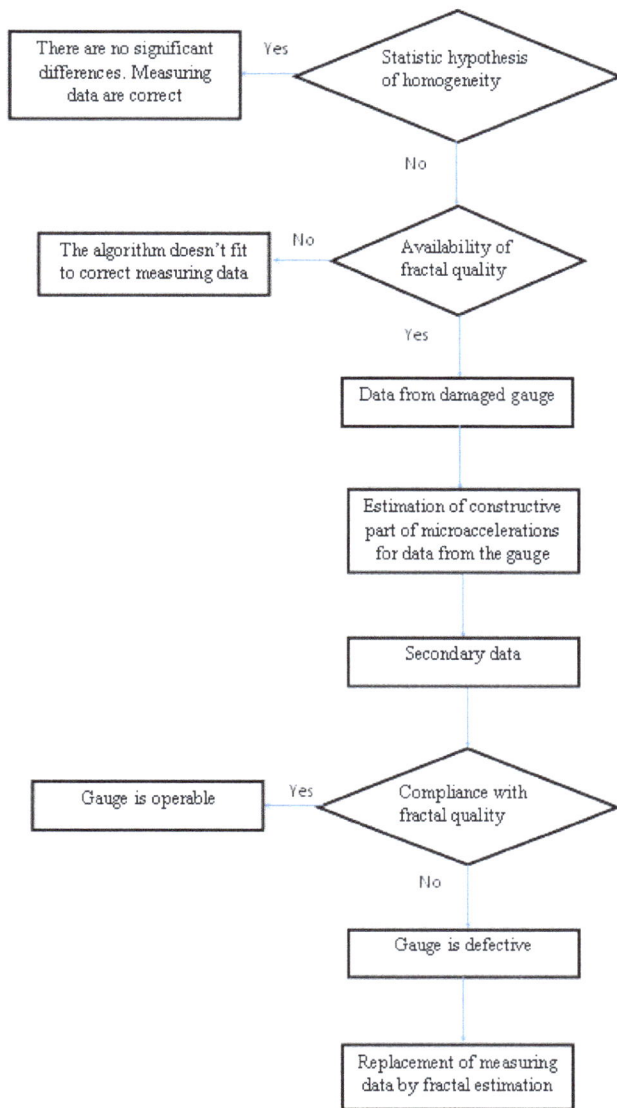

Figure 9. Algorithm 4.2.

4.4. The Problem to Improve Accuracy of Data in Case of False Run of Gauges

Description: False run of one or few gauges was uncovered by efficiency control of measuring equipment. It necessitates improvements of accuracy of data from the time period.

Urgency: In case some gauge are defective it is necessary to raise data accuracy to analyze facilities for realization of technological processes, uncover gravity-sensitive characteristics of this processes and also investigate effect of different factors on microacceleration field in the indoor environment of space laboratory.

Algorithm: Majority inspection [15] method is usually applied to raise accuracy of data in case of on or few gauges are damaged. This method is based on creating of additional measuring line by approximation of measured data. To algorithm to solute problem 3.3 precede solution of this

problem. After this it is necessary to:

1. Estimate median of data on the basis of their measurement.

2. Make functional dependence.

3. Check compliancy between accuracy of approximation and prescribed accuracy.

4. Apply generated model as measuring data of all defective gauges.

This algorithm is shown in figure 11.

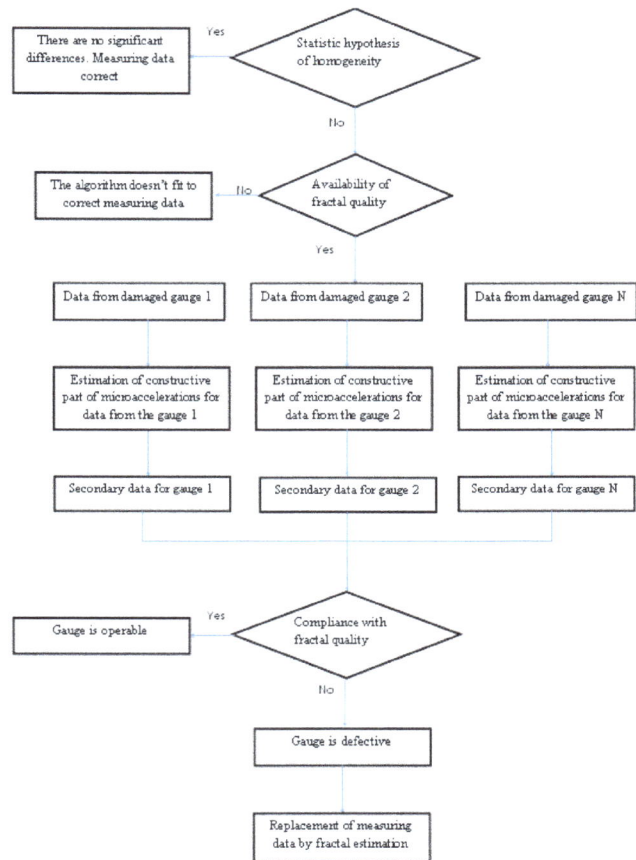

Figure 10. Algorithm 4.3.

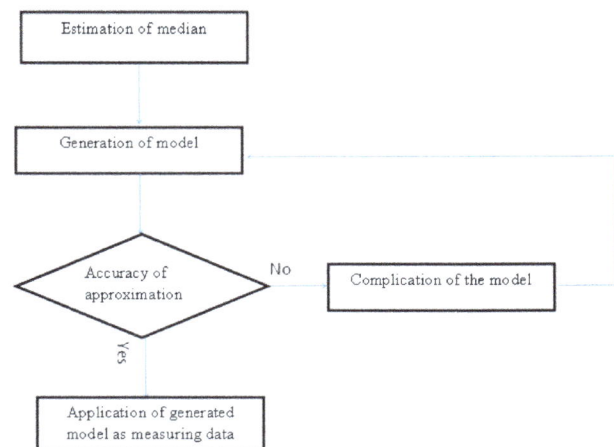

Figure 11. Algorithm 4.4.

5. Conclusion

In the article are shown four different problems. They are concerned with recovering of lost measuring data, efficiency check of gauges and improvement of accuracy of measurements in case some gauges are damaged.

Reliable data are necessary for qualitative analyses of microacceleration field in the indoor environment of space laboratory.

If constructive part of microaccelerations is dominant the fractal quality of this part should be applied for effective solution of these problems. In case data comply with fractal quality they can be used for estimation of microaccelerations. Otherwise accuracy of measuring data can be improved by majority inspection method. Moreover the data can be replaced by fractal model.

The problems and algorithms which are shown in the article can be applied either for development of new space laboratories and gauges or for realization of space projects to conduct gravity-sensitive processes in space.

References

[1]   A.V. Sedelnikov The Problem of Microaccelerations: from Comprehension up to Fractal Model, (Moscow, Russian Academy of Sciences: The Elected Works of the Russian School, 2012), p. 277.

[2]   R.G. Owen, D.I. Jones, A.R. Owens and A.A. Robinson "Integration of a microgravity isolation mount within a Columbus single rack," Acta Astronautica, vol. 22, 1990, pp. 127-135.

[3]   V.V. Sazonov, S. Yu. Chebukov, V.I. Abrashkin, A.E. Kazakova and A.S. Zaytsev "The analysis of low-frequency microaccelerations onboard FOTON-11 artificial satellite," Space Research, vol. 39, No 4, 2001, pp. 419-435.

[4]   S.A. Nikitin, V.I. Polezhayev and V.V. Sazonov "About measurement quasistatic microacceleration components onboard an artificial satellite by means of the convection sensor," Space Research, vol. 39, No 2, 2001, pp. 179-187.

[5]   V.S. Zemskov, M.R. Raukhman and V.P. Shalimov "Gravitational sensitivity of fusions at cultivation of crystals of InSb:Te Bridzhmen's by methods and bestigelny zonal melting in the conditions of a microgravity," Space Research, vol. 39, No 4, 2001, pp. 375-383.

[6]   V.S. Zemskov, M.R. Raukhman and E.A. "Kozitsyna Features of crystallization of multicomponent alloys in the conditions of zero gravity," Physics and chemistry of processing of materials, No 5, 1985, pp. 44-49.

[7]   V.S. Zemskov, M.R. Raukhman and V.P. Shalimov On the way to understanding of processes of cultivation from fusions of crystals of semiconductors in zero gravity on spacecrafts (Moscow, Institute of metallurgy and materials science of A.A. Baykov of 60 years, ELIS, Editions Lyakishev N. P., 1998), pp. 295-317.

[8]   S.F. Konovalov, Yu.A. Ponomarev, D.V. Maiorov, V.P. Podchezercev and A.G. Sidorov "Hybrid microelectromechanical gyroscopes and acceleration gages," Science and Education, No 10, 2011, pp. 1-23.

[9]   A.N. Boiko, A.V. Zavodyan and B.M. Simonov "Micromechanical accelerometers: modeling of elements of a design and production," Electronics: Science, Technology, Business, No 8, 2009, p.100-103.

[10] A.V. Sedelnikov and A.A. Kireeva "Alternative solutions to increase the duration of microgravity calm period on board the space laboratory," Acta Astronautica, vol. 69, 2011, pp. 480–484.

[11] A.V. Sedelnikov "Fractal quality of microaccelerations," Microgravity Scienes and Technology, vol. 24, No 5, 2012, pp. 345–350.

[12] A.V. Sedelnikov "The usage of fractal quality for microacceleration data recovery and for measuring equipment efficiency check," Microgravity Scienes and Technology, vol. 26, No 5, 2014, pp.327–334.

[13] A.I. Belousov and A.V. Sedelnikov "Probabilistic estimation of fulfilling favorable conditions to realize the gravity-sensitive processes aboard a space laboratory," Russian Aeronautics, vol. 56, No 3, 2013, p. 297–302.

[14] A.I. Belousov and A.V. Sedelnikov "Problems in formation and control of a required microacceleration level at spacecraft design, tests, and operation," Russian Aeronautics, vol. 57, No 2, 2014, pp. 111–117.

[15] S.V. Zhernakov "Algorithms of control and diagnostics of aviation gas-turbine engine in the conditions of onboard realization on the basis of technology of neural networks," The Bulletin of the Ufa State Aviation Technical University, vol. 14, No 3(38), 2010, pp. 42-56.

22

Natural Convection in an Anisotropic Non-Darcy in Differentially Heated Porous Cavity Under G-Jitter

P. Ghosh[*], **S. Tuteja**

Department of Mechanical Engineering Indian Institute of Technology(Banaras Hindu University), Varanasi, Uttar Pradesh, India

Email address:

pradyumna_ghosh@rediffmail.com (P. Ghosh)

Abstract: Natural convection in a square porous cavity under sinusoidal g-jitter has been studied for hydro dynamically and thermally anisotropic porous media. The difference with the homogeneous porous media under sinusoidal g-jitter with the anisotropic porous medium under sinusoidal g-jitter is the circulation pattern change. Fluid flow aligns with the porosity distribution. An effort has also been made to understand the non-Darcy effect for the above mentioned problem. It has been observed that at very low velocities, results from the porous media following Darcy's model and Forchheimer's equation (non-Darcy model) closely resemble each other. Velocity and pressure behave in a sinusoidal fashion with the same frequency as with the gravitational acceleration. Last but not the least an effort has also been made to understand the behaviors of average Nusselt number in the above mentioned situations.

Keywords: Anisotropic, G-Jitter, Forchheimer's Equation, Non-Darcy, Porous Media, Sinusoidal

1. Introduction

Microgravity indicates low gravity where the main gravity ranges between 10^{-1} to 10^{-5} m/s^2. It was considered that space environment was completely weightless. All onboard objects (Space Shuttle and the International Space Station) experience low amplitude broadband perturbed acceleration, or g-jitter, where there is a non-zero steady (metastable) part and an oscillatory part (random and constructive) [1,2]. This perturbed residual acceleration is called "g-jitter. Experiments have been carried out in space, to find out the influence of the g-jitter.

Porous media is used in many areas of applied science and engineering: filtration, mechanics (soil mechanics, rock mechanics), engineering (petroleum engineering, construction engineering), geosciences (hydrogeology, petroleum geology, geophysics), biology and biophysics, material science, etc. Natural convection through porous media under g jitter has emerged as a subject of interest with interesting applications biological and material science.

Thermal convection in a layer heated from below for the study of non-linear fluid dynamics and the transition to turbulence have been studied by Busse (1978) [3]. Natural convection in porous medium heated from below was investigated by Cheng (1978) [4]. Biringen and Danabasoglu (1990) [5] solved fully nonlinear time-dependent Boussinesq equations for g-jitter in a two-dimensional rectangular cavity with an aspect ratio of 2 at Ra = 1.775×105 and at a Prandtl number (Pr) = 0.007 (liquid germanium). They specified the critical value of ω above which the system experiences transition from convective temperature fields to a conductive one.

A detailed flow field and heat transfer analysis under g-jitter with a zero-mean base gravity parallel to the applied thermal gradient inside the enclosure has been studied by Hirata et al. (2001) [6] which also includes the analysis of onset of convection. Natural convection in an isotropic Darcy square porous cavity under constant gravity has been studied. The result is validated with a similar study conducted by Ghosh and Ghosh [2009] [7].

However, there isn't any experimental or computational study for natural convection in a anisotropic square cavity where g-jitter is applied perpendicular to the thermal gradient. Moreover, Natural convection in a Non-Darcy medium under g-jitter has not been studied. In many applications, porous materials are anisotropic due to preferential orientation for solid bundles. A detailed analysis on natural convection in a heat generating anisotropic porous medium is yet to appear in the literature. The paper deals with the study of flow behaviour in isotropic and anisotropic porous medium in which fluid is subjected to g-jitter applied in a particular direction. Further, Non Darcy model has to be applied to the porous medium, and its effect has to be studied. Moreover an effort has also been made in a Non Darcy porous medium following Forchheimer's equation has also been analysed.

2. Transport Equations and Numerical Solution Methodology

A two-dimensional square cavity has been modelled and hexahedral mesh is generated for discretization of the domain. Figure 1 shows the schematic diagram of the domain to be analysed. G-jitter is applied parallel to Y-axis. Top and bottom walls parallel to X-axis are adiabatic. Left and right walls are isothermal with different temperatures applied to them. It is assumed that isothermal walls are perfect conductors of heat and that the square cavity is completely filled with homogeneous porous media saturated with Newtonian fluid. Different temperatures at the walls create density differences in the fluid which produces buoyancy and drives convection. Number of computational cells has been kept at 2500 after performing the grid independence tests.

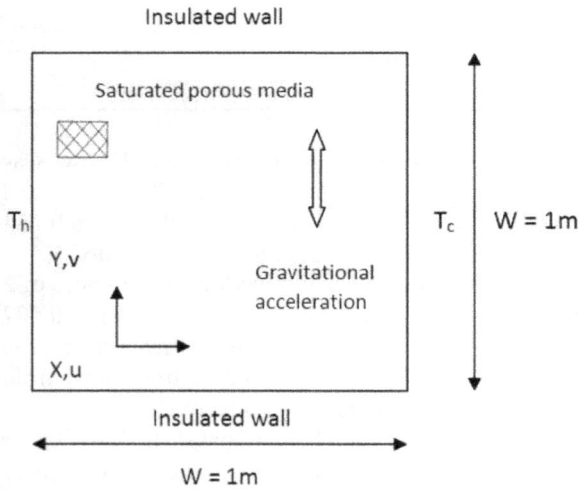

Figure 1. The CFD domain.

In a porous medium following the Darcy model, the x directional pore velocity (u) and y directional pore velocity (v) can be presented as follows,

$$u = \frac{K}{\mu}\left(-\frac{\partial p}{\partial x}\right); \quad v = \frac{K}{\mu}\left(-\frac{\partial p}{\partial y} + \rho g\right)$$

Using Darcy's law the dimensionless transport equations under sinusoidal g-jitter are as follows. Heat transfer through the porous medium is represented, subject to the assumption of thermal equilibrium between the medium and the fluid flow, by the energy equation below.

All asterisked quantities and θ in this paper are in dimensionless form

$$\frac{\partial u^*}{\partial x^*} + \frac{\partial v^*}{\partial y^*} = 0; \quad u^* = -\frac{\partial p^*}{\partial x^*}; \quad v^* = -\frac{\partial p^*}{\partial y^*} + Ra\,\mathrm{mod}\,\theta\sin\left(\omega^* t^*\right);$$

$$v^* = -\frac{\partial p^*}{\partial y^*} + Ra\,\mathrm{mod}\,\theta\sin\left(\omega^* t^*\right)$$

$$\omega^* = \omega\frac{\sigma W^*}{\alpha_e}; \quad p^* = \frac{p}{\rho\left(\frac{\vartheta}{W}\right)^2}$$

Where,

$$u^* = \frac{u}{v_0}; \quad v^* = \frac{v}{v_0}; \quad t^* = t\left(\frac{\alpha_e}{W^2\sigma}\right); \quad x^* = \frac{x}{W}; \quad y^* = \frac{y}{W};$$

$$Da = \frac{K}{\omega^2}; \quad \mathrm{Pr} = \frac{\vartheta}{\alpha_e}; \quad \theta = \frac{T_h - T_C}{2\Delta T}; \quad \alpha_e = \frac{k_{eff}}{\rho_f c_{pf}}; \quad v_0 = \frac{\alpha_e}{W};$$

$$k_{eff} = sk_f + (1-s)k_s; \quad Ra\,\mathrm{mod} = \frac{gk\beta\Delta TW}{\alpha_e\,v};$$

$$\Delta T = T_h - T_C; \quad T_0 = T_C; \quad \theta^* = \frac{T - T_0}{\Delta T}; \quad \sigma = \frac{s\rho_f c_{pf} + (1-s)\rho_s c_s}{\rho_f c_{pf}}.$$

The $\sigma = \frac{s\rho_f c_{pf} + (1-s)\rho_s c_s}{\rho_f c_{pf}}$. Non-Darcy porous layer follows Forchheimer's equation. According to Joseph et al. (1982) [8] the appropriate modification to Darcy's equation is to replace it by:

$$\nabla p = -\frac{\mu}{K}\vec{v} - c_F\sqrt{K}\rho_f|\vec{v}|\vec{v},$$

where c_F is a dimensionless form-drag constant.

This is the Forchheimer's equation and the last term is call the forchheimer's term. Using Forchheimer's equation the dimensionless transport equations under sinusoidal g-jitter are as follows.

$$\frac{\partial u^*}{\partial x^*} + \frac{\partial v^*}{\partial y^*} = 0; \quad u^* = -\frac{\partial p^*}{\partial x^*}; \quad v^* = -\frac{\partial p^*}{\partial y^*} + Ra\,\mathrm{mod}\,\theta\sin\left(\omega^* t^*\right);$$

$$\frac{\partial p^*}{\partial y^*} = -\frac{1}{Da\cdot\mathrm{Pr}}v^* - \frac{1}{\sqrt{Da}\cdot\mathrm{Pr}^2}v^{*2} + Ra\,\mathrm{mod}\,\theta\sin\left(\omega^* t^*\right).$$

The energy equation remains the same as in the Darcy model:

$$v^* = -\frac{\partial p^*}{\partial y^*} + Ra\,\mathrm{mod}\,\theta\sin\left(\omega^* t^*\right)$$

Air and aluminium has been used as the fluid and porous solid respectively. Incompressible ideal gas model is used for varying air density. The base microgravity ($10^{-6}g_0$) [1,2] has been determined by solving transport equations numerically.

The source term added to the y direction momentum equation is shown below:

$$s = \rho g \sin(\omega t).$$

An unsteady segregated solver is used and second order implicit time marching scheme is used. Second order upwind scheme is used for momentum and energy discretization. The Body Force- Weighted scheme is applied for pressure discretization. The SIMPLE algorithm is applied as a standard scheme for pressure- velocity coupling. Fixed time stepping method is used for time marching. Residual normalization method is used for reaching the convergence criterion.

3. Isotropic Darcy Square Porous Cavity Under Constant Gravity

Natural convection in an isotropic Darcy square porous cavity under constant gravity has been studied. The result is validated with a similar study conducted by Ghosh and Ghosh (2009) [7]. Results are presented for Ramod = 100. Fluid circulation is observed and the velocity contour plot is shown in the figure 2.

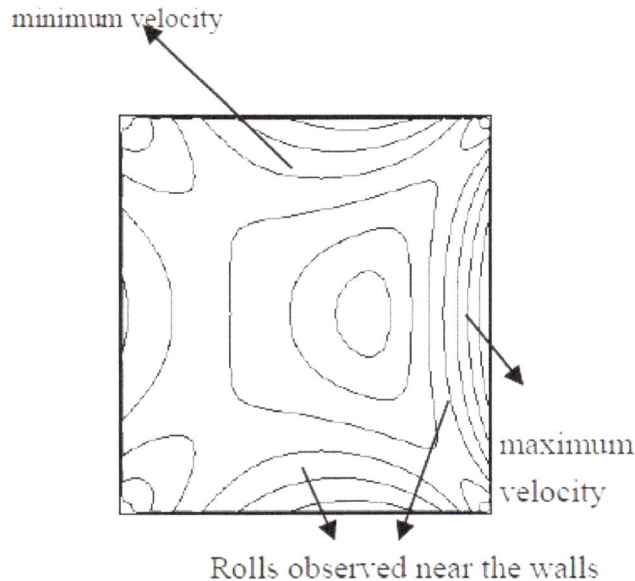

Figure 2. Velocity contours in porous cavity under constant gravity.

The centre of rotation is near the cold wall towards the right of the cavity centre. Greater convection effects are seen near the cold wall resulting in greater velocities. Rolls are observed near the walls. These rolls are due to the thermal diffusivity difference of wall material, fluid and solid matrix.

The maximum and minimum velocities of $3.41 \cdot 10^{-9}$ m/s and $4.26 \cdot 10^{-10}$ m/s are observed in the entire porous cavity. In the mid plane of the cavity at the upper half, flow is from hot side to cold wall and in the lower half the flow is from cold side to hot wall, which clearly indicate a clockwise rotation.

Table 1 and fugure 3 shows the average Nusselt numbers

calculated in the present study and in the study conducted by Ghosh and Ghosh (2009) [5].

Table 1. Comparison of average Nusselt number with numerical results by Ghosh and Ghosh (2009)[5] at constant gravity (sin(ωt) =1).

Name	Ramod = 100	Ramod = 1000
Present prediction	3	13,2
Ghosh and Ghosh (2009)	3,14	13,80

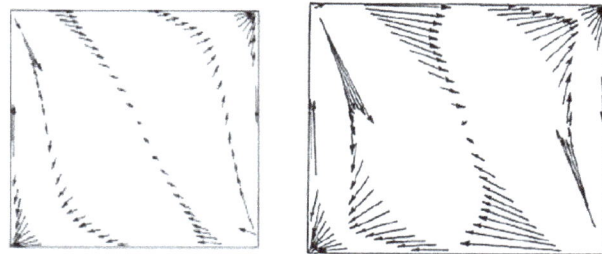

Figure 3. Velocity vectors in porous cavity under constant microgravity (10μm/s²) in present model (left) and simulations (right) performed by Ghosh and Ghosh (2009) [5].

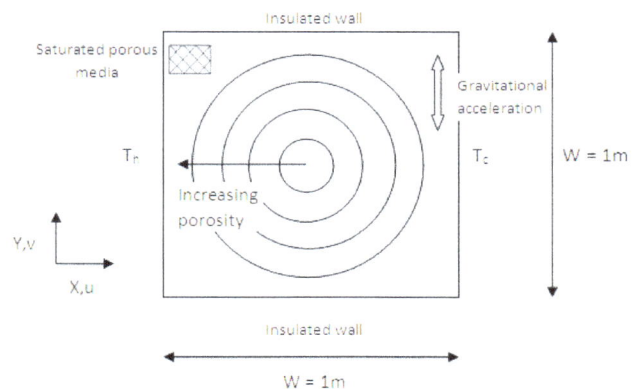

Figure 4. Anisotropic porous medium with five different porosity values in different concentric circular regions.

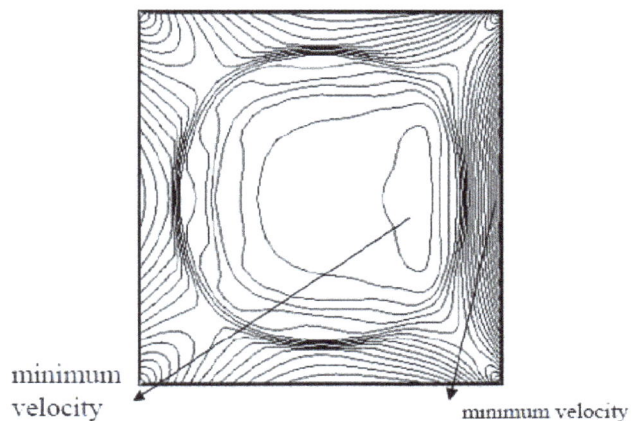

Figure 5. Velocity contour in the anisotropic porous cavity under constant gravity.

4. Anisotropic Darcy Porous Layer Under G-Jitter

Natural convection in the anisotropic(different porosity in different circular region) square porous cavity under g-jitter for Ramod =100, and $\omega = \pi/50$ has been studied. Gravity peaks at every 25 s and reverses its direction every 50 s. Figure 4 shows the anisotropic porous medium with five different porosity values in different concentric circular regions. Porosity increases from 0.6 in the inner circle to 0.8 towards the walls. Figure 5 shows the velocity contours for natural convection in the anisotropic porous medium under constant gravity.

Fig. 6 show velocity vectors at Ramod = 100 and $\omega = \pi/50$ at different instances of time.

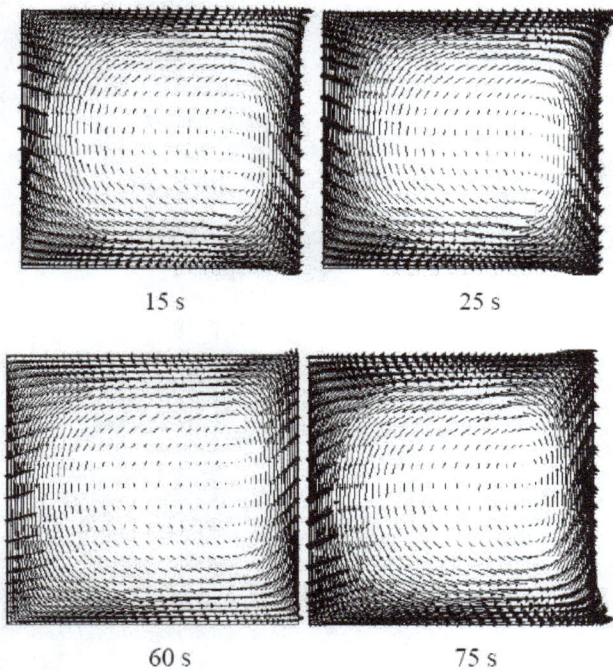

15 s   25 s

60 s   75 s

Figure 6. *Vectors in the anisotropic porous cavity for Ramod = 100, $\omega = \pi/50$.*

The maximum and minimum velocities are $6.2 \cdot 10^{-9}$ m/s and $5.2 \cdot 10^{-10}$ m/s. The flow field is divided in zones of different velocities. The region with maximum porosity has maximum velocity. Though the centre has the lowest porosity, it does not have the lowest velocity. The region with the minimum velocity is towards the right of the centre near the cold wall. The highest velocity is seen near the cold wall.

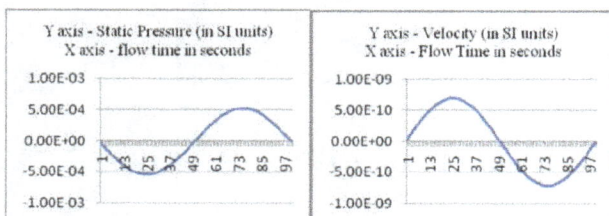

Figure 7. *Static Pressure (N/m²) and velocity (m/s) at point (0.8, 0.4).*

Under g-jitter fluid behaviour is periodic with the same time period as that of the gravitational acceleration. The velocity and pressure behaviour is same for all points in the porous media. Fluid circulation changes from clockwise to anticlockwise direction at t=50 s and fluid velocity peaks at 25 s and 75 s. Fig 7 shows that the fluid velocity and pressure at point 3 behave in a sinusoidal manner.

The temperature remains constant at a point in the porous media at all instances of time. The magnitudes of the average Nusselt number at the hot wall remains constant with time. Table 2 shows the magnitude of average Nusselt number in the anisotropic porous medium. The average Nusselt number in the case with circular porosity function is greater than in earlier discussed cases. This indicates that convection effect is more dominant than that in the earlier discussed cases.

Table 2. *Average Nusselt number for oscillating gravity, $\omega = \pi/50$.*

Parametr	Ramod = 100	Ramod = 1000
Average Nu	3,6	14,1

5. Isotropic Non-Darcy Square Porous Cavity Under G-Jitter

In this case, the isotropic porous medium follows the Forchheimer's equation. As observed from Fig 8, the velocity contour is similar to that in the isotropic Darcy case.

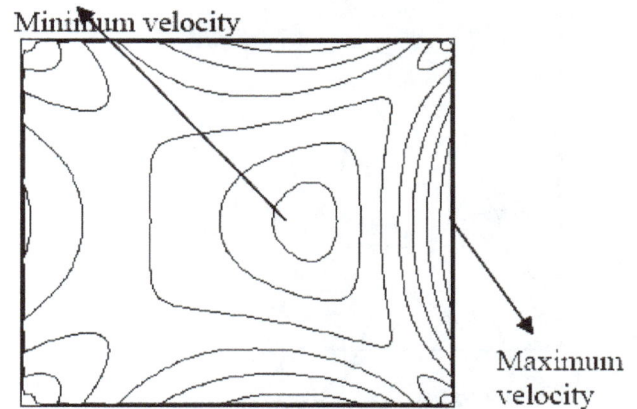

Figure 8. *Velocity contour in the Non-Darcy porous cavity under constant gravity.*

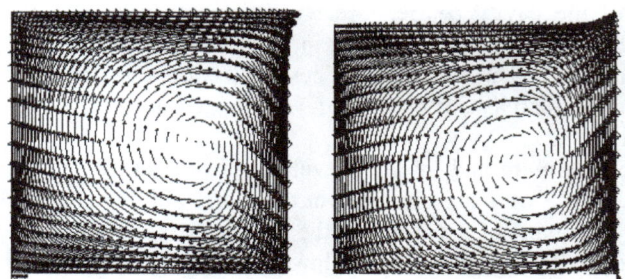

Figure 9. *Velocity vectors in the Non-Darcy porous cavity for Ramod = 100, $\omega = \pi/50$.*

The maximum and minimum velocities are $3.41 \cdot 10^{-9}$ m/s and $7.9 \cdot 10^{-11}$ m/s which are comparable with those in the Darcy case. There is no significant difference between the Darcy and Non-Darcy cases. Fluid behaviour is periodic with the same time period as that of the gravitational acceleration. Fluid circulation changes from clockwise to anticlockwise direction at t=50 s and fluid velocity peaks at 25 s and 75 s (fugure 9).

Under g-jitter, the temperature remains constant at a point in the porous media at all instances of time. The magnitudes of the average Nusselt number at the hot wall remains constant with time. Table 3 shows the magnitude of average Nusselt number in the Non Darcy porous medium. Average Nusselt number magnitude indicates that convection effect is more dominant than that in the earlier discussed cases. This is due to the extra convective term added in the momentum equation.

Table 3. *Average Nusselt number for oscillating gravity, $\omega = \pi/50$.*

Parametr	Ramod = 100	Ramod = 1000
Average Nu	3,5	14,3

6. Conclusion

Natural convection in anisotropic porous media under g-jitter has been studied for four different cases. In all the cases, fluid circulation is seen with different centers of circulation as described in the results. Velocity and pressure behave in a sinusoidal fashion with the same frequency as with the gravitational acceleration. The average Nusselt number remains constant at all instances of time under g-jitter. At the hot wall, anisotropic medium with a circular porosity function has a higher average Nusselt number magnitude, while anisotropic medium with porosity increasing in the x directions has a lower average Nusselt number magnitude as compared to the isotropic medium. The average Nusselt number at the hot wall in the Non Darcy porous layer is more than in the Darcy porous layer. This is due to the inertial term in the Forcheimer equation which is not present in the Darcy model equation. Natural convection in an isotropic Non-Darcy porous medium under g-jitter closely resembles that in the Darcy porous medium.

References

[1]   A.V. Sedelnikov "Classification of microaccelerations according to methods of their control," Microgravity Scienes and Technology, vol. 27, No 3, 2015, pp.327–334.

[2]   A.V. Sedelnikov "The usage of fractal quality for microacceleration data recovery and for measuring equipment efficiency check," Microgravity Scienes and Technology, vol. 26, No 5, 2014, pp.327–334.

[3]   F.H. Busse "Non-linear properties of thermal convection," Rep. Prog. Phys., No 41, 1978, pp. 1929-1967.

[4]   P. Cheng "Heat transfer in geothermal systems," Adv. Heat Transfer, No 14, 1978, pp. 1-105.

[5]   S. Biringen and G. Danabasoglu "Computation of convective flow with gravity modulation in rectangular cavities," J. Thermophys, No 4, 1990, pp. 357-365.

[6]   K. Hirata, T. Sasaki and H. Tanigawa "Vibrational effects on convection in a square cavity at zero gravity," J. Fluid Mech., vol. 45, No 4, 2001, pp. 327-344.

[7]   P. Ghosh and M.K. Ghosh "Streaming flows in differentially heated square porous cavity under sinusoidal g-jitter,"Int. J. Therm. Sc., No 48, 2009, 514-520.

[8]   D.D. Joseph, D.A. Nield and G. Papanicolaou "Nonlinear equation governing flow in a saturated porous medium," Water Resour. Res., vol. 18, No 4, 1982, pp. 1049-1052.

Permissions

List of Contributors

Lucian Milea and Doina Moraru
Solaris Consult S.R.L., Bucharest, Romania

Florin Lazo, Elteto Zoltan and Monica Dascalu
Centre for New Electronic Architecture, Research Institute for Artificial Intelligence, Bucharest, Romania

Eduard Franti
Centre for New Electronic Architecture, Research Institute for Artificial Intelligence, Bucharest, Romania

Micromachined Structures, Microwave Circuits and Devices Laboratory, National Institute for Research and Development in Microtechnologies (IMT), Bucharest, Romania

Suzana Cismas
Department of Biotechnology, University of Agricultural Sciences and Veterinary Medicine, Bucharest, Romania

Yassen El-Sayed Yassen and Ahmed Sharaf Abdelhamed
Mechanical Power Engineering, Faculty of Engineering, Port Said University, Port Said, Egypt

A. G. Syromyatnikov
St. Petersburg University, Department of Physics, Universitetskaya Nab., St. Petersburg, Russia

Sergey Orlov
Petrozavodsk State University, Petrozavodsk, Russia

Ahmed Farouk AbdelGawad
Professor of Computational Fluid Mechanics, Mech. Eng. Dept., Umm Al-Qura Univ., Makkah, Saudi Arabia

Talal Saleh Mandourah
Mech. Eng. Dept., Umm Al-Qura Univ., Makkah, Saudi Arabia

Mihail Stefan Teodorescu, Lucian Milea, Dan Coroama and Doina Moraru
Faculty of Electronics and Telecommunications, Politehnica University of Bucharest, Bucharest, Romania

Monica Dascalu
Faculty of Electronics and Telecommunications, Politehnica University of Bucharest, Bucharest, Romania

Center for New Electronic Architecture, Research Institute for Artificial Intelligence, Bucharest, Romania

Anca Plavitu
Center for New Electronic Architecture, Research Institute for Artificial Intelligence, Bucharest, Romania

Faculty - Exact Sciences and Engineering, Hyperion University, Bucharest, Romania

Eduard Franti
Center for New Electronic Architecture, Research Institute for Artificial Intelligence, Bucharest, Romania

National Institute for Research and Development in Microtechnologies, Bucharest, Romania

Yassen El-Sayed Yassen and Ahmed Sharaf Abdelhamed
Mechanical Power Engineering, Faculty of Engineering, Port Said University, Port Said, Egypt

Abdurrhman A. Alroqi and Weiji Wang
Department of Engineering and Design, University of Sussex, Brighton, UK

Ahmed Fayez EL-Saied, Mohamed Hassan Gobran and Hassan Zohier Hassan
Mechanical Power Eng. Dept., Zagazig University, Zagazig, Egypt

Alexander Degtyarev, Irina Vorobiova and Anatoliy Sheptun
Yuzhnoye SDO, Dnepropetrovsk, Ukraine

Qiu Hu
China Center for Resource Satellite Data & Application, Beijing 100830, China

Eyad Amen Mohamed, Muhammad Naeem Radhwi and Ahmed Farouk Abdel Gawad
Mech. Eng. Dept., College of Eng. & Islamic Archit., Umm Al-Qura Univ., Makkah, Saudi Arabia

N. I. Klyuev
Department of Mathematical Modeling in Mechanics, Samara State Aerospace University, Samara, Russia

Yang Yang and Xiaokui Yue
School of Astronautics, Northwestern Polytechnical University, Xi'an, China

Andrew G. Dempster
School of Electrical Engineering & Telecommunications, University of New South Wales, Sydney, Australia

Yong Li and Chris. Rizos
School of Civil & Environmental Engineering, University of New South Wales, Sydney, Australia

Ahmed Farouk Abdel Gawad
Professor of Computational Fluid Mechanics, Mech. Eng. Dept., Umm Al-Qura Univ., Makkah, Saudi Arabia

Arezu Jahanshir
Buein Zahra Technical University, Department of Eng. Physic, Qazvin, Iran

Atef Mohamed Alm-Eldien and Gamal Hafaz
Mech. Power Eng. Dept., Faculty of Eng., Port Said University, Port Said, Egypt

Ahmed Farouk Abdel Gawad
Professor of Computational Fluid Mechanics, Mech. Eng. Dept., Umm Al-Qura Univ., Makkah, Saudi Arabia

Mohamed Gaber Abd El Kreim
Gas Turbine Maintenance Engineer, East Delta Electricity Production Company, Ismailia, Egypt

Sherif A. Mohamed, Ibrahim S. Taha and Mahmoud G. Morsy
Department of Mechanical Engineering, Faculty of Eng., Assuit University, Assuit, Egypt

Hany A. Mohamed
Mech. Eng. Department, Faculty of Eng., Al Taif University, Al Taif, Saudi Arabia

Mahmoud S. Ahmed
Faculty of Industrial Education, Sohag University, Sohag, Egypt

V. M. Kotov
Republican State Enterprise "National Nuclear Center of the Republic of Kazakhstan", Kurchatov, Republic of Kazakhstan

A. V. Sedelnikov
Departament of Space Mechanical Engineering of Samara State Aerospace University, Samara, Russia

K. I. Potienko
Faculty of Computer Information Technologies and Automatic Equipment, Student of Donetsk National Technical University, Donetsk, Ukraine

P. Ghosh and S. Tuteja
Department of Mechanical Engineering Indian Institute of Technology(Banaras Hindu University), Varanasi, Uttar Pradesh, India

Index